ADVANCED LECTURES IN MATHEMATICS (ALM)

www.intlpress.com/alm

ALM 1: Superstring Theory
ALM 2: Asymptotic Theory in Probability and Statistics with Applications
ALM 3: Computational Conformal Geometry
ALM 4: Variational Principles for Discrete Surfaces
ALM 6: Geometry, Analysis and Topology of Discrete Groups
ALM 7: Handbook of Geometric Analysis, No. 1
ALM 8: Recent Developments in Algebra and Related Areas
ALM 9: Automorphic Forms and the Langlands Program
ALM 10: Trends in Partial Differential Equations
ALM 11: Recent Advances in Geometric Analysis
ALM 12: Cohomology of Groups and Algebraic K-theory
ALM 13: Handbook of Geometric Analysis, No. 2
ALM 14: Handbook of Geometric Analysis, No. 3
ALM 15: An Introduction to Groups and Lattices: Finite Groups and Positive Definite Rational Lattices
ALM 16: Transformation Groups and Moduli Spaces of Curves
ALM 17: Geometry and Analysis, No. 1
ALM 18: Geometry and Analysis, No. 2
ALM 19: Arithmetic Geometry and Automorphic Forms
ALM 20: Surveys in Geometric Analysis and Relativity
ALM 21: Advances in Geometric Analysis
ALM 22: Differential Geometry: Under the Influence of S.-S. Chern
ALM 23: Recent Developments in Geometry and Analysis
ALM 24: Handbook of Moduli, Volume I
ALM 25: Handbook of Moduli, Volume II
ALM 26: Handbook of Moduli, Volume III
ALM 27: Number Theory and Related Areas
ALM 28: Selected Expository Works of Shing-Tung Yau with Commentary, Volume I
ALM 29: Selected Expository Works of Shing-Tung Yau with Commentary, Volume II
ALM 30: Automorphic Forms and L-functions
ALM 31: Handbook of Group Actions, Volume I
ALM 32: Handbook of Group Actions, Volume II
ALM 33: Introduction to Modern Mathematics
ALM 34: Some Topics in Harmonic Analysis and Applications
ALM 35: The Legacy of Bernhard Riemann After One Hundred and Fifty Years, Volumes I and II

Advanced Lectures in Mathematics
Volume 3

Computational Conformal Geometry

Xianfeng David Gu
Shing-Tung Yau

International Press
www.intlpress.com

高等教育出版社
HIGHER EDUCATION PRESS

Advanced Lectures in Mathematics, Volume 3
Computational Conformal Geometry

Xianfeng David Gu (State University of New York, Stony Brook)
Shing-Tung Yau (Harvard University)

2010 Mathematics Subject Classification. 68U05

ISBN: 978-1-57146-324-1

Paperback re-issue 2016. Originally published in 2008 under ISBN 978-1-57146-171-1 (casebound).

Printed in the United States of America.

Supplemental materials are available online at
www.intlpress.com/alm3

Preface

Conformal geometry is in the intersection of many fields in pure mathematics, such as Riemann surface theory, differential geometry, algebraic curves, algebraic topology, partial differential geometry, complex analysis, and many other related fields. It has a long history in pure mathematics, and is an active field in both modern geometry and modern physics. For example, the conformal fields in super string theory and modular space in theoretic physics are research areas with very fast developments.

Recently, with the rapid development of three dimensional digital scanning technology, computer aided geometric design, bio-informatics, and medical imaging, more and more three dimensional digital models are available. The need for effective methods to represent, process, and utilize the huge amount 3D surfaces has become urgent. Digital geometric processing emerges as an inter-disciplinary field, combining computer graphics, computer vision, visualization, and geometry.

Computational conformal geometry plays an important role in digital geometry processing. It has been applied in many practical applications already, such as surface repairing, smoothing, de-noising, segmentation, feature extraction, registration, re-meshing, mesh spline conversion, animation, and texture synthesis. Especially, conformal geometry lays down the theoretic foundation and offers rigorous algorithms for surface parameterizations. Computational conformal geometry is also applied in computer vision for human face tracking, recognition, expression transfer; in medical imaging, for brain mapping, virtual colonoscopy, data fusion; in geometric modeling for constructing splines on manifolds with general topologies.

The fundamental reason why conformal geometry is so useful lies in the following facts:

- Conformal geometry studies conformal structure. All surfaces in daily life have a natural conformal structure. Therefore, the conformal geometric algorithms are very general.
- Conformal structure of a general surface is more flexible than Riemannian metric structure and more rigid than topological structure. It can handle large deformations, which Riemannian geometry cannot efficiently deal with; it preserves a lot of geometric information during the deformation, whereas topological methods lose too much information.

- Conformal maps are easy to control. For example, the conformal maps between two simply connected closed surfaces form a 6-dimensional space, therefore by fixing three points, the mapping is uniquely determined. This fact makes conformal geometric methods very valuable for surface matching and comparison.
- Conformal maps preserve local shapes, therefore it is convenient for visualization purposes.
- All surfaces can be classified according to their conformal structures, and all the conformal equivalent classes form a finite dimensional manifold. This manifold has rich geometric structures, and can be analyzed and studied. In comparison, the isometric classes of surfaces form an infinite dimensional space, which is really difficult to deal with.
- Computational conformal geometric algorithms are based on elliptic partial differential equations, which are easy to solve and the process is stable. Therefore, computational conformal geometry methods are very practical for real engineering applications.
- In conformal geometry, all surfaces in daily life can be deformed to three canonical spaces: the sphere, the plane, or the disk (the hyperbolic space). In other words, any surface admits one of the three canonical geometries: spherical geometry, Euclidean geometry, or hyperbolic geometry. Most digital geometric processing tasks in three dimensional space can be converted to the task in these two dimensional canonical spaces.

The major goals for writing this book are twofold. First, we want to introduce the beautiful theories of conformal geometry to general audiences, and make the elegant conformal structures better appreciated. The major concepts in conformal geometry are profound and abstract, which mainly existed in the imaginations of professional mathematicians. Our conformal geometric methods can compute those concepts explicitly on all kinds of surfaces in daily life, and display them using modern computer graphics and visualization technologies. Therefore, the students can see them, sense them, and accumulate intuition. Professional mathematicians can design experiments and use computers to help their exploration.

Furthermore, we would like to introduce the practical conformal geometric algorithms, and make them easily accessible for computer scientists and engineers. Therefore, the whole book is written to use less abstract mathematical reasoning, but more intuitive explanations and hands on experience. Major concepts and theorems are visualized by figures and computational algorithms are given. Students can implement the algorithms by themselves and see the abstract concepts represented as data structures on computers and create the images reflecting various geometric structures.

The book has two parts. The first part focuses on the theoretical foundations. It covers algebraic topology, differential exterior calculus, differential geometry, Riemann surface theory, surface Ricci flow, and general geometric structures. All of this knowledge is required for doing research in computational conformal geometry. Most of these topics are elementary, and some advanced topics are briefly touched with thorough references.

The second part focuses on computational algorithms, and is completely written in computer science language. It covers the computational algorithms for surfaces, which can be easily generalized to 3-manifolds. Then the algorithms on computing conformal structures for surfaces using various methods are explained in detail. Finally, algorithms for computing hyperbolic structure, and projective structure using Ricci flow method are examined. All algorithms are accompanied by pseudo-code, which is extremely easy to convert to programming language. We hope students can build the software system from scratch, and follow the book to implement various algorithms. The algorithms described in the book have already been applied in industrial applications.

The major content of the book is summarized from our research projects during the last several years. This textbook has been taught in graduate level courses in the Mathematics Department at Harvard University and the Computer Science Department at the State University of New York at Stony Brook. The theory part takes one semester, the computer science part takes one semester. The problem sets and programming exercises are valuable for students to improve their understanding and build their practical skill for developing geometric processing software. More teaching materials, coding samples and geometric surface data sets are available from the authors by request. [1]

The first author is very grateful to all professors in the *Center for Visual Computing* at Stony Brook: Arie Kaufman, Hong Qin, Dimitris Samaras, Klaus Mueller and all faculty members in the Computer Science Department at Stony Brook University and the Computer Information Science and Engineering Department at the University of Florida. The first author wants to thank Steven Gortler, Hugues Hoppy, John Snyder, Julie Dorsey, Leonard McMillan, who led him to the graphics field; Tom Sederberg, Ralph Martin, Shi-Min Hu, Jörg Peters, who led him to the geometric modeling field; Tony Chan, Paul Thompson and Baba Vemuri, who led him to medical imaging field.

Special thanks to Harry Shum, Baining Guo, Dinesh Manocha, Ming Lin, Helmult Pottman, Leif Kobelt for their consistent support and guidance. The first author also wants to thank Craig Gostman, Alla Sheffa, Peter Schröder, Mathieu Desbrun, Bruno Lévy, Myung-Soo Kim, Cindy Grimm, Seungyong Lee, and Kun Zhou for collaborations and support.

The first author also wants to thank the following mathematicians for fruitful collaborations, inspiring discussions, and valuable advices: Feng Luo, Kefeng Liu, Huai-dong Cao. Special thanks for Xiaowei Wang and Wei Luo for the collaborations in applied mathematics.

Special thanks to Yalin Wang, who helped us to conduct the most advanced research in brain mapping, and adapted conformal geometric methods for real industrial applications. Special thanks to Song Zhang, who developed the 3D scanner with high accuracy and high speed, and used our conformal geometric technology for motion capture applications. Special thanks to Junho Kim, who developed an efficient random access mesh compression algorithm, Euclidean Ricci flow and volumetric harmonic forms. The first author also appreciates all the students in the Center for Visual Computing at Stony Brook.

Special thanks to Lance Cong, Alex Mohr, Luke Farrer, and Susan Frank, who contribute to the artwork of this beautiful textbook. Special thanks to Ralph Martin, Feng Luo and Joe Marino, who helped proof reading the manuscript.

[1] Supplemental materials are available online at www.intlpress.com/alm3

Last but not least, we also want to thank both of our families, without their supports, this book could not be accomplished.

Stony Brook, New York,
Summer 2007

David Gu
Shing-Tung Yau

Contents

Part II Algorithms

1

Introduction

Computational conformal geometry is an interdisciplinary field, combining modern geometry theories from pure mathematics with computational algorithms from computer science.

From a theoretical point of view, computational conformal geometry has deep roots in mathematics and physics. In mathematics, it is the intersection of many fields, such as algebraic topology, differential geometry, Riemann surface, harmonic analysis, and so on. It also has a close relation with many physics fields, such as electromagnet field in electrodynamics, elasticity deformation in mechanics, heat diffusion in thermal dynamics, and modular space theory in super string theory. This field has many open problems of fundamental importance. For example, the space formed by all Riemann surfaces sharing the same topology is called the Modular space, which is a finite dimensional manifold. The topology and geometry of the modular space are not well understood even today. The theoretical research in conformal geometry is very active in pure mathematics and theoretical physics.

From a practical point of view, computational conformal geometry offers many powerful tools to handle a broad range of geometric problems. With the development of three dimensional scanning technologies, 3D shapes in real life can be easily acquired. The cutting edge 3D scanning technology can capture dynamic geometry, such as human expressions as fast as more than 180 frames per second. This huge amount of real time geometric data need to be processed by algorithms with high efficiency and robustness to process. The basic geometric processing tasks include shape representation, geometric compression, surface repairing, shape de-noising and smoothing, surface stitching and merging, meshing and re-meshing, and mesh-spline conversion. High level shape operations include surface classification, shape comparison, surface matching and recognition, shape manipulation, and many others. Computational conformal geometric methods can handle most of the above tasks directly or indirectly. The power comes from the following fact:

All surfaces in real life can be deformed to three canonical shapes: the sphere, the plane and the disk. The deformation preserves angles and is determined by a small number of control parameters, such as several landmarks. Therefore, all geometric problems in three dimensional Euclidean space $\mathbb{R}^3$ *can be converted to two dimensional problems on the plane.*

The major purpose of this textbook is to give a thorough introduction of computational conformal geometry, from the basic theoretic foundations, the profound and beautiful mathematic structures, to practical algorithms and state-of-the-art geometric processing techniques.

This textbook is useful for students majoring in mathematics who are interested in conformal geometry. Most general mathematics books focus on a single subject, but in order to fully understand conformal geometry, one needs the knowledge from many fields in mathematics. This is the only book which covers algebraic topology, differential geometry, Riemann surface, harmonic maps, and even Ricci flow. The content is self-contained; a general reader familiar with calculus can teach himself by using this book and penetrate the frontier quickly. Another uniqueness is that the textbook is visual. All of the important concepts are visualized using our conformal geometric algorithms and graphics techniques. For example, the mathematicians before our generation knew the existence of holomorphic differentials on surfaces, but they could not imagine them on real surfaces (such as the surface of the Stanford bunny as shown in Fig. 1.1). In this book, you will see the holomorphic differentials on all kinds of surfaces in real life. Another prominent feature of this book is that the major theoretic proofs are constructive, because our goal is to use computer to really calculate the conformal structures of surfaces. From this point of view, this book raises many open questions for mathematicians to convert abstract conceptual geometric beings to constructive algorithms and prove the rigor of them. For example, holomorphic quadratic differentials play a fundamental role in quasi-conformal maps, but so far, there is no rigorous algorithm to explicitly compute them on discrete surfaces.

Fig. 1.1. Visualization of a holomorphic differential form on the Stanford bunny surface.

If you major in computer science, this book can be very practical and useful. If you seek more of hands-on experience, you can start from the second part of the book right away, and follow the receipts of the algorithms step by step to build your own software from scratch. Then you can adapt the algorithm for your own applications. The content of the book is self-contained, by implementing the algorithms in the book, you can easily build many useful tools. However, in order to understand why the algorithms work and how they are invented, you need to study the first part of the book. If you want to do research to improve the algorithms and invent novel methods, it is helpful to study the theoretic part. We believe that for computer science majors, the best way to learn abstract theory is through implementing algorithms and playing with shapes first, then they can get the real feeling and deep understanding of the dry, obscure mathematics. Seeing is believing, one can easily grasp the essence of conformal mapping by looking at Fig. 1.2 without knowing the concept of Riemannian metric. The methods in the book are valuable for research in graphics, geometric modeling, vision, visualization, medical imaging, scientific computing and many other engineering fields. Most of the algorithms have been applied in industry today and are protected by patents.

Fig. 1.2. Visualization of conformal mappings.

1.1 Overview of Theories

Conformal geometry studies the *conformal structure* of general surfaces. Conformal structure is a natural structure of all surfaces in real life. Riemannian metric is a structure to measure the lengths of curves on the surface, the areas of domains on the surface and the intersection angles between curves. Conformal structure is a structure to only measure the intersection angle between two curves on the surface. Topological structure gives the neighborhood information. Roughly speaking, *Conformal structure is more rigid than topological structure and more flexible than Riemannian metric. Conformal geometry is between topology and Riemannian geometry.* We will get a better understanding of the above statement through further discussion.

1.1.1 Riemann Mapping

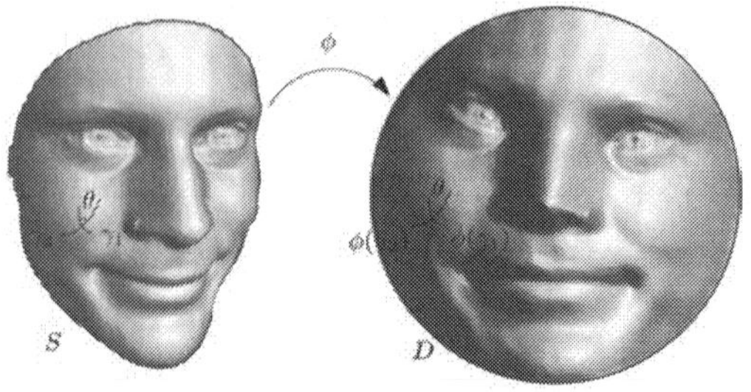

Fig. 1.3. A Riemann mapping from a human face to the unit disk, the mapping is angle-preserving.

A *conformal map* between two surfaces preserves angles. The *Riemann mapping* theorem states that any simply connected surface with a single boundary (a topological disk) can be conformally mapped to the unit disk. As shown in Fig. 1.3, a human face S is a topological disk and mapped to the unit disk D by a conformal mapping $\phi : S \to D$. Suppose γ_1, γ_2 are two arbitrary curves on the face surface S, ϕ maps them to $\phi(\gamma_1), \phi(\gamma_2)$. If the intersection angle between γ_1, γ_2 is θ, then the intersection angle between $\phi(\gamma_1)$ and $\phi(\gamma_2)$ is also θ. γ_1 and γ_2 can be chosen arbitrarily. Therefore, we say ϕ is conformal, meaning angle-preserving.

The conformality can be visualized using *texture mapping* techniques in computer graphics. Fig. 1.4 illustrates the idea. A texture refers to an image on the plane. First, a conformal mapping from the face surface (a) to the unit disk (b) is established. Then we cover the disk with a texture image, and then pull back the image onto the surface. In this way, the mapping can be directly visualized. If the texture is a checker board, then the pull back map glues the checker board to the face surface. Because the map is conformal, all the right angles of the corners of the checkers are preserved on the human face as shown in (c). If we replace the texture by a circle packing pattern, then planar circles are mapped to circles on the surface. The tangency relation among circles are preserved as shown in (d).

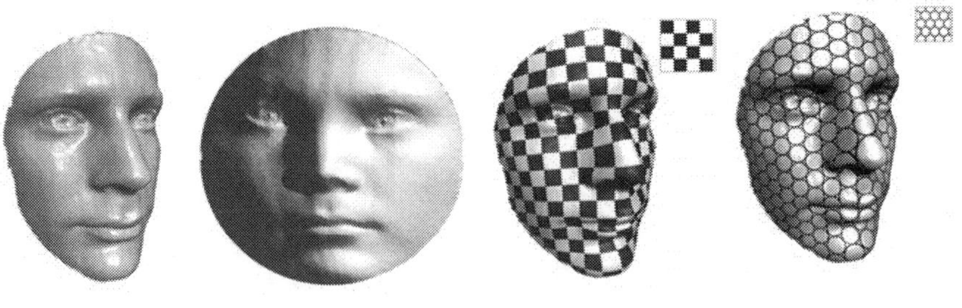

(a) Face surface (b) Map to disk (c) Checker texture (d) Circle packing texture

Fig. 1.4. Visualization of conformality using texture mapping in computer graphics.

Fig. 1.5 shows the conformal mappings of a multi-holed annulus. The planar domains are a circular disk with circular holes, an annulus with concentric circular arcs, and a strip with slits. The planar domains are determined by the conformal structure of the original surface.

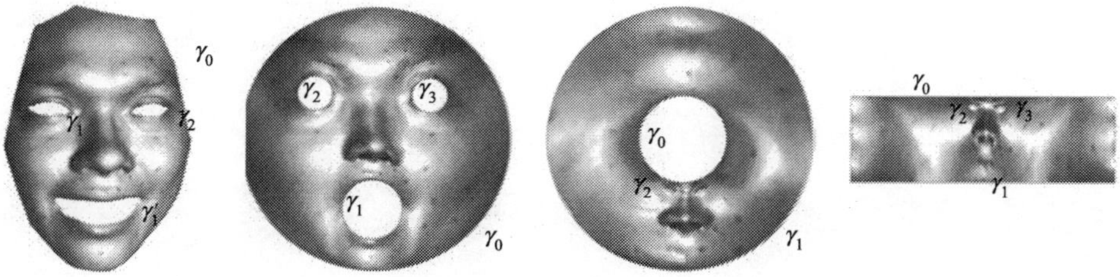

Fig. 1.5. Conformal mappings of a multi-holed annulus.

1.1.2 Riemann Uniformization

The conformal map between two planar domains is the conventional analytic function, or holomorphic function. From this point of view, conformal mappings are the generalization of holomorphic functions, and Riemann surfaces are the generalization of the complex plane. All surfaces in real life are real surfaces. The derivative of an analytic function is called a *holomorphic differential*. Holomorphic differentials can be defined on surfaces directly. They can be visualized using the same technique as the visualization of conformal mapping. By integrating the holomorphic differentials, we can locally map the surface to the plane. The mapping is conformal and is visualized by checker board texture mapping.

Fig. 1.6 shows the holomorphic differentials on animal surfaces. Fig. 1.7 shows the holomorphic differentials on the surface of Michelangelo's David.

Fig. 1.6. Visualization of holomorphic differentials on animal surfaces.

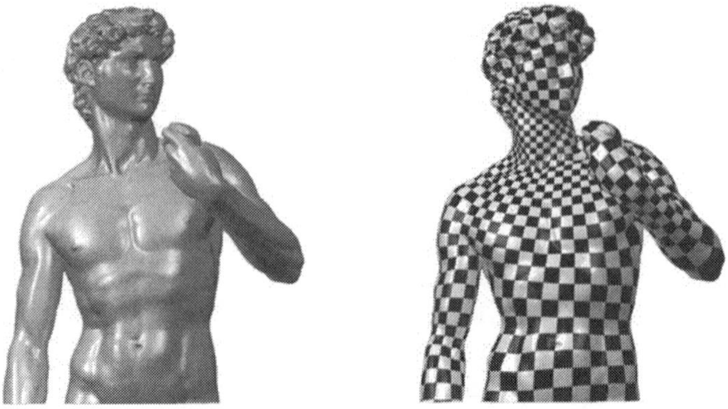

Fig. 1.7. Visualization of holomorphic differentials on Michelangelo's David.

Holomorphic functions, meromorphic functions, and holomorphic, meromorphic differentials on the surface form special groups. The group structures are governed by *Riemann-Roch theorem*, which is a profound theory connecting geometry, topology, and partial differential equation.

Riemann uniformization theorem states that all surfaces in real life can be conformally mapped to one of three canonical shapes: the unit sphere, the Euclidean plane, and the hyperbolic space. Namely, all surfaces admit one of the three canonical geometries: spherical, Euclidean or hyperbolic geometry. It can also be interpreted as all surfaces admit a canonical Riemannian metric, which is conformal to the original Riemannian metric and induces constant Gaussian curvature, which is $+1$, 0 or -1. Fig. 1.8 illustrates the uniformization theorem. For closed surfaces without any handles as shown in the first column, they can be conformally mapped to the unit sphere. Closed surfaces with one handle can be periodically mapped to the plane. For example, the whole kitten surface in the middle column of the figure can be conformally mapped to a parallelogram on the plane; the repetition of the parallelogram forms a tiling of the whole plane. The third column shows an exotic bottle, which has two handles. The surface has no self-intersection and is embedded in the Euclidean space. It can be conformally periodically mapped to the unit disk, which represents the hyperbolic space. The whole surface is mapped to a hyperbolic octagon, the repetition of the octagon forms a tiling of the whole hyperbolic space.

1.1.3 Shape Space

Surfaces can be classified using conformal geometry. Two surfaces are conformally equivalent if they can be conformally mapped to each other. It is challenging to verify if two surfaces are conformally equivalent. Roughly speaking, for closed surfaces with one handle, if the shapes of the parallelograms are similar, then they are conformally equivalent. The same result holds for surfaces with more handles. For closed surfaces with two handles, if their hyperbolic octagons are congruent in the hyperbolic space, then they are conformally equivalent.

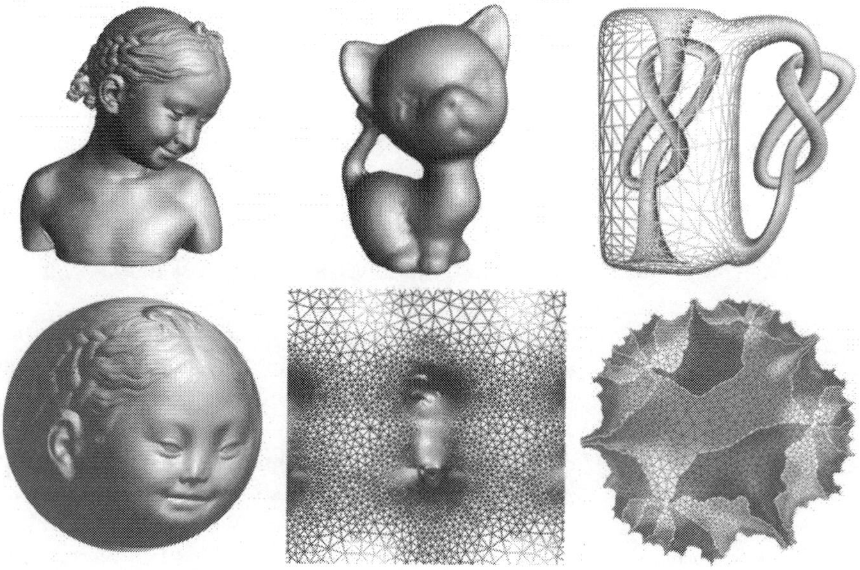

(a) $\chi > 0$ (b) $\chi = 0$ (c) $\chi < 0$

Fig. 1.8. Uniformization Theorem: all surfaces with Riemannian metric can be conformally embedded onto one of three canonical spaces: sphere, plane and hyperbolic space.

The conformal equivalence classes form a finite dimensional space, which is called the *modular space*, which is the space of shapes, and therefore, *shape space*. The universal covering space of the modular space is the *Teichmüller space*, which has simpler topology and is easier to handle. Each point in the shape space represents a shape, and each curve represents a deformation process. A lot of active research are being conducted today to fully understand the topological and geometric structures of the shape space.

Surface classification using conformal structures is described in [1]. The methods for computing general geometric structures are in [2] and [3]. A shape space application using Ricci flow is described in [4].

1.1.4 General Geometric Structure

A surface cannot be covered by one coordinate system. In general, we can find a collection of open sets to cover the surface and map each open set to the plane. We then use the planar coordinates as the local coordinates for the corresponding open set. Such a kind of collection of local coordinate systems is called an *atlas* of the surface. One point on the surface may be covered by multiple local coordinates. The transformation from one local coordinate system to another is called the *transition map* or *coordinates change*.

Suppose X is a topological space, G is the transformation group of X. An (X,G) *structure* is an atlas, such that the local coordinates are in X and the transition functions are in G. For example, a spherical structure is an atlas, where all of the local coordinates are on the sphere and all of the transition functions are rotations. The three frames in Fig.

1.8 can be interpreted as the visualization of spherical structure, Euclidean structure and hyperbolic structure respectively.

According to Klein's Erlangen program, different geometries study the invariants under different transformation groups. For example, let X be the Euclidean plane. If G is rigid motion, then the geometry is Euclidean geometry, and the invariants are lengths, angles, area etc. If G is an affine transformation group, then the geometry is the affine geometry, and the major invariants are the ratio for three points on a line and the parallelism between two lines. If G is the real projective transformation group, then the corresponding geometry is real projective geometry. The major invariant is the cross ratio for four points on a line.

If a surface admits an (X,G) structure, then the corresponding geometry can be defined on the surface directly. For example, in the automobile industry and mechanics engineering fields, surfaces are represented as splines, which are piecewise rational polynomials. Conventional splines are defined on the Euclidean plane and constructed based on affine invariants. The fundamental problem in the computer aided geometric design (CAGD) field is to construct splines defined on arbitrary surfaces. If the surface admits an *affine structure*, then splines can be defined on the surface without any difficulty. Unfortunately, very few surfaces admit affine structure (Fig.1.9). This fact causes intrinsic difficulty for applications in the CAGD/CAD field. But fortunately, all surfaces admit *real projective structure*. (Fig.1.10) How to construct splines based on real projective geometry is an active research area in geometric modeling today.

Like conformal structure, the shape space of all conformal structures has rich topological and geometric properties. The understanding of the shape space of all (X,G) structures is also widely open.

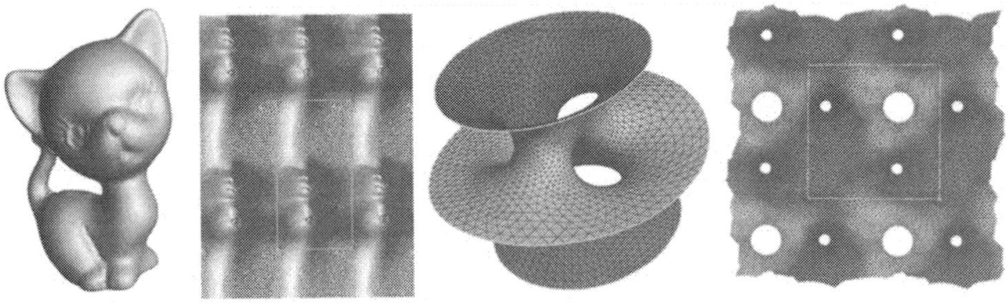

(a) (b) (c) (d)

Fig. 1.9. Visualization of the affine structures of two genus one surfaces.

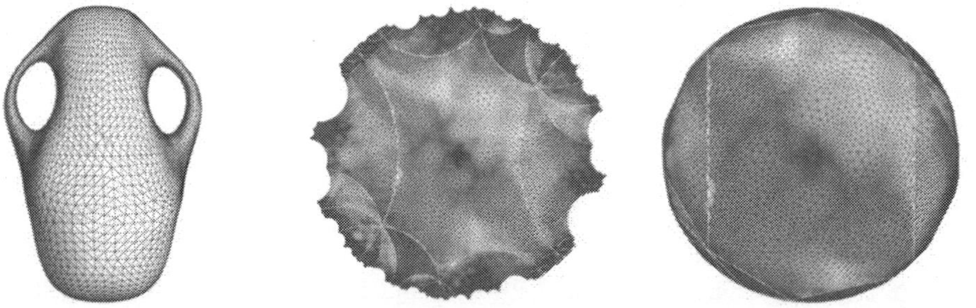

Fig. 1.10. Visualization of the hyperbolic structure and the real projective structure of a genus two vase model.

1.2 Algorithms for Computing Conformal Mappings

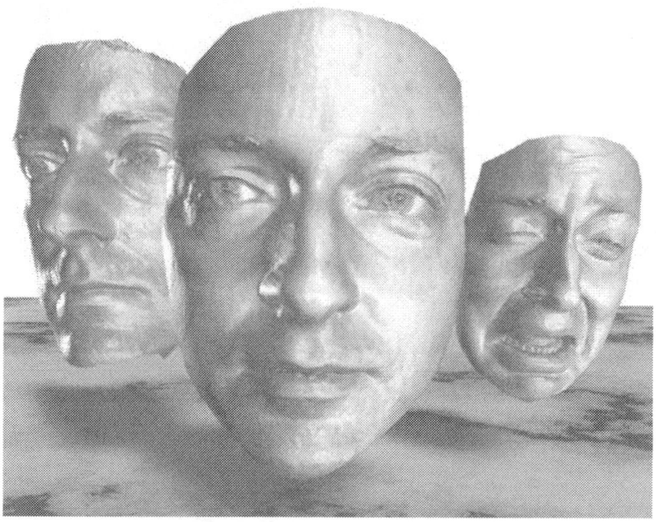

Fig. 1.11. Face geometries with different expressions of the same person, scanned using a real time high speed high resolution scanner. Courtesy of Geometric Informatics Inc.

Recently, $3D$ scanning technology is developing extremely fast. Fig. 1.11 shows several facial surfaces with different expressions of the same person, scanned by a scanner based on the phase-shifting method. The scanning speed is as fast as 180 frames per second, and each frame has 640×480 samples. The scanner can capture dynamic facial expressions in real time. It is challenging to process the huge amount of geometric data efficiently and robustly. Conformal geometry offers powerful tools to tackle this problem. The main strategy is to use conformal mappings to transform 3D surfaces to canonical 2D domains, and convert 3D geometric problems to 2D ones.

We summarize the major algorithms for computing conformal mappings of surfaces with various topologies.

Harmonic Maps for Topologoical Disks

The harmonic map between a simply connected surface with a single boundary and a convex planar domain can be computed by solving the Dirichlet problem. First we fix the boundary on a convex planar curve, and compute the interior by minimizing harmonic energy. The Euler-Lagrange equation of the critical point of the harmonic energy is the Laplace equation. By using the finite element method, the Laplace equation is formulated as a symmetric positive definite linear system. The problem is a linear problem. The energy optimization can be performed using the conjugate gradient method efficiently. Fig. 1.12 shows a harmonic map between a human face and a rectangle on the plane. The algorithm is explained in detail in Chapter 10.

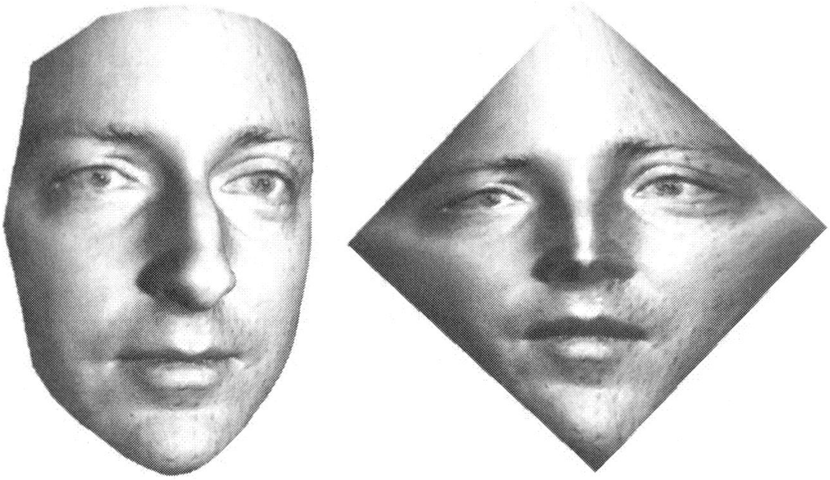

Fig. 1.12. Harmonic maps for topological disks.

Harmonic Maps for Topological Spheres

The harmonic map between a topological sphere and the canonical unit sphere is automatically conformal. The computational algorithm is based on the non-linear heat diffusion process. We first construct a degree one map, such as the Gauss map, then compute the Laplacian of the map, and update the map along the negative direction along the tangential component of the Laplacian. Because of the projection to the tangential space, the heat diffusion process becomes non-linear. Different solutions differ by Möbius transformations. Therefore, normalization conditions are necessary. Fig. 1.13

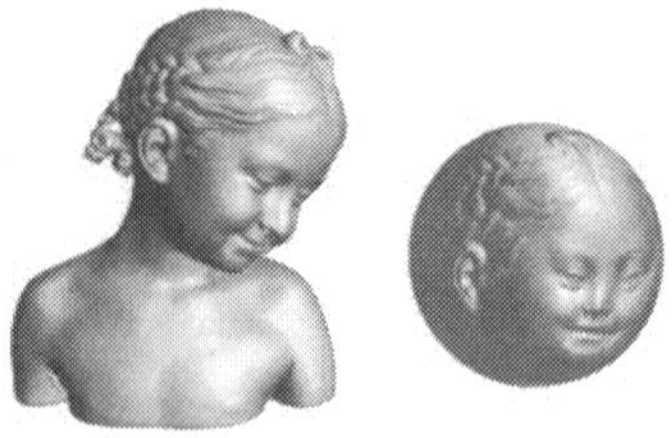

Fig. 1.13. Harmonic maps for topological spheres.

shows one example of conformal mapping of a topological sphere. Details can be found in the Chapter 10.

Riemann Mappings of Topological Disks (Fig. 1.14)

Harmonic maps between a topological disk and the unit disk may not necessarily be conformal. We compute the double covering of the topological disk, which is a topological sphere, then compute a conformal map between the doubled surface and the unit sphere, such that each copy of the topological disk is mapped to a hemisphere. Then we use stereographic to project the unit sphere onto the whole plane, the lower hemisphere is mapped to the unit disk. This induces the mapping from the surface to the unit disk, and the map is conformal.

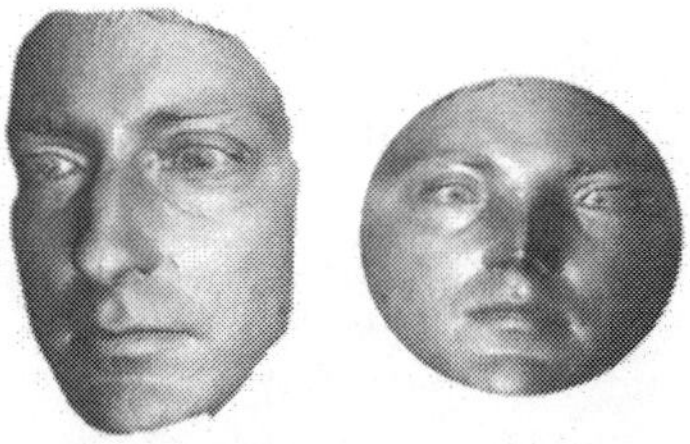

Fig. 1.14. Riemann mapping.

Conformal Mappings with Free Boundaries

Conformal mappings with free boundaries can be achieved by discrete approximation of the Beltrami equation (Fig. 1.15). A special case is the Cauchy-Riemann equation. The advantage of this method is that it is linear and efficient. The disadvantage is that there is less control of the boundaries. It mainly handle genus zero surfaces. The mapping results may have self-overlapping. Extra constraints can be added to enhance the mapping result, such as feature point constraints. The chapter 10 explains the details.

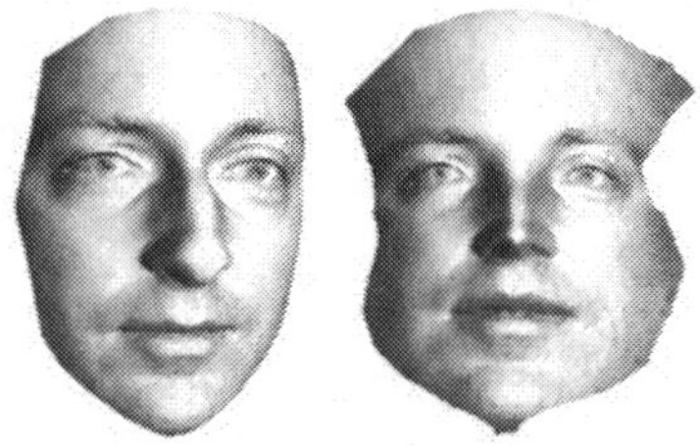

Fig. 1.15. Solving Beltrami equation using free boundary condition.

Conformal Mapping of Multi-Holed Annuli to Disk with Circular Holes

If the input surface is an annulus with multiple holes, the conformal mapping is more complicated to compute. It can be mapped onto the unit disk with circular holes (Fig. 1.16). In each homotopy class of degree one mappings, such kind of mapping exists and is unique. We apply the Euclidean Ricci flow method to construct such kind of mapping. The centers of the circles and the radii of the circles are conformal invariants and determined by the geometry of the input surface. The chapter 12 explains the details.

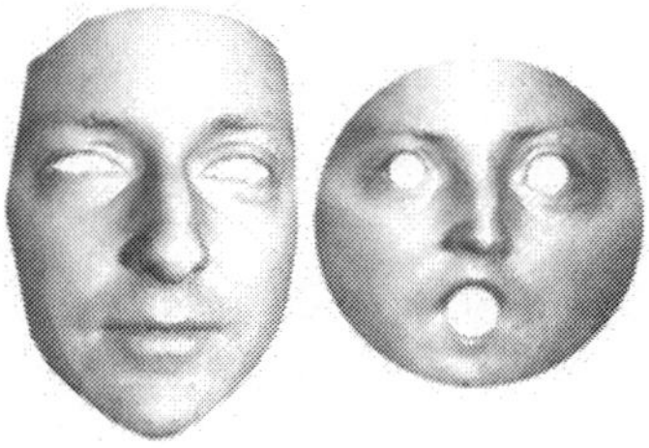

Fig. 1.16. Conformal mapping between a multi-holed annulus to a unit disk with circular holes.

Holomorphic 1-Forms

All metric surfaces are Riemann surfaces, which admit special complex differential forms, *holomorphic 1-forms*. The group of the holomorphic 1-forms has special structure; the generators can be explicitly calculated. A holomorphic 1-form has zero points and the number of zero points is equal to the absolute value of the Euler number. In the neighborhood of normal points, holomorphic 1-form induces conformal maps between the neighborhood and the complex domains. Iso-parametric curves through zero points can be used to segment the surface. Fig. 1.17 illustrates a conformal texture mapping induced by a holomorphic 1-form. The chapter 11 explains the details.

Fig. 1.17. Conformal mapping induced by a holomorphic 1-form.

Holomorphic 1-Forms for Affine Structure

General geometric structures on the surfaces refer to the atlases, such that all of the local coordinate changes belong to the special transformation groups. Most popular spline schemes are constructed based on affine invariants. Therefore they can be generalized to be defined on the surfaces with affine geometric structure.

Unfortunately, the existence of an affine atlas depends on the topology of the surface. If the surface is with boundaries, or the surface is a closed torus, then it admits an affine atlas. In general cases, extraordinary points have to be introduced. Conventional subdivision surfaces are splines defined on the surfaces with extraordinary points. Holomorphic 1-forms naturally induce affine structures on the surfaces. The extraordinary points are the zero points of the holomorphic 1-form, as shown in Fig. 1.18. The centered octagons are zero points. The number of zero points equals the Euler number of the surface. This paves the way to defining various planar splines on general surface domains. The chapter 11 explains the details.

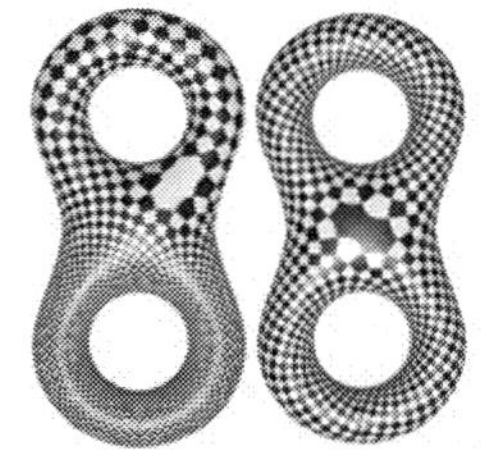

Fig. 1.18. Affine structure induced by holomorphic 1-forms.

Conformal Mapping of Multi-Holed Annuli to Annulus with Concentric Circular Arcs

Special holomorphic 1-forms can be constructed on a multi-holed annulus, such that the whole surface is mapped to an annulus with concentric circular arcs. Two boundaries are mapped to the inner and the outer boundaries of the annulus, other boundaries are mapped to the slits. Computing such holomorphic 1-forms is a linear problem, and the most difficult part is to find harmonic 1-forms. The chapter 11 explains the details. Fig. 1.19 shows the conformal mapping of a three-holed face surface (the mouth is open). The target domain is a unit disk with two concentric circular arcs. The exterior boundary of the face is mapped to the outer circle of the annulus. The mouth boundary is mapped to the inner circle. The boundaries of the eyes are mapped to the two circular slits. Then we conformally mapped the annulus to the rectangle. The outer and inner circles are mapped to parallel lines and the boundaries of eyes are mapped to horizontal slits.

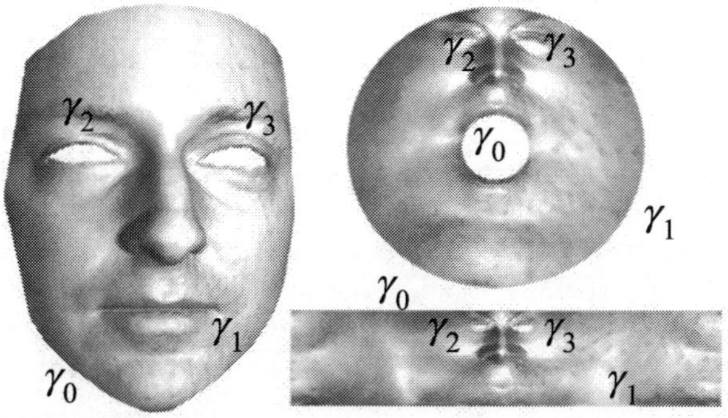

Fig. 1.19. Conformal mapping between a multi-holed annulus and an annulus with concentric circular arcs and a rectangle with slits.

Euclidean Ricci Flow for Genus One Surface

Euclidean Ricci flow method computes special metrics of the surface conformal to the original metric with prescribed target curvature. For a genus one closed surfaces, we set the target Gaussian curvature to be zero everywhere, and compute the flat metric, which is conformal to the original induced Euclidean metric. The universal covering space of the surface can be isometrically embedded on the plane. Fig. 1.20 shows one example. The kitten surface is of genus one. The universal covering space is embedded on the plane. The rectangle is a fundamental polygon. The details can be found in the chapter 12.

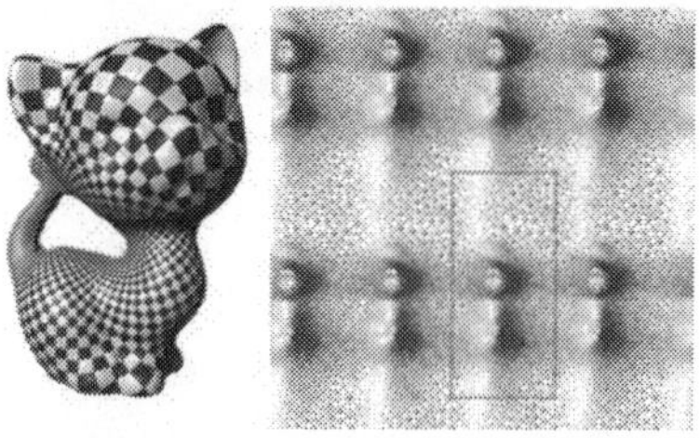

Fig. 1.20. Conformal flat metric of a genus one surface, computed using Euclidean Ricci flow.

Hyperbolic Ricci Flow for High genus Surface

For high genus surfaces, there exists a unique Riemannian metric, which is conformal to the original Riemannian metric, and induces a constant Gaussian curvature everywhere. This constant is -1. Such kind of metric can be computed using hyperbolic Ricci flow. The universal covering space of the surface can be isometrically embedded on the hyperbolic space. Fig. 1.21 demonstrates the embedding of the universal covering space of a genus two surface on the Poincaré model of hyperbolic space. The details can be found in the chapter 12.

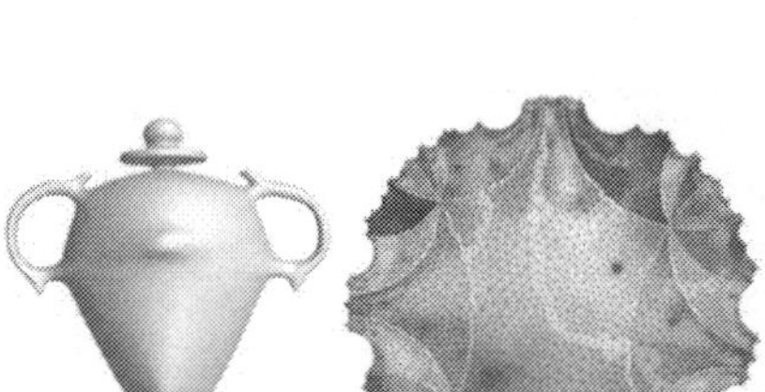

Fig. 1.21. Conformal hyperbolic mapping of a genus two surface.

Hyperbolic Ricci Flow for Real Projective Structure

All surfaces admit a special atlas such that all local coordinate changes are real projective transformations. Such a kind of atlas is called a real projective structure of the surface. Real projective atlases can be computed using the hyperbolic Ricci flow method. Fig. 1.22 demonstrates the computing result. First, we compute the conformal hyperbolic metric of the surface. Then, we embed its universal covering space on the Poincaré model of the hyperbolic space. Finally, we transform the Poincaré model to the Klein model, where all of the rigid motions are real projective transformations. This embedding induces a real projective atlas of the surface. The details can be found in the chapter 12.

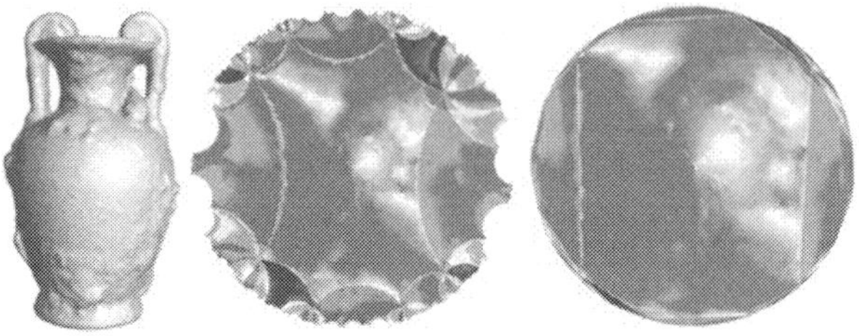

Fig. 1.22. Real projective structure of a genus two surface.

Conformal Metric Designed by a Prescribed Curvature

The conformal metrics and the curvatures of a surface are essentially of one-to-one correspondence. The conformal metric can be computed using a prescribed curvature on the surface using Euclidean Ricci flow method. Fig. 1.23 shows one example. The input surface is a topological disk. It is mapped to the planar domains specified by the curvature on the boundaries. The curvature of interior points are zero everywhere. The conformal mapping induced by the metric is fully controlled by the prescribed curvatures. It is also possible to concentrate all the curvature of a surface with arbitrary topology to a single point. The details can be found in the chapter 12.

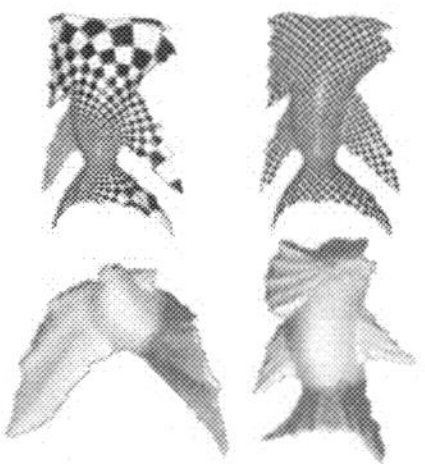

Fig. 1.23. Conformal flat metrics are designed by the target curvature.

1.3 Applications

Computational conformal geometric methods are valuable for a broad range of application in geometric modeling, computer graphics, computer vision, visualization, medical imaging, scientific computing and many other engineering fields.

Computational conformal geometric methods have the following merits:

- *General* All surfaces in daily life are Riemann surfaces and admit conformal structures.

- *Flexible* All surfaces, no matter how complicated the shape is, can be conformally deformed to one of three canonical shapes. For example, all genus zero surfaces can be mapped to the unit sphere, preserving the conformal structures. It is impossible for them to be mapped to the sphere preserving the Riemannian metric structure. Therefore, conformal geometry is more flexible than Riemannian geometry. In practice, conformal geometric methods are suitable to efficiently handle large deformations.
- *Rigid* The conformal mappings between a topological disk and the unit disk as shown in Fig. 1.4 are infinite, but all such mappings form a 3-dimensional group. Therefore, by fixing the images of 3 boundary points, the conformal map is uniquely determined. In general, conformal mappings are controlled by finite parameters.
- *Stable* Computing conformal structure is equivalent to solving some elliptic partial differential equations on surfaces. The solutions of elliptic partial differential equations are with infinite smoothness, with low requirements for the continuity of the input surface. The solution is very robust to the change of initial conditions. Therefore, in practice, we can simply approximate the surface using polygonal meshes, and approximate functions using piecewise linear ones. The approximate solutions are insensitive to the noise and converge fast.
- *Intrinsic* Conformal mappings are determined by the Riemannian metric, independent of the embedding. Many shape deformations in nature preserve the metric, such as the surface of human skin during the limb motion. The conformal structures are invariant during the motion. In the discrete case, the approximate solutions depend on the edge lengths of the mesh, not the positions of the vertices. The solutions are not dependent on the triangulation of the mesh.
- *Shape Preserving* Conformal mappings preserves angles; locally, they are just scaling transformations. Therefore, features in 3D are kept on the 2D image. This is useful for visualization purposes. In practice, general geometric programs are extremely difficult to debug because of the intricate pointer relations. Conformal geometric codes are relatively easy to debug, because it is very easy to tell whether the mapping is shape preserving or the texture mapping is harmonic.
- *Universal* According to conformal geometry, all surfaces admit one of three geometries: spherical, Euclidean or hyperbolic geometry. Therefore, a large class of geometric problems related to surfaces can be solved by one of them.
- *Finite Dimensional* Conformal geometry offers an effective ways to classify shapes. For shapes sharing the same topology, there are infinitely many conformally equivalent classes, but the shape space of these classes is finite dimensional. Most conformal geometric beings, such as holomorphic differentials, Riemannian mappings, and Fuchsian groups are finite dimensional. In practice, it is enough to find the generators of the groups. Therefore, it is easy to deal with.
- *Simple* Conformal geometric methods reduce 3D problems to 2D ones to drastically simplify the problems. On the other hand, in differential geometry, the representations of differential operators heavily depends on the choice of local coordinates. Conformal geometry offers a canonical coordinate system, the so called iso-thermal coordinates. Most of differential operators have the simplest format under these coordinates. This simplifies the theoretic deduction and programming.
- *Easy to implement* In engineering applications, the concepts from differential geometry, such as Gaussian curvature and principle directions are very sensitive to local

noises. Conformal geometric concepts, such as holomorphic forms, harmonic maps are very robust to the noises. Therefore, it is easy to implement and debug conformal geometric algorithms for beginners.

On the other hand, conformal geometry ignores Riemannian metric and extrinsic geometric information, such as the principle curvatures. In practice, we need to combine it with other algorithms from other geometries, especially differential geometry and Riemannian geometry. In the following, we briefly browse some direct applications of conformal geometric methods.

1.3.1 Computer Graphics

Conformal geometry has numerous applications in computer graphics, including surface parameterization, mesh repairing, texture mapping and synthesis, surface re-meshing, mesh matching, mesh-spline conversion, geometric morphing, efficient rendering, animation, and many more.

Surface Parameterizations

In computer graphics, surface parameterizations refer to the process of mapping the surface onto 2D planar domains. The mapping will unavoidably introduce distortions, which can be further classified to *area distortion* and *angle distortion*. It is impossible to eliminate both area distortion and angle distortion, unless the shape is flat. Conformal geometric methods can completely eliminate angle distortions.

The basic process is to construct a conformal map from the surface to the plane. In practice, we can add other requirements for the mapping. For example, we can specify feature points and landmarks on the surface, and require them be to mapped to some user-defined positions on the plane; We can design the mapping such that some interested parts on the surface are enlarged on the image plane; We can require the area distortion of the mapping to be as uniform as possible.

Fig. 1.24 illustrates the process. We first construct a conformal mapping to map the Stanford bunny surface shown in (b) and (c) to the planar domain shown in (d), which is called the texture domain. In order to verify the conformality of the parameterization, we texture map the bunny with a checker board as shown in (a). In order to improve the parameterization quality on the two ears, we have carefully designed the mapping. The left part and the right part of the bunny are mapped to two rectangles. We preserve the normal information on the planar image (d). So we can see the correspondence between the surface and the image. We can recognize the body parts from the image directly, such as the eyes, the nose, the ears, etc. Locally, the shapes are well preserved. Globally, the sizes of different parts are changed drastically. This also demonstrates that conformal mapping locally can be treated as scalings. The area distortions can be predicted by the Gaussian curvature of the surface. The relation between the area distortion and the Gaussian curvature can be accurately formulated and solved using the surface Ricci flow method.

The global conformal surface parameterization algorithms based on Riemann surface theory can be found in [5]. Algorithms for computing conformal structures are explained

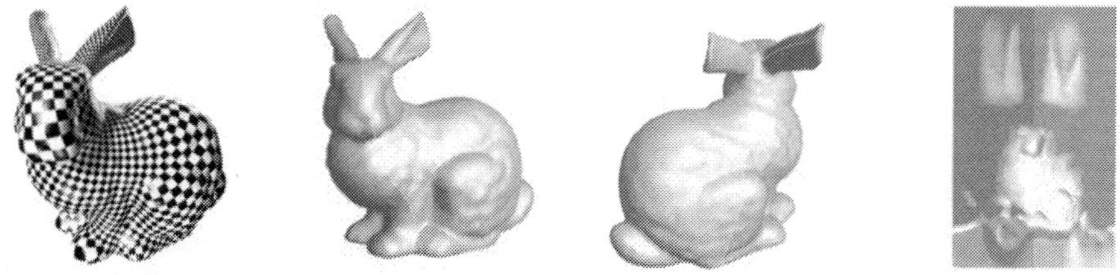

(a) Texture mapped bunny (b) Front view (c) Back view (d) Planar image

Fig. 1.24. Conformal parameterization of the Stanford bunny model.

in the first author's thesis [6], [7] and [8]. Optimal conformal surface parameterizations based on conformal structure is described in [9].

Texture Mapping

In computer graphics, surfaces are approximated by triangular meshes (polygonal surfaces, each face is a triangle), which can be supported by graphics hardware directly. The rendering efficiency of the hardware depends on the resolution of the mesh. For real time applications, usually low resolution meshes are preferred. Small geometric details and material properties are modeled as texture images. The parameterization process maps each vertex to the planar domain, and obtains its 2D coordinates, which are called the texture coordinates. Then graphics hardware will glue the texture to the meshes using the texture coordinates. Fig. 1.25 illustrates the process.

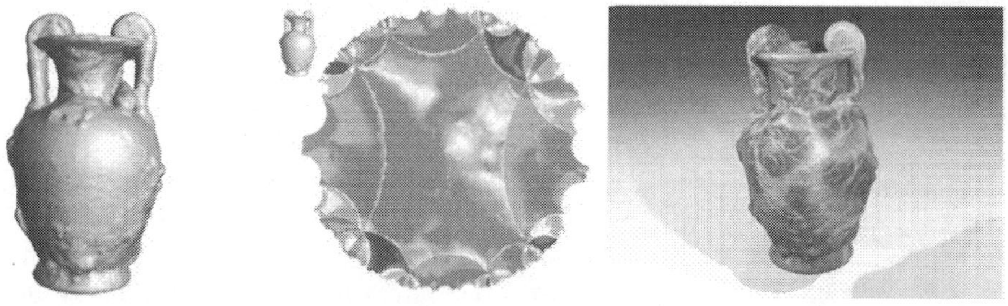

(a) Geometry (b) Parameterization (c) Texture mapping

Fig. 1.25. Texture mapping using hyperbolic parameterization.

Fig. 1.26 shows texture mappings of the genus one kitten surface. Different textures are illustrated at the upper right corners. The conformal parameterization is demonstrated by the checker board texture mapping on the left.

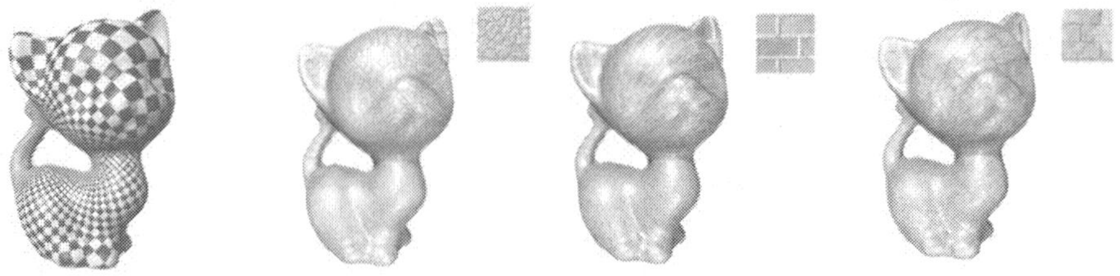

Fig. 1.26. Conformal texture mapping of a genus one closed surface.

Texture Synthesis

Fig. 1.27 demonstrates a texture synthesis application of conformal parameterization. Given a small patch of texture sample, the texture synthesis process generates a large texture to cover the whole surface. Instead of generating a new texture in 3D on the surface directly, we generate the texture on the 2D parameter plane. By using conformal parameterization, the local shape of the texture will not be distorted. Therefore, the synthesis algorithm solely focuses on the size of the local shape. It is much more efficient to generate uniform textures. The work on uniform texture synthesis based on conformal structure is explained in [10].

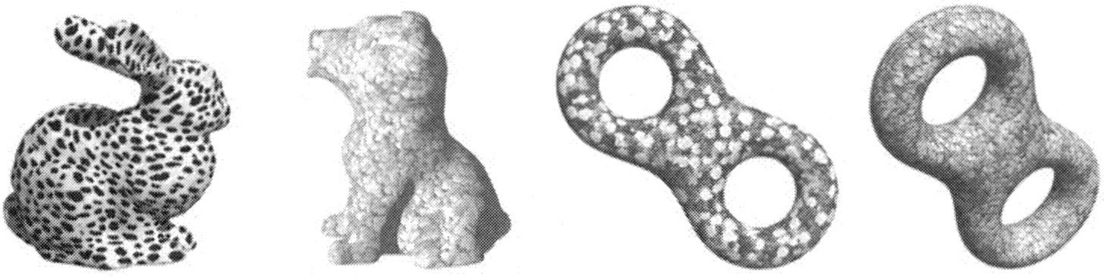

Fig. 1.27. Texture synthesis using conformal parameterization, courtesy of Klaus Mueller [10].

Geometric Morphing

Fig. 1.28 shows a geometric morphing process using the conformal mapping technique. The Stanford bunny surface S is a topological sphere, which can be conformally mapped

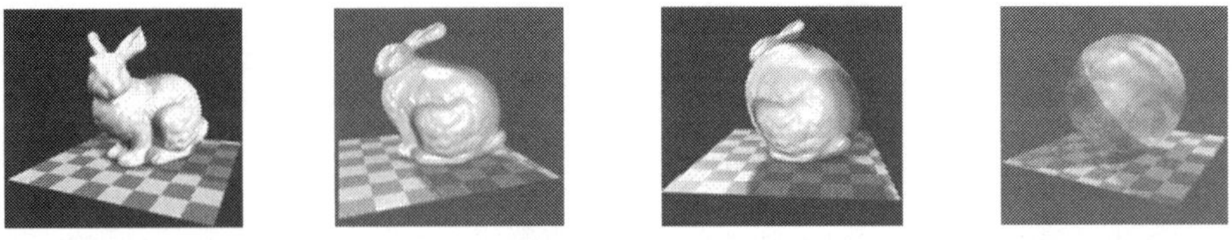

Fig. 1.28. Geometric morphing of the Stanford bunny surface to the unit sphere.

to the unit sphere $\mathbb{S}^2$; we denote the mapping as $\phi : S \rightarrow \mathbb{S}^2$. Then we can construct a family of surfaces to deform from the bunny to the sphere. The simplest morphing is the linear combination of the initial shape and the target shape. More complicated morphing sequences can be constructed similarly. The conformal mapping is computed using the non-linear harmonic map method.

Re-meshing and Geometry Images

Surfaces are represented as meshes in computer graphics. In order to convert them to spline format in geometric modeling, it is highly desirable to re-tessellate the triangular meshes to quad meshes, because most popular spline schemes are based on tensor-product. We first parameterize the triangular mesh onto the planar region, and use regular grids to tessellate the planar image of the surface. This induces the tessellation of the original surface and converts it to a quad mesh. Fig. 1.29 demonstrates one example for re-meshing a triangular mesh to a quad mesh.

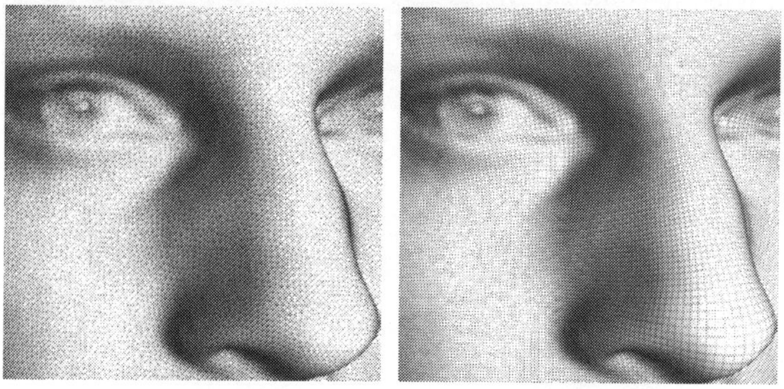

(a) Original mesh (b) Quad mesh

Fig. 1.29. Surface remeshing using conformal parameterization.

General meshes have both connectivity information of the triangulations and geometric information represented as the coordinates of the vertices. After re-meshing, the quad-mesh connectivity is regular. It is not necessary to encode the connectivity any more; we just record the coordinates of the vertices. We can color encode the coordinates and represent a surface as an image, which is called geometry image. There are two pipelines in graphics hardware, one handles meshes and the other handles texture images. Geometry image unifies both geometry and texture, which has the potential to simplify the graphics hardware. Geometry images can be applied to efficient rendering. Fig. 1.30 shows a geometry image for Michelangelo's David head model. The details of geometry image are presented in [11].

1.3.2 Computer Vision

Conformal geometry has been applied in computer vision for surface matching, shape comparison, shape classification, geometric analysis and tracking.

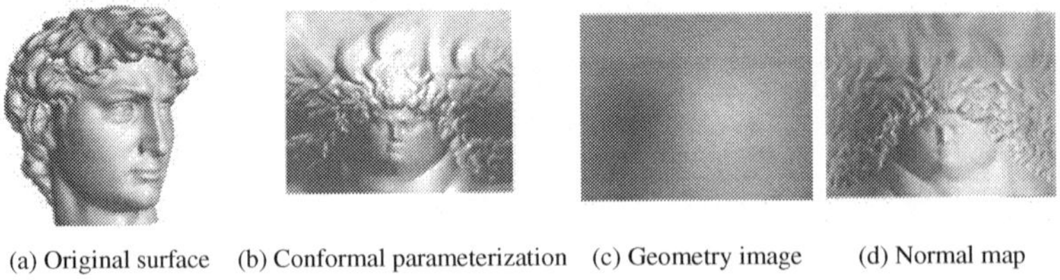

Fig. 1.30. Geometry image of Michelangelo's David head model.

Surface Matching

Surface matching is a fundamental task for computer vision, graphics and medical imaging. Fig. 1.31 shows the basic idea of using conformal parameterizations to convert 3D matching problems to 2D ones. Suppose S_1 and S_2 are two surfaces in $\mathbb{R}^3$. $\phi_1 : S_1 \to D$ and $\phi_2 : S_2 \to D$ are conformal mappings to map surfaces to the canonical planar domain. $\bar{f} : D \to D$ is a map from D to itself, this is a 2D matching process. Then

$$f = \phi_2^{-1} \circ \bar{f} \circ \phi_1, \quad S_1 \to S_2,$$

is the desired 3D matching.

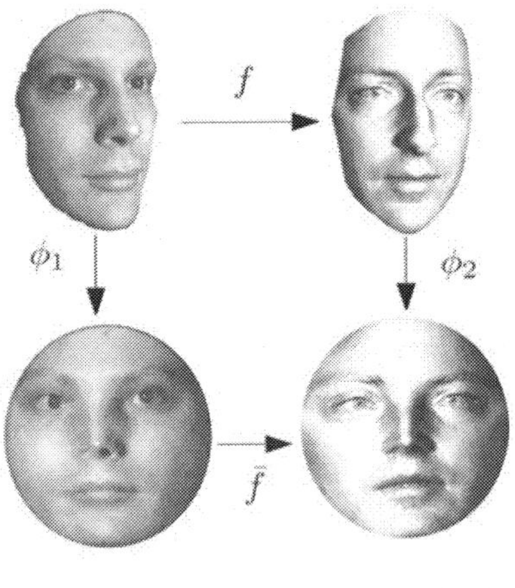

Fig. 1.31. Surface matching using conformal mapping.

If S_1 and S_2 are similar to each other in terms of their geometries, then their conformal structures are close to each other. Under some appropriate boundary conditions, the images of corresponding feature points will be close to each other on the planar images. In Fig. 1.31, we can see that the images of the nose tips of two surfaces are very close to each other on the 2D plane. The images of the eye corners are also close.

Fig. 1.32 demonstrates the fact that isometric deformation preserves the conformal structure. The original surface in (a) is a plastic mask, which can only be bent and can hardly be stretched. We deform it to get another surface shown in (c). We use a 3D scanner to acquire their shapes, denoted as S_1 and S_2. Then we use conformal map $\phi_1 : S_1 \rightarrow \mathbb{R}^2$ and $\phi_2 : S_2 \rightarrow \mathbb{R}^2$ with the only constraints that the images of all the boundaries are circles. We do not specify the centers and radii of the images of the boundaries, they are calculated automatically by our conformal geometric algorithms. Their planar images are shown in (b) and (d), which are identical. Then the 2D map $\bar{f}$ is the identity of the two hole annulus, the 3D map $f = \phi_2^{-1} \circ \bar{f} \circ \phi_1$, which is exactly the isometric deformation. This example shows that conformal geometric methods can recover isometric maps automatically. Therefore, for the purpose of surface matching, conformal geometric methods reduce the dimensionality and recover isometric maps, furthermore, they can handle surfaces with arbitrary topologies.

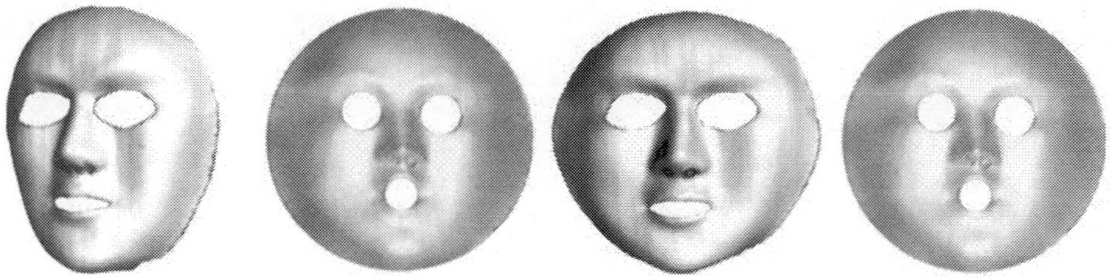

(a) Original surface (b) Planar image of (a) (c) Deformed surface (d) Planar image of (c)

Fig. 1.32. Isometric deformation from (a) to (c) preserves conformal structures, their planar images of conformal mappings are consistent shown in (b) and (d). Courtesy of Dimitris Samaras [12].

In Fig. 1.33, we matched two genus two surfaces and two genus six surfaces using the hyperbolic Ricci flow method. We visualize the matching by transferring the textures from the domain surfaces to the range surfaces. Fig. 1.34 shows the surface matching between two genus two surfaces by a geometric morphing.

Surface matching with exact feature alignment is in [13]. 3D surface matching, recognition and stitching using conformal geometry are described[14], [15].

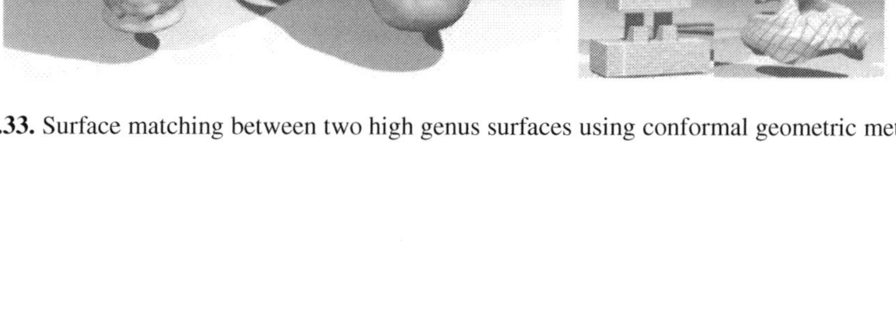

Fig. 1.33. Surface matching between two high genus surfaces using conformal geometric methods.

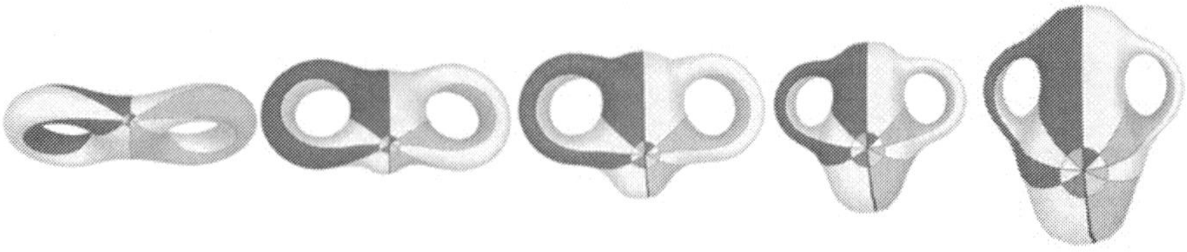

Fig. 1.34. Visualization of surface matching by surface morphing.

Surface Tracking

Tracking dynamic shapes with large deformation is a challenging task. By using conformal geometric methods, 3D surfaces are mapped onto 2D domains, and dynamic surfaces can be consistently matched. Major feature points are aligned and tracked. Fig. 1.35 illustrates one such an example. We match a generic face mesh to real facial data sets acquired using a high speed 3D scanner.

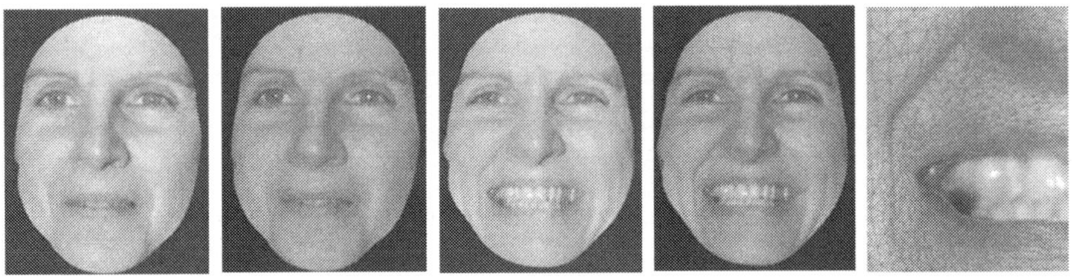

(a) (b) (c) (d) (e)

Fig. 1.35. Snapshots from a Tracking Sequence of Subject A: (a) Initial data frame. (b) Initial tracked frame. (c) Data at the expression peak. (d) Tracked data at the peak. (e) Close-up at the peak. Courtesy of Dimitris Samaras [12].

We performed tracking on four subjects performing various expressions for a total of twelve sequences of 250 – 300 frames each (at 30Hz). Each frame contains approximately 80k 3D points, whereas the generic face mesh contains 8k nodes. Our technique tracks very accurately even in the case of topology change. Details of high resolution tracking of non-rigid 3D motion are presented in [12].

Surface Stitching

Fig. 1.36 demonstrates the alignment and stitching of two 3D surfaces undergoing non-rigid deformations. 3D faces are captured using 3D scanners. Each face has approximately 80k 3D points with both shape and texture information available. The subjects were not asked to keep their head and facial expression still during the 3D face scanning.

An important property of conformal mappings is that they can map a 3D shape to a 2D domain in a continuous manner with minimized local angle distortion. This implies that conformal mappings are not sensitive to surface deformations. For example, since the stretching distortion between two faces of the same person with different expressions

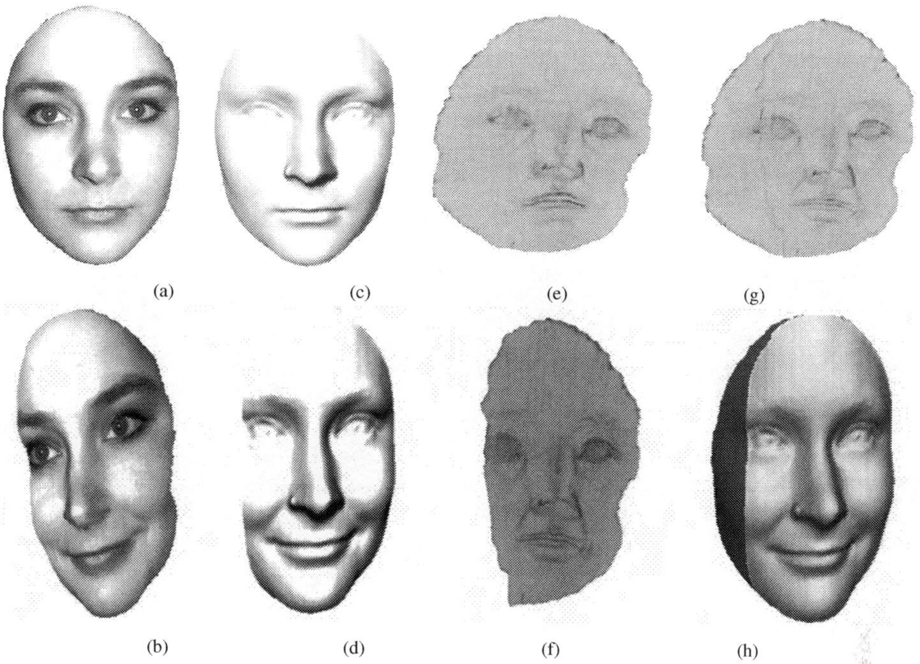

Fig. 1.36. An example of surface alignment and stitching: (a,b) Two original 3D faces with texture in different poses and deformations. (c,d) Original 3D faces without texture. (e,f) The conformal mapping images of the faces. (g) The aligned planar images of the two faces. (h) The resulting 3D face by stitching a part of (c) into (d). Courtesy of Dimitris Samaras [15].

is not very big, their planar images are very similar. Therefore, matching on the planar images of conformal mappings is more reliable and accurate than direct matching in 3D. For partial surface matching, extra constraints are needed, such as feature points, feature lines, or area distortion factors.

Shape Space

Surfaces can be classified by conformal equivalence. For closed surfaces with one handle (genus one), all conformal classes form a 2-dimensional space, namely, each conformal class can be represented by 2 real parameters. For genus $g > 1$ closed surfaces, all of the conformal equivalent classes form a $(6g - 6)$-dimensional space, which is called the Teichmüller space. The *Teichmüller coordinates* of a surface can be explicitly computed as the fingerprint of the shape, which can be applied to geometric database indexing and shape comparison purposes.

Given a pair of topological pants (a topological annulus with two holes), we can compute a unique hyperbolic metric and embed it in the hyperbolic space. The Teichmü ller coordinates of it are the geodesic lengths of three boundaries under the hyperbolic metric. Fig. 1.37 shows three such kind of surfaces and their embedding in the hyperbolic space, which is called the fundamental polygon. If the fundamental polygons are congruent in the hyperbolic space, the corresponding surfaces are conformally equivalent. It is easy to see that every surface in the figure is not conformally equivalent to any other one.

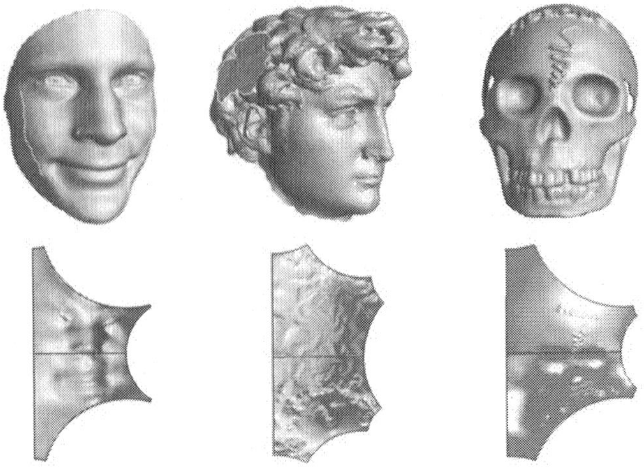

Fig. 1.37. The coordinates of topological annuli with 2 holes in the shape space are the lengths of their boundaries under the hyperbolic metric. The shape of their fundamental polygon indicates the shape space coordinates.

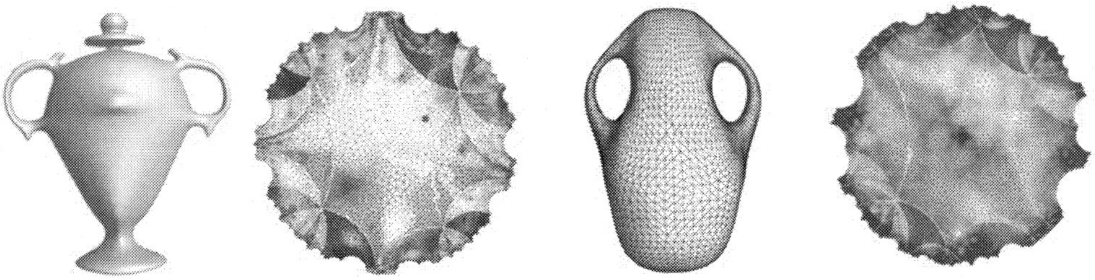

Fig. 1.38. The shape of the fundamental polygon determines the conformal class of the surface. The two genus two surfaces in the figure are not conformal equivalent.

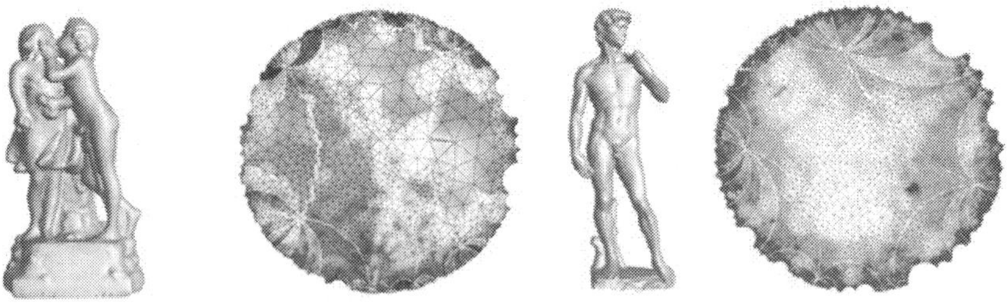

Fig. 1.39. The shape of the fundamental polygon determines the conformal class of the surface. The two genus three surfaces in the figure are not conformally equivalent.

A similar idea can be applied to surfaces with more complicated topologies. Fig. 1.38 shows the fundamental polygons of two genus two surfaces in the hyperbolic space. Because these hyperbolic polygons are not congruent, there is no conformal mapping between the two surfaces (in the given homotopy class). Fig. 1.39 shows the fundamental polygons of two genus three surfaces in the hyperbolic space, one is the sculpture model, the other is Michelangelo's David. They are not conformal equivalent either. The details for computing the fundamental polygon and the shape space coordinates are explained in detail later. From our experience, it is very rare to find two high genus surfaces in real life are conformal equivalent, unless one is obtained by the other by an isometric transformation. We believe shape classification using conformal invariants has the potential to index large scale geometric database.

1.3.3 Geometric Modeling

The surfaces obtained by 3D scanners are represented as point clouds. After geometric processing, triangular meshes are constructed. In geometric modeling fields, surfaces are usually represented as piecewise polynomials or rational polynomials with higher order continuity, called splines. The conformal geometric method is a useful tool to convert meshes to splines. For the purpose of generalizing splines from the planar domain to the manifold domain, a special atlas needs to be constructed using conformal geometric methods.

Manifold Splines

In graphics, surfaces are approximated by piecewise linear polygonal meshes, which are with C^0 continuity. In automobile and manufacturing industries, the requirements for the continuity of surface representations are much higher. In general, a surface representation with C^2 continuity is highly desirable. For example, in order to ensure the continuity of a mirror reflected image on the car body, the car body surface should be at least C^2. Therefore, surfaces are approximated by piecewise polynomials, or rational polynomials.

Conventional spline schemes are defined on planar domains and based on affine invariants, such as barycentric coordinates. Therefore, conventional splines are based on affine geometry on the plane. One of the most important properties of a spline scheme is *parametric affine invariance*. This means if we transform the parameters of a spline surface by an affine transformation and preserve the control points of the spline surface, then the shape of the surface does not change.

As we discussed before, if a surface admits an affine structure, then conventional splines can be defined on it. Fig. 1.40 demonstrates the key components of manifold spline. M is the domain surface, the spline will be defined on M. U_α, U_β are two open sets on M, $\phi_\alpha : U_\alpha \to \mathbb{R}^2$, $\phi_\beta : U_\beta \to \mathbb{R}^2$ map the open sets to the parameter planes to establish local coordinates. Any point $p \in U_\alpha \cap U_\beta$ has two sets of local coordinates, the coordinate transition function is given by $\phi_{\alpha\beta} : \phi_\alpha(U_\alpha) \longrightarrow \phi_\beta(U_\beta)$. The transition function $\phi_{\alpha\beta}$ is an affine map from $\mathbb{R}^2$ to itself.

Then we can define conventional splines on the parameter domains $\phi_\alpha(U_\alpha)$ of the local chart (U_α, ϕ_α) , $\phi_\beta(U_\beta)$ of the local chart (U_β, ϕ_β). C_α, C_β are the control nets, which are consistent for the overlapping regions. Then by the parametric affine invariance property, the shapes of the spline surfaces defined on the overlapping region are identical. In Fig. 1.40, the green patch on F has two definitions, one is on $\phi_\alpha(U_\alpha)$, the other is on

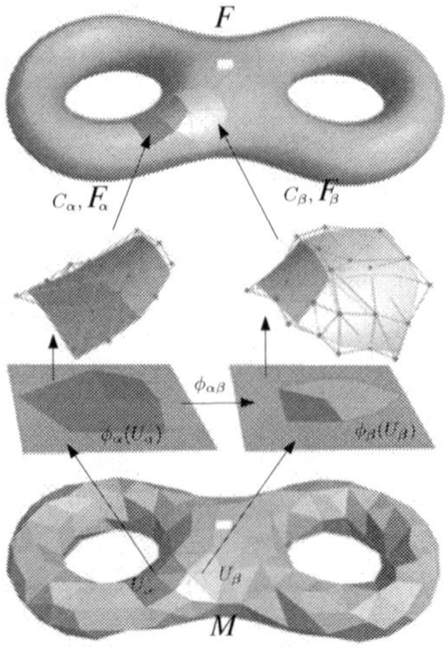

Fig. 1.40. Framework of manifold spline.

$\phi_\beta(U_\beta)$. Because the parametric transformation between $\phi_\alpha(U_\alpha)$ and $\phi_\beta(U_\beta)$ is affine, and the two control nets are identical, the two definitions are therefore consistent and induce the same shape. By this way, we can define conventional splines on each local chart and keep the consistencies of the control nets for the overlapping regions. Then the spline surface is coherently defined on the domain manifold M, and the resulting spline surface F has the desired continuity.

Fig. 1.41. Manifold spline for a genus three surface. Curtesy of Hong Qin [16].

Therefore, the key is to construct an affine atlas for the domain surface. Any surface with boundaries admits an affine structure and only genus one closed surfaces admit affine structure. In the figure, we punch two holes on the genus two surface, and construct an affine atlas using holomorphic differential forms and construct manifold splines. Holomorphic differential forms induce an affine atlas covering the whole surface except several zero points. The affine atlas can be used to construct the manifold splines. Fig. 1.41 shows a manifold triangular B-Spline defined on a genus three surface. Fig. 1.42 shows a manifold T-Spline defined on Michelangelo's David head model and the Buddha model.

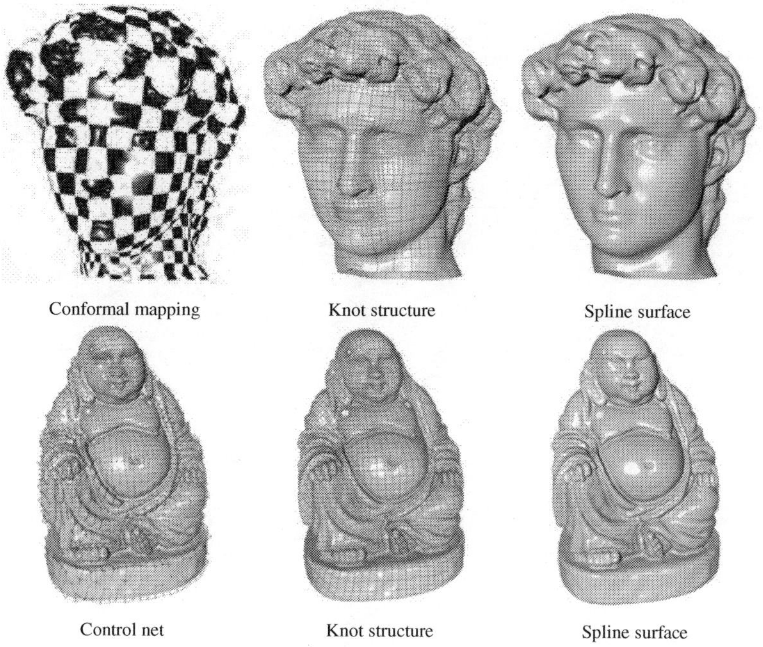

Fig. 1.42. Manifold Splines for Michelangelo's David Head and the Buddha model. Courtsey of Hong Qin and Ying He [16, 17, 18].

The theoretic framework of manifold splines is established in [19] and [16]. Then the theory is applied to generalize many spline schemes on manifolds, such as manifold T-Spline [17], triangular B-Spline [20], and polycube spline [21]. We construct manifold splines with single singularity [18], which reaches the theoretic limit.

1.3.4 Medical Imaging

With the rapid development of medical imaging technologies, a vast amount of medical imaging data are available today. In order to fuse medical images acquired from different modalities, extract surfaces or volumes, register, fuse and compare different geometric data sets, conformal geometric algorithms have been developed and proven to be valuable for real applications.

Conformal Brain Mapping

Brain imaging technology has accelerated the collection and databasing of brain maps. Computational problems arise when integrating and comparing brain data. The cortex surface of a brain is highly convoluted and the anatomical structures vary from person to person. One way to analyze and compare brain data is to map them into a canonical space while retaining geometric information on the original cortex surface as far as possible.

Cortical surfaces are of genus zero and therefore can be conformally mapped onto the unit sphere. All such conformal mappings differ by Möbius transformations of the sphere, which form a 6-dimensional group. For genus zero closed surfaces, harmonic maps are also conformal. We can obtain a conformal mapping by optimizing the harmonic energy. Further constraints are added to ensure that the conformal map is unique. Empirical tests on magnetic resonance imaging (MRI) data show that the mappings preserve angular relationships, are stable in MRIs acquired at different times, and are robust to differences in data triangulation, and resolution. Fig. 1.43 shows the conformal brain mapping of a real human cortical surface.

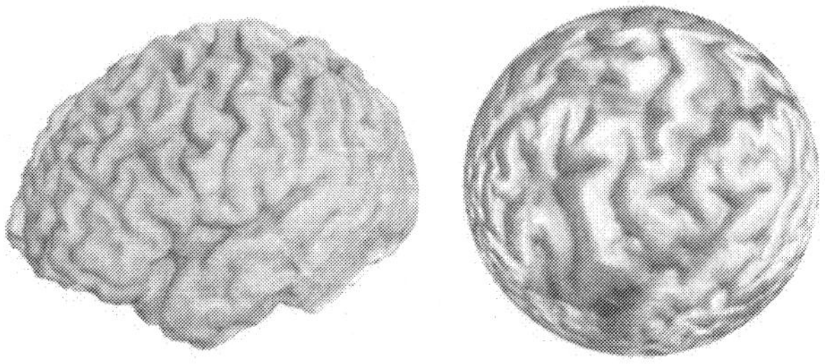

(a) Brain cortex surface (b) Conformal spherical brain mapping

Fig. 1.43. Conformal Brain Mapping. Courtesy of Tony Chan, Paul Thompson and Yalin Wang [22].

Conformal brain mapping using nonlinear heat diffusion is introduced in [22] and [23]. Spherical conformal mapping is presented in [24]. Conformal brain mapping based on Riemann surface structure is explained in [25].

Conformal Virtual Colon Flattening

Virtual colonoscopy uses computed tomographic (CT) images of a patient's abdomen and a virtual fly-through visualization system that allows the physician to navigate within a 3D model of the colon searching for polyps, the precursors of cancer. Virtual colonoscopy has been successfully demonstrated to be more convenient and efficient than the real optical colonoscopy. However, because of the length of the colon, inspecting the entire colon wall is time consuming, and prone to errors. Moreover, polyps behind folds may be hidden, which results in incomplete examinations.

Virtual dissection is an efficient visualization technique for polyp detection, in which the entire inner surface of the colon is displayed as a single 2D image. This can be visualized using checkerboard texture mapping. The colon surface is reconstructed from CT

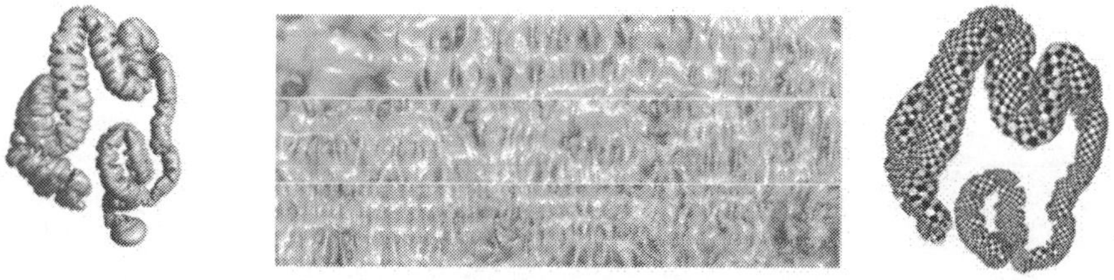

(a) Colon surface reconstructed from CT

(b) Conformal colon flattening images

(c) Conformal parameterization

Fig. 1.44. Conformal colon flattening. Courtesy of Arie Kaufman [26].

images. We developed a method for colon flattening by computing the conformal structure of the surface, and map the whole colon surface onto a rectangle. The flattening is angle preserving, so the shape of the polyps is preserved. This can be visualized using checkerboard texture mapping. It is convenient for doctors to visually locate the polyps. The colon surface is a cylinder and can be conformally mapped to the plane periodically, where the boundaries are mapped to parallel lines and each period is a rectangle. We cut the rectangle into three segments. The mapping is conformal, therefore all of the geometric details of the colon surface, such as the muscle structure, are preserved on the flattened image (Fig. 1.44). Details of conformal virtual colon flattening can be found in [26].

Further Readings

The following textbooks include the theories for algebraic topology [27], [28]; differential geometry [29] [30]; Riemann surface theory and Teichmüller theory [31], [32], [33], [34]; Geometric structure [35]; circle packing [36] and [37]; Ricci flow [38]. Advanced topics on harmonic maps on general manifolds are covered in [39]. Advanced topics on conformal metric deformation are in [40].

The research in computational conformal geometry is a very active field in recent years. Hundreds of research articles have been published in computer graphics, computer vision, visualization, geometric modeling, and other engineering fields. More than one hundred related research works can be found in the survey papers [41] and [42]. In the following chapters, we will give more references on related topics.

Part I Theories

2

Homotopy Group

2.1 Algebraic Topological Methodology

The ultimate goal of algebraic topology is to study topological problems using algebraic methods. Basically, special groups are associated with a space, and the topological properties of the space are reflected by the structures of the groups. It is just like taking a photo of the space, the negative of the photo is algebraic. The photo filters out some information, but preserves the most crucial topological information of the space. In the language of category, algebraic topological method is a functor, which maps the category of $\{topological\ space,\ topological\ map\}$ to the category of $\{group,\ homomorphism\}$.

The most important algebraic structures associated with a topological space are homology, cohomology, and homotopy group (fundamental group). We briefly explain the major ideas of this method.

Suppose a small insect is crawling on a surface. The insect cannot leave the surface and jump into the 3-dimensional space. Locally, the insect just sees a small region of the surface. Then how can the insect tell if the surface is a sphere or a donut? For us humans, it is trivial, because we have the experience of the third dimension and can "see" the tunnel or the handle of the surface. But can we ask the similar question of our universe? How do we know if the whole space-time of our universe has no voids, tunnels, or handles? How can we fully understand the topology of our universe? We cannot jump out of the whole space-time, we can not "see" the universe from outside. From this point of view, us human beings are not superior than the insect on the surface. If you are a computer scientist, you can imagine that the CPU of your computer is the insect, which has no perception of the 3rd dimensional world, you need to design an algorithm to enable your CPU to detect if the surface is a topological sphere or a topological donut.

The father of topology, Poincaré was the first person who realized the profoundness of the problem and originated the subject of topology. Suppose the insect is highly intelligent and with a very good memory. It can crawl on the surface to form a loop path, then it can walk along another loop, which is close to the original one but shrunken a little bit, therefore shorter. The insect repeats this process, keeps shrinking the loop. Eventually, there are two possible situations: first, the insect finds a loop, which can not shrink any more, then this loop must surround a tunnel or a handle; second, the loop shrinks to a point. Suppose all loops on the surface can shrink to a point, then the surface must be a genus zero surface; otherwise, the surface must have at least one handle. This method is the exact intuition behind a homotopy group. Therefore, in order to verify if a closed surface is a sphere, one can ask

Can all closed loops on the surface shrink to a point?

When Poincaré generalized the idea to the three-dimensional spaces, he raised the famous Poincaré conjecture:

If all closed loops in a closed 3-dimensional space can shrink to a point, is the space topologically equivalent to the 3-dimensional sphere?

After more than one hundred years of tremendous effort, mathematicians eventually proved the conjecture, which looks so natural and intuitive. The most powerful tool to tackle the problem is Hamilton's Ricci flow. In the later part of this book, we will introduce the theory of Ricci flow on surfaces, and develop a series of computational algorithms to compute the conformal structure of surfaces with all kinds of topologies.

Later, Poincaré generalized the idea of the homotopy group to the homology group. If a closed loop can shrink to a point, then it must be the boundary of a surface patch (a topological disk). It is obvious that the boundary of any surface patch must be closed. (The boundary of a surface patch may have several connected components, for example, the boundary of a cylinder is formed by two circles.) Then he believed that

The difference between the closed loops and the boundary loops conveys the topological information of a surface.

This is the exact intuition behind homology group. We can formulate the above idea rigorously using modern algebraic language, the greatest advantage is that

Algebra can be computed using a modern computer.

Therefore it can convert a human's mental labor for solving topological problems to sheer computational complexity of a computational algorithm.

A homotopy group is much more complicated than a homology group in general. The intrinsic reason is that a homotopy group is non-abelian (the product of an Abelian group is commutative), while a homology group is abelian. But a homotopy group reflects much more topological information than those of a homology group. For example, we can differentiate two spaces with the same homology groups by their homotopy groups.

Both homotopy and homology groups are geometric. For the surface case, they can be easily visualized or imagined. The cohomology group is hard to image, but it is more useful in topology. From a computational point of view, cohomology is much easier to compute. The intuition behind a cohomology group is trickier to explain. Suppose a function is defined on a surface, and its gradient field is a tangent vector field, then the circulation around any closed loop is zero, namely, the curl of the vector field is zero everywhere. We can ask the inverse problem: whether any curl-free vector field on the surface is a gradient field of some function.

The difference between the curl-free vector fields and the gradient fields conveys the topological information of the surface.

Cohomology is a powerful tool. Many geometric algorithms can be applied to a local region on the surface. When we extend them to the whole surface, usually we will encounter some global topological obstruction. Such an obstruction can be precisely formulated using the language of cohomology. Therefore in the studying of global geometric problems, cohomology plays a crucial role.

2.2 Surface Topological Classification

Topology considers the global properties of a space. We can imagine the surface is made of rubber. We can stretch and compress but not tear the surface, and consider those preserved properties.

Genus

For example, we have a sculpture of a girl as shown in the first picture in Fig. 2.1. Imagine the sculpture is made of elastic rubber, we can blow it up to become a sphere, as shown in the second picture in Fig. 2.1. Suppose we have another plastic sculpture, a kitten model, as shown in the first picture of Fig. 2.2, it can be deformed elastically to a donut in the second picture. By no means, we can deform the girl sculpture to the donut without tearing the surface. In this case, we can see that the girl sculpture surface is topologically equivalent to the sphere; the kitten sculpture surface is topologically equivalent to the donut. It is obvious that the donut is not equivalent to the sphere, because the donut has one "handle" and the sphere has no handle.

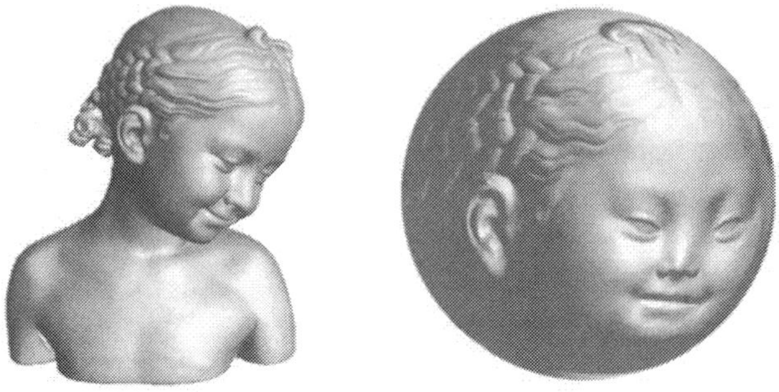

Fig. 2.1. Topological deformation between two genus zero surfaces.

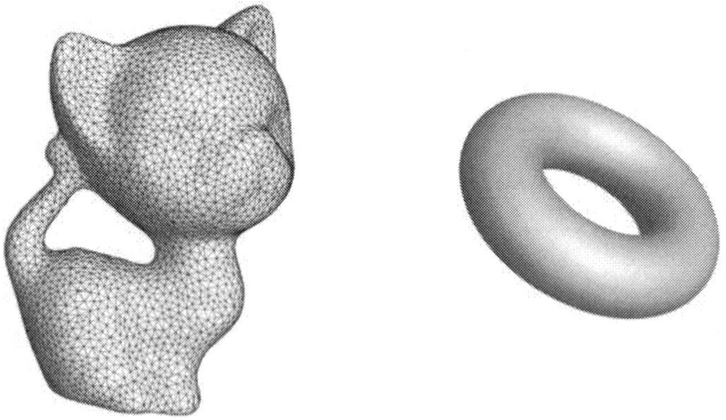

Fig. 2.2. Topological deformation between two genus one surfaces.

More complicated examples are given in Fig. 2.3, where the two surfaces in the same column are topologically equivalent and share the same number of handles. The number of handles of a surface is called the "genus" of the surface, which is the major topological invariant.

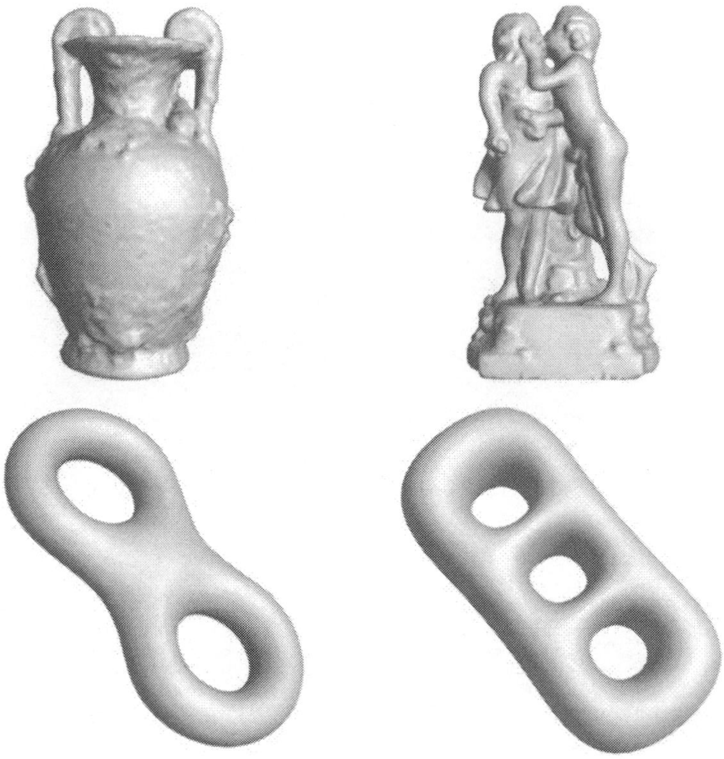

genus 2 genus 3

Fig. 2.3. Topological invariant: genus of surfaces.

Boundary

All previously illustrated surfaces are closed, namely, without boundaries. Imagine a small insect is crawling on a surface, at each point, it can see only a local region of the surface. From the insect's perspective, the local surface looks like a part of plane, and it has two degree of freedoms to move around. Therefore, we say the surface is of two dimensions. The insect has no concept of the third dimension, it can never jump off of the surface, so it cannot see the handle.

Suppose the surface is not closed, it has "holes" on it. When the insect travels to the vicinity of the edge, it can notice it. If the insect crawls along the boundary, it has one degree of freedom. That means the dimension of the boundary of a two-dimensional surface is one.

A surface with boundaries is called an *open surface*. The number of boundaries is also a topological invariant. We cannot deform a closed sphere to a disk without splitting or tearing it. In Fig. 2.4, we show the surface of a human face, and deform it to a disk on the plane. It is obvious that this surface has one boundary, and it is topologically equivalent to a planar disk.

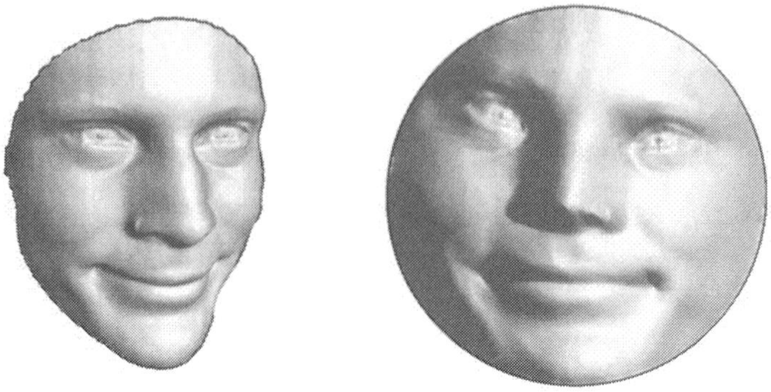

Fig. 2.4. A human face surface is with one boundary and topologically equivalent to a planar disk.

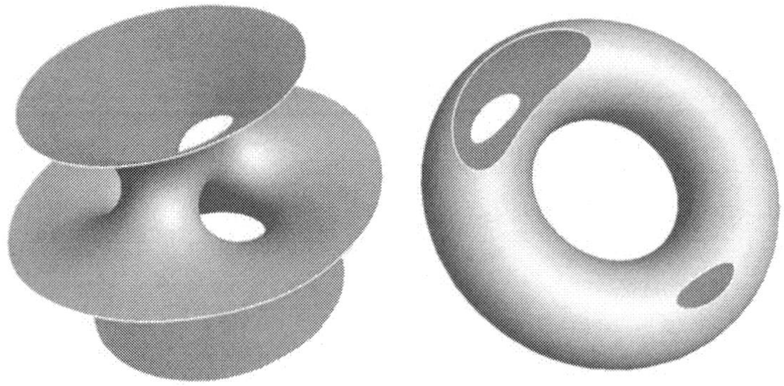

Fig. 2.5. A minimal surface is topologically equivalent to a torus with three holes.

If surfaces are with more complicated topologies, it is not so obvious to see the topological equivalence relation. The surface in Fig. 2.5, which is called Costa's minimal surface, was discovered in 1982 by Celso Costa. If we fix the boundaries of the surface in our three-dimensional Euclidean space $\mathbb{R}^3$, then Costa's minimal surface is with the minimal area. This surface is of genus one with three boundaries. It is topologically equivalent to the surface in the right column in the same figure. This example shows that if two surfaces are topologically equivalent, then they share the same the number of boundaries and the genus.

Orientability

There is another topological invariant, called "orientability". Intuitively, if a closed surface is embedded in $\mathbb{R}^3$, then it separates $\mathbb{R}^3$ into two parts. One is finite, called the interior, the other is infinite, called the exterior. Then the surface has two sides, one is inside, the other is outside. We say a surface is *orientable*, if we can differentiate its inside from its outside.

The most common non-orientable surface is the Möbius band, which is formed by joining the ends of a rectangle with a twist (180 degree). Fig. 2.6 depicts a Möbius band. A cylinder can be formed in the same way without the twist. Then it is easy to see that the cylinder has two boundaries, but the Möbius band has only one boundary. Furthermore, if an insect is crawling on the cylinder, without crossing the boundary, it cannot go from one side to the other. However, if the insect is crawling on a Möbius band, then it can touch the same point from both "sides" without crossing the boundary.

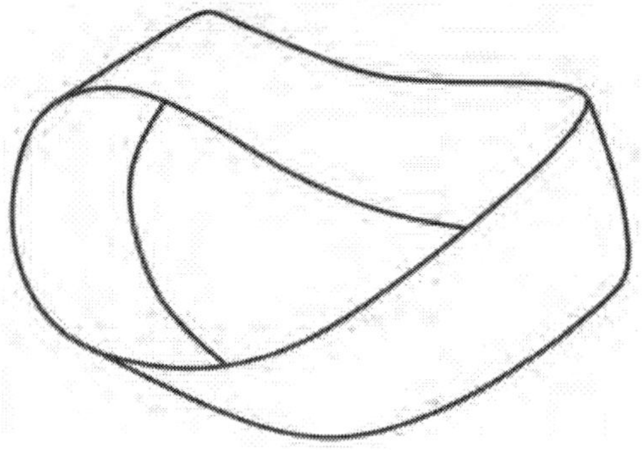

Fig. 2.6. A Möbius band is a non-orientable surface with one boundary.

Orientability is another important topological property; if two surfaces are topologically equivalent, both of them are either orientable or non-orientable. Non-orientable surfaces are rare in daily life. General surfaces are the boundary of volumes, therefore it naturally has two sides, inside and outside. In mathematics and physics, there are a lot of natural spaces which are non-orientable. For example, the space of all the lines through the origin in $\mathbb{R}^3$ is non-orientable, and the space of all rotations of $\mathbb{R}^3$ is non-orientable.

In the rest of the book, we mainly study the surfaces embedded in $\mathbb{R}^3$, therefore we assume all the surfaces are orientable, unless we explicitly state otherwise.

Surface Topological Classification

Two simple surfaces can be glued together to form a more complicated one. Basically, we can remove a small topological disk from each surface, and glue the two left pieces along the boundaries of the removed disks.

Definition 2.1 (Connected Sum). *The connected sum $S_1 \# S_2$ is formed by deleting the interior of disks $D_i \subset S_i$ and attaching the resulting punctured surfaces $S_i - D_i$ to each other by a homeomorphism*

$$h : \partial D_1 \to \partial D_2,$$

where ∂D_i represents the boundary of D_i, so

$$S_1 \# S_2 = (S_1 - D_1) \cup_h (S_2 - D_2).$$

All surfaces can be decomposed to the connected sum of simple surfaces, as shown in Fig. 2.7; the building blocks are tori for orientable surfaces, and the crosscaps for unorientable surfaces.

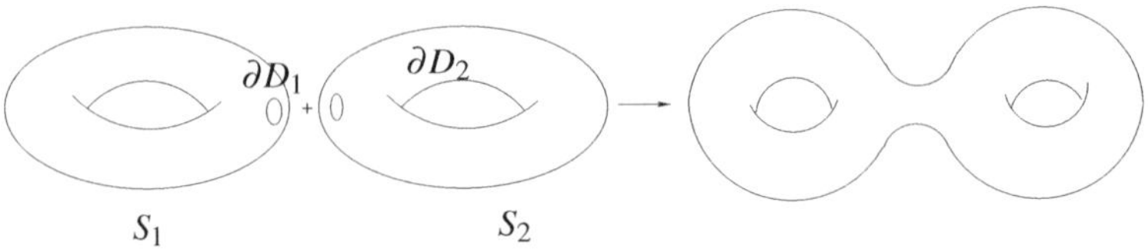

Fig. 2.7. Connected sum.

Definition 2.2 (Projective Plane). *All straight lines through the origin in $\mathbb{R}^3$ form a two-dimensional manifold, which is called the projective plane $\mathbb{R}P^2$.*

A straight line intersects the unit sphere at two antipodal points. Therefore a projective plane can be obtained by identifying two antipodal points of the unit sphere. A projective plane with a hole is called a *crosscap*.

The basic topological structures of all surfaces are described in the following theorem.

Theorem 2.3 (Classification Theorem for Surfaces). *Any closed connected surface is homeomorphic to exactly one of the following surfaces: a sphere, a finite connected sum of tori, or a sphere with a finite number of disjoint discs removed and with crosscaps glued in their place. The sphere and connected sums of tori are orientable surfaces, whereas surfaces with crosscaps are unorientable.*

Any closed surface S is the connected sum

$$S = S_1 \# S_2 \# \cdots \quad S_g.$$

If S is orientable and of genus g, then each S_i is a torus T^2. If S is unorientable and of genus g, then S_i is a projective plane $\mathbb{R}P^2$.

2.3 Homotopy of Continuous Mappings

Intuitively, two continuous maps $f_0, f_1 : M \to N$ are said to be homotopic if there is an intermediate family of continuous maps $f_t : M \to N$, for $0 \leqslant t \leqslant 1$ which vary continuously with respect to t.

Definition 2.4 (Homotopy). *Two continuous maps $f_0, f_1 : M \to N$ are said to be* homotopic *if there is a continuous map $F : M \times I \to N$ such that $F(\cdot,0) = f_0$ and $F(\cdot,1) = f_1$. The map F is called a* homotopy *between f_0 and f_1, denoted as $f_0 \cong f_1$ or $F : f_0 \cong f_1$. For each $t \in [0,1]$, we denote $F(\cdot,t)$ by $f_t : M \to N$, where f_t is a continuous map.*

It is easy to verify that the relation $\cong$ on the set of continuous maps from M to N is an equivalence relation. We can use the concept of homotopic maps to classify topological spaces.

Suppose that K is a subset of M and that f_0 and f_1 are two continuous maps from M to N, we say that f_0 and f_1 are *homotopic relative* to K if there is a homotopy $F : M \times I \to N$ such that $F(p,t) = f_0(p), \forall p \in K, \forall t \in I$.

Definition 2.5 (Homotopy Equivalence). *Two topological spaces M and N are homotopy equivalent or of the* same homotopy type *if there exist continuous maps $f : M \to N$ and $g : N \to M$, such that*

$$g \circ f \cong id_M : M \to M, \quad f \circ g \cong id_N : N \to N.$$

The maps f and g are called the homotopy equivalences.

Suppose $N \subset M$ is a *strong deformation retract* of M if there is a homotopy $F : M \times I \to N$, such that for all $p \in M$, $F(p,0) = p$, for all $p \in N$, $t \in I$, $F(p,t) = p$, and $F(p,1) \in N, \forall p \in M$. A topological space M and its strong deformation retract are of the same homotopy type. For example, an open disk can retract to a point, and an open annulus can retract to a circle.

2.4 Homotopy Group

A map $f : [0,1] \to M$ from the unit interval to a topological space M is called a *path* in M. If f and g are two paths in M with $f(1) = g(0)$, then the *product* of f and g is a path $f \cdot g$, which is defined as

$$f \cdot g(t) = \begin{cases} f(2t), & 0 \leqslant t \leqslant \frac{1}{2}, \\ g(2t-1), & \frac{1}{2} \leqslant t \leqslant 1. \end{cases}$$

Definition 2.6 (Homotopic Paths). *Two paths f, g in M are said to be* equivalent, *if f and g are homotopic relative to $\{0,1\}$, denoted as $f \cong g$.*

On a surface, two closed curves on a surface are homotopic to each other, if they can deform to each other without leaving the surface, as shown in Fig. 2.8.

$\cong$ is an equivalent relation on the set of paths in M. We denote the equivalence class of a path γ by $[\gamma]$. We can define the product of equivalence classes of paths as

$$[f][g] = [f \cdot g].$$

It can be easily verified that the product of path classes is well defined, namely, it is independent of the representative path among the equivalence class. We can also easily verify that the multiplication of equivalence classes of paths is associative

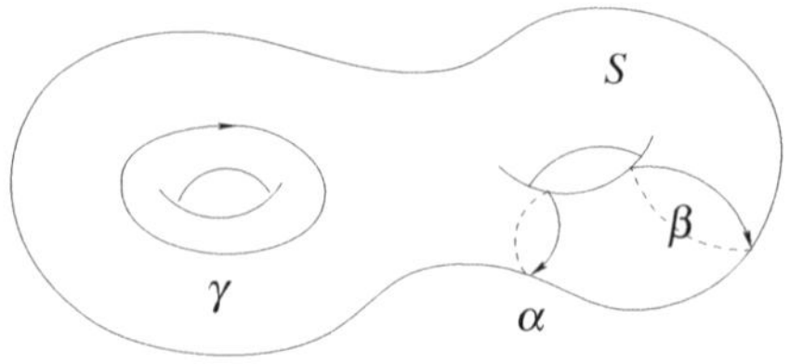

Fig. 2.8. α is homotopic to β, not homotopic to γ.

$$([f][g])[h] = [f]([g][h]).$$

We say f is a *closed path (or a loop) based at* p, if $f(0) = f(1) = p \in M$. We define $\varepsilon_p : I \to M$ as the constant path, that is $\varepsilon_p(t) = p$. Then it is obvious that

$$[f][\varepsilon_p] = [f] = [\varepsilon_p][f].$$

We further define the inverse of a path $f^{-1}(t) = f(1-t)$. Then it is obvious

$$[f][f^{-1}] = [\varepsilon_p] = [f^{-1}][f].$$

We denote the set of equivalence classes of closed paths based at $p \in M$ by $\pi(M,p)$. From the above discussion, we see that $\pi(M,p)$ form a group, which is called the *fundamental group* or the *homotopy group* of M.

Let $p,q \in M$. If there is a path γ from p to q, then groups $\pi(M,p)$ and $\pi(N,q)$ are isomorphic, $u_\gamma : \pi(M,p) \to \pi(M,q)$,

$$u_\gamma[g] = [\gamma \cdot g \cdot \gamma^{-1}]. \tag{2.1}$$

Therefore, we usually omit the base point for path connected spaces.

2.5 Homotopy Invariant

Let $\phi : M \to N$ be a continuous map. Then if $[\gamma] \in \pi(M,p)$, then $[\phi\gamma]$ is a well-defined element of $\pi(N,\phi(p))$, therefore, the map

$$\phi_* : \pi(M,p) \to \pi(N,\phi(p))$$

is well defined by

$$\phi_*[\gamma] = [\phi\gamma]. \tag{2.2}$$

$\phi_*([f][g]) = \phi_*[f \cdot g] = [\phi(f \cdot g)] = [\phi f \cdot \phi g] = [\phi f][\phi g] = \phi_*[f]\phi_*[g]$. Therefore ϕ_* is a homomorphism of groups, which is called the induced homomorphism. If $\phi : M \to N$ is a homeomorphism, then $\phi_* : \pi(M,p) \to \pi(N,\phi(p))$ is an isomorphism. Suppose $\phi : M \to N$, $\psi : N \to K$ are continuous maps. Then $(\psi\phi)_* = \psi_*\phi_*$. The homomorphism id_* induced by the identity map $id : M \to M$ is the identity homomorphism on $\pi(M,p)$.

The fundamental group offers a tool to convert topology to algebra. The process has the following characteristics.

1. For each topological space, a fundamental group is defined.
2. For each continuous map between topological spaces, a homomorphism between groups is induced.
3. The composition of continuous maps induces the composition of the induced homomorphisms.
4. The identity map induces the identity homomorphism.
5. A homeomorphism induces an isomorphism.

In category language, the fundamental group is a functor from the topological category {topological spaces, continuous maps} to the algebraic category {groups, homomorphisms}. We can learn partial information of topology by algebraic means.

Theorem 2.7. *Let $\phi, \psi : M \to N$ be continuous mappings between topological spaces M and N. Let $F : \phi \cong \psi$ be a homotopy. If $f : I \to N$ is the path from $\phi(p)$ to $\psi(p)$, given by $f(t) = F(p,t)$, then the homomorphisms*

$$\phi_* : \pi(M,p) \to \pi(N,\phi(p))$$

and

$$\psi_* : \pi(M,p) \to \pi(N,\psi(p))$$

are related by $\psi_* = u_f \phi_*$, where u_f is the isomorphism from $\pi(N,\phi(p))$ to $\pi(N,\psi(p))$ defined by the path f (in equation 2.1).

Namely, the following diagram is commutative:

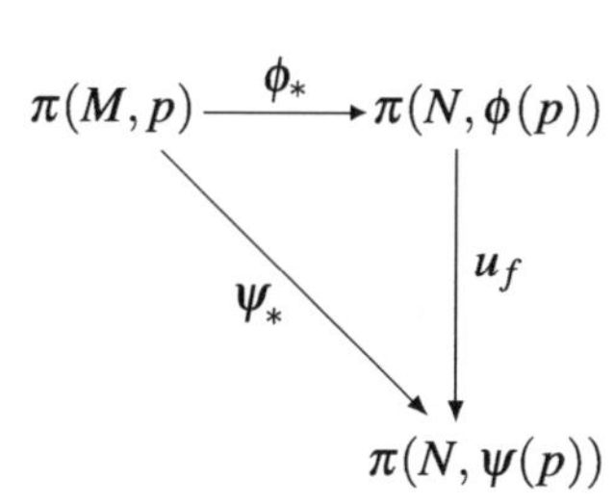

Furthermore, we can talk about the homotopy equivalent spaces.

Theorem 2.8. *If $\phi : M \to N$ is a homotopy equivalence, then $\phi_* : \pi(M,p) \to \pi(N,\pi(p))$ is an isomorphism for any $p \in M$.*

Hence, topological spaces with the same homotopy type have the isomorphic fundamental groups. If a space can shrink to a point, then its fundamental group is trivial. In general, a space and its strong deformation retract share the same fundamental group.

2.6 Covering Spaces

Definition 2.9 (Covering Space). *Let $\mathbf{p} : \tilde{M} \to M$ be a continuous map, p is onto and for all $q \in M$, there is an open neighborhood U of q such that*

$$\mathbf{p}^{-1}(U) = \cup_{j \in J} U_j,$$

for some collection $\{U_j, j \in J\}$ *of subsets of* $\tilde{M}$*, satisfying* $U_j \cap U_k = \emptyset$ *if* $j \neq k$*, and with* $\mathbf{p}|_{U_j} : U_j \to U$ *a homeomorphism for each* $j \in J$. $\mathbf{p} : \tilde{M} \to M$ *is a* covering.

Namely, locally a covering map is a homeomorphism, so all of the local properties of M can be "lifted" to $\tilde{M}$.

Definition 2.10 (Lift). *Suppose* $p : \tilde{N} \to N$ *is a covering, and* $f : M \to N$ *is a continuous map. Then a* lift *of* f *is a continuous map* $\tilde{f} : M \to \tilde{N}$ *such that* $p \circ \tilde{f} = f$.

In other word, the following diagram is commutative:

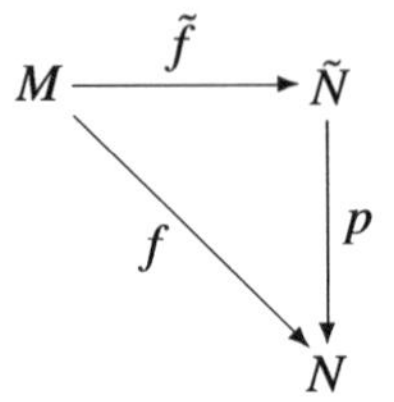

Globally, $\tilde{M}$ and M have different topologies. If $\tilde{M}$ has simpler topology, then $\tilde{f}$ is easier to study than f. In general, the map f may have different lifts. Suppose $\tilde{f}_1, \tilde{f}_2 : M \to \tilde{N}$ are two lifts of $f : M \to N$. If M is connected, then if $\tilde{f}_1(q) = \tilde{f}_2(q)$ for some $q \in M$, then $\tilde{f}_1 = \tilde{f}_2$.

Definition 2.11 (Deck Transformation). *Suppose* $p : \tilde{M} \to M$ *is a covering. Then an automorphism* $\tau : \tilde{M} \to \tilde{M}$ *is called a* deck transformation *if*

$$\mathbf{p} \circ \tau = \mathbf{p}.$$

A deck transformation makes the following diagram commute.

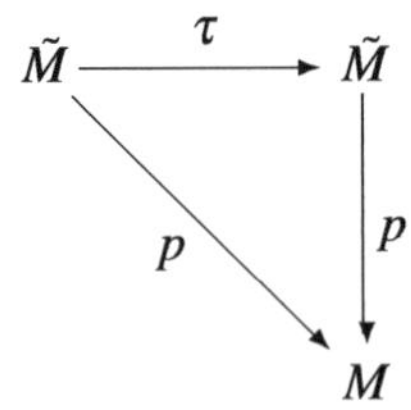

All of the deck transformations form a group $Deck(\tilde{M})$, the *deck transformation group*. M is homeomorphic to the quotient space

$$\tilde{M}/Deck(\tilde{M}) \cong M.$$

Definition 2.12 (Fundamental Domain). *A closed subset $D \in \tilde{M}$ is called a* fundamental domain *of the $Deck(\tilde{M})$, if $\tilde{M}$ is the union of conjugates of D,*

$$\tilde{M} = \bigcup_{\tau \in Deck} \tau D,$$

and the intersection of any two conjugates has no interior.

The fundamental groups and deck transformation groups satisfy the following relation.

Theorem 2.13 (Fundamental Group of Covering Space). *Suppose M is path connected, $\mathbf{p} : \tilde{M} \to M$ is a covering. Then $\mathbf{p}_* \pi(\tilde{M}, \tilde{q})$ is a normal subgroup of $\pi(M,q)$, where $p(\tilde{q}) = q$,*

$$\pi(M,q)/\mathbf{p}_*\pi(\tilde{M},\tilde{q}) \cong Deck(\tilde{M}).$$

On the other hand, for any normal subgroup H of $\pi(M,q)$, there exists a covering $\mathbf{p} : \tilde{M} \to M$, such that $\mathbf{p}_*\pi(\tilde{M})$ is isomorphic to H. Especially, e is a normal subgroup of $\pi(M,q)$, therefore there exists a covering $\mathbf{p} : \tilde{M} \to M$, such that $\tilde{M}$ is simply connected. Such a kind of covering space is universal, and extremely useful in practice.

Definition 2.14 (Universal Covering). *Suppose $\mathbf{p} : \tilde{M} \to M$ is a covering. If $\tilde{M}$ is simply connected ($\pi(\tilde{M}, \tilde{q}) = e$), then the covering is a* universal covering.

Theorem 2.15 (Universal Covering Space for Surfaces). *The universal covering spaces of orientable closed surfaces are the sphere (genus zero), plane (genus one), and disk (high genus).*

Fig. 2.9 shows the universal covering spaces of all orientable surfaces.

Suppose $q \in M$. Then there is a one-to-one correspondence between the pre-images of q, $p^{-1}(q) \subset \tilde{M}$, and the elements in $\pi(M,q)$,

$$\phi : p^{-1}(q) \to \pi(M,q).$$

First we fix a point $\tilde{q}_0 \in p^{-1}(q)$, for any $\tilde{q}_k \in p^{-1}(q)$, we can find a path $\tilde{\gamma} : I \to \tilde{M}$ connecting $\tilde{q}_0$ and $\tilde{q}_k$. Then the projection of $\tilde{\gamma}$ is a loop on M. We match $\tilde{q}_k$ with the homotopy class of that loop, namely

$$\phi(\tilde{q}_k) = [p(\tilde{\gamma})].$$

In order to show that ϕ is well defined, we need to show that the homotopy class $[p(\tilde{\gamma})]$ is independent of the choice of $\tilde{\gamma}$. This is because since $\tilde{M}$ is simply connected, any two paths connecting $\tilde{q}_0$ and $\tilde{q}_k$ are homotopic, and their projections are also homotopic.

Any nontrivial closed loop γ on M is lifted to an open curve $\tilde{\gamma}$ on $\tilde{M}$. The homotopy class of γ is determined by the starting and ending points of $\tilde{\gamma}$, as shown in Fig. 2.10.

Suppose we have a surface embedded in $\mathbb{R}^3$, γ is a loop on the surface. We would like to find the shortest loop in the homotopy class $[\gamma]$. This is a challenging problem. By lifting γ to the universal covering space $\tilde{M}$, the problem is converted to find the shortest path, which is much easier to handle. A detailed algorithm can be found in [43]. Fig. 2.11 shows one such an example, the shortest loop on a genus one kitten surface is lifted to the shortest path on its universal covering.

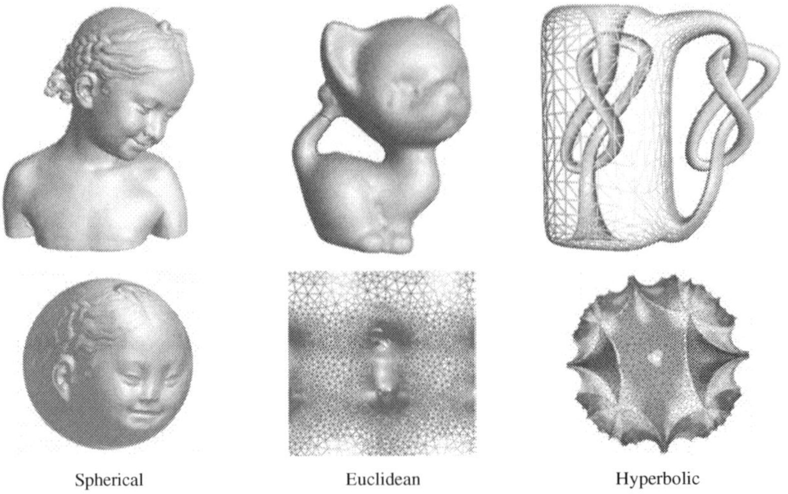

Fig. 2.9. The universal covering spaces of all closed orientable surfaces are sphere, plane or the disk.

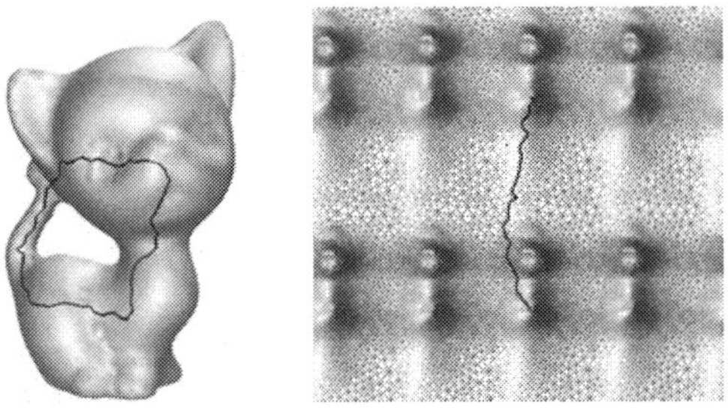

Fig. 2.10. A loop on the surface is lifted to a path on the universal covering.

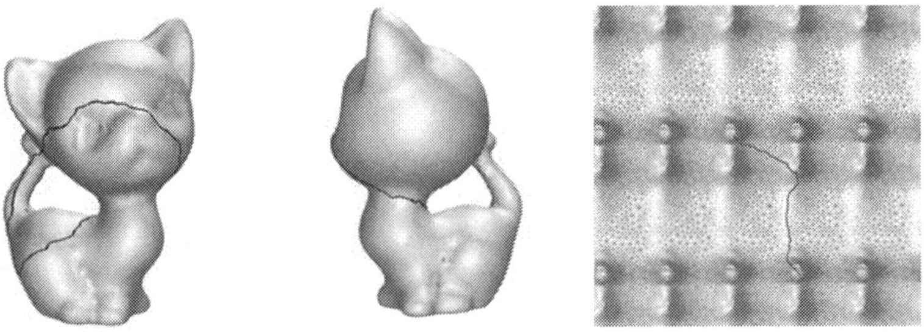

Fig. 2.11. Shortest loop on the surface is lifted as the shortest path on the universal covering.

2.7 Group Representation

Suppose $S = \{x_1, x_2, \cdots, x_n\}$ is a set of non-commutative symbols, using these symbols to form words,

$$w = x_{i_1}^{e_1} x_{i_2}^{e_2} \cdots x_{i_n}^{e_n},$$

where $x_{i_k} \in S$ and $e_k \in \{+1, -1\}$. A reduced word contains no pair $x_i x_i^{-1}$ or $x_i^{-1} x_i$. Using juxtaposition of reduced words as a law of composition, then the set of reduced words form a group. The empty word acts as the identity. The group is the *free group generated by S*. Suppose R is a set of words in S, two words w_1, w_2 are equivalent if w_1 can be obtained by finite steps of the following operations: inserting or deleting a word in R, or $x_i x_i^{-1}$ or $x_i^{-1} x_i$. All the equivalent word class form a group using juxtaposition as the composition. The group is denoted as $\langle S; R \rangle$. S is called the *generator*, R is called the *relation*.

Given two groups $\langle S_1; R_1 \rangle$ and $\langle S_2; R_2 \rangle$, by symbolic manipulation, one can verify whether they are isomorphic. But in general, the computation is NP-hard.

2.8 Seifert-van Kampen Theorem

Suppose a topological space $M = U_1 \cup U_2$, both U_1 and U_2 are path connected with $U_1 \cap U_2$ non-empty and path connected. Let $\phi_1, \phi_2, \psi_1, \psi_2$ denote the inclusion maps as shown in the following diagram:

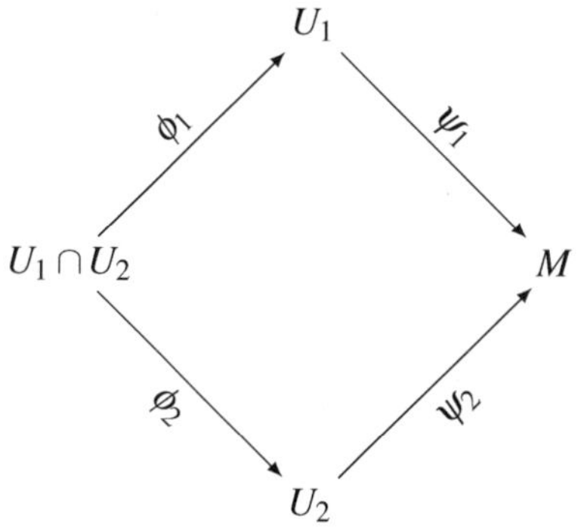

select a base point $q \in U_1 \cap U_2$. Then the following diagram of homomorphisms are commutative:

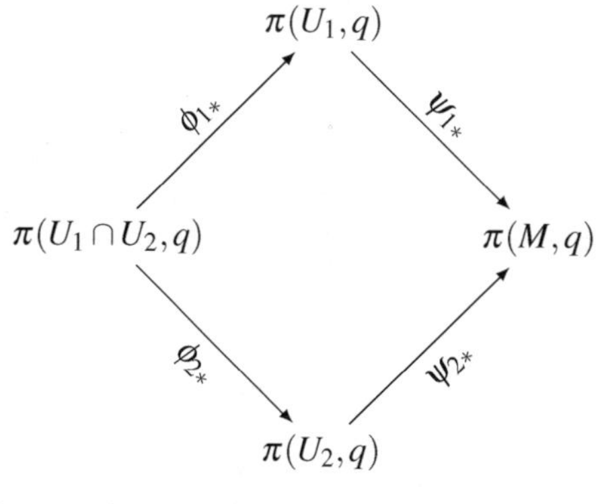

Suppose that the fundamental groups are

$$\begin{aligned}\pi(U_1\cap U_2,q) &= \langle S;R\rangle,\\ \pi(U_1,q) &= \langle S_1;R_1\rangle,\\ \pi(U_2,q) &= \langle S_2;R_2\rangle.\end{aligned}$$

Let R_S denote the following set of words:

$$R_S = \{(\phi_{1*}s)(\phi_{2*}s)^{-1}, \forall s\in S\}$$

Theorem 2.16 (Seifert-van Kampen). *$\pi(M,q)$ is isomorphic to the group defined by the generators $S_1\cup S_2$ and the relations $R_1\cup R_2\cup R_S$, $\langle S_1\cup S_2;R_1\cup R_2\cup R_S\rangle$.*

This is the important tool needed to compute homotopy groups of general spaces. In the following, we compute the fundamental group of general closed surfaces.

Definition 2.17 (Canonical Fundamental Group Basis). *A fundamental group basis $\{a_1,b_1,a_2,b_2,\cdots,a_g,b_g\}$ is canonical, if*

1. *a_i and b_i intersect at the same point p.*
2. *a_i and a_j, b_i and b_j only touch at p.*

The surface M can be sliced along a set of canonical basis and form a simply connected patch D, the fundamental domain. The fundamental domain is with the boundary

$$a_1b_1a_1^{-1}b_1^{-1}a_2b_2a_2^{-1}b_2^{-1}\cdots a_gb_ga_g^{-1}b_g^{-1}.$$

shown in Fig. 2.12.

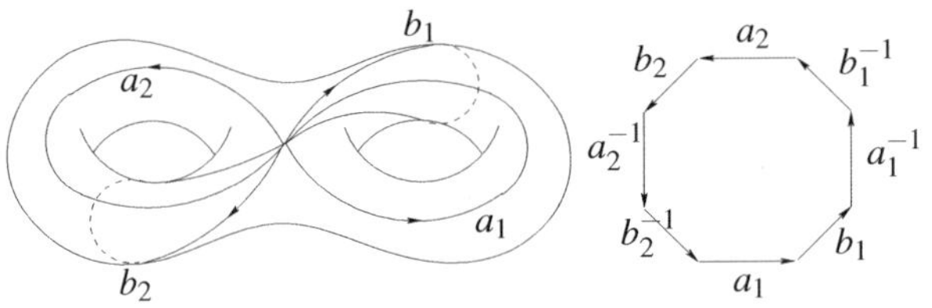

Fig. 2.12. Canonical basis of fundamental group $\pi_1(M,p_0)$.

Now let q be an interior point of D, let U_1 be M with q removed, and let U_2 be the interior of D. Then $M=U_1\cup U_2$, $U_1\cap U_2$ can retract to a circle S^1. The strong deformation retract of U_1 is the $2g$ loops a_i,b_i's sharing a common point. By using Seifert-van Kampen theorem, it is obvious that

$$\pi(U_1) = \langle\{a_1,b_1,a_2,b_2,\cdots,a_g,b_g\},\emptyset\rangle$$

and $\pi(U_2)=e$. Let γ be the loop surrounding the point q in $U_1\cap U_2$. Then

$$\pi_{1*}([\gamma]) = a_1b_1a_1^{-1}b_1^{-1}a_2b_2a_2^{-1}b_2^{-1}\cdots a_gb_ga_g^{-1}b_g^{-1}$$

and

$$\pi_{2*}=e,$$

therefore,

$$\pi(M,q) = \langle \{a_1,b_1,a_2,b_2,\cdots,a_g,b_g\};\{a_1b_1a_1^{-1}b_1^{-1}a_2b_2a_2^{-1}b_2^{-1}\cdots a_gb_ga_g^{-1}b_g^{-1}\}\rangle.$$

For general surfaces, the following theorem holds.

Theorem 2.18 (Fundamental Groups of General Surfaces). *The fundamental group of the surface M,*

$$M = \mathbb{S}^2 \# mT^2 \# n\mathbb{R}\mathrm{P}^2$$

is the group with generators

$$a_1,b_1,a_2,b_2,\cdots,a_m,b_m,f_1,f_2,\cdots,f_n$$

and one relation

$$a_1b_1a_1^{-1}b_1^{-1}a_2b_2a_2^{-1}b_2^{-1}\cdots a_mb_ma_m^{-1}b_m^{-1}f_1^2f_2^2\cdots f_n^2 = e.$$

Problems

2.1. Fundamental Group of Graph
(a) What is the fundamental group of a tree T?
(b) Design an algorithm to compute the fundamental group $\pi(G)$ of a graph G.

2.2. Fundamental Group of Surfaces
Suppose S is a sphere with q disk removed, glued with q crosscaps,
(a) Prove that the fundamental group of S has q generators $f_1,f_2,\cdots,f_q$.
(b) Prove that the generators have the relation

$$f_1^2f_2^2\cdots f_q^2 = 1.$$

2.3. Covering Space
(a) Construct two different covering spaces of a torus.
(b) If we remove a disk from the torus, construct the universal covering space of the torus with a hole.

2.4. Universal Covering
(a) Let S be a multi-holed annulus, construct the universal covering of S.
(b) If S is of genus g and with b boundaries, construct its universal covering.

2.5. Deck Transformation
Suppose S is a genus two surface,
(a) Construct a two sheet covering space, $(\tilde{S},\mathbf{p})$.
(b) Construct a deck transformation

$$\tilde{f}:\tilde{S}\to\tilde{S},$$

such that $\tilde{f}$ is not an identity map.

2.6. Structures on the Covering Space
Suppose S is a topological space, $(\tilde{S}, \mathbf{p})$ is its universal covering space, show that
(a) if S is a manifold, so is its universal cover $\tilde{S}$;
(b) if S is a Riemann surface, so is its universal cover $\tilde{S}$, and π is a holomorphic map;
(c) if S is a Riemannian manifold, so is its universal cover $\tilde{S}$, and π is an isometric map.

2.7. Fundamental Group of Double Covering
Suppose S is a surface with b boundaries and g handles. We double cover S to get $(\tilde{S}, \mathbf{p})$.
(a) Compute the fundamental group of S, $\pi(S)$.
(b) Represent the fundamental group of $\tilde{S}$ using $\pi(S)$.

2.8. Knot
The figure at the beginning of this chapter shows a knot surface. The surface separates $\mathbb{R}^3$ into two parts, the inside volume I and the outside volume O.
(a) Compute its fundamental group.
(b) Compute the fundamental group of I.
(c) Compute the fundamental group of O.
(d) If the knot is a common torus, which of the above three fundamental groups will be different?

2.9. Topological Equivalence
Prove that if two surfaces have the same fundamental group, then they are homeomorphic to each other.

2.10. Milnor's Theorem on Surfaces
Suppose S is a compact surface. Prove the following proposition. If S has nonnegative Gaussian curvature, then its fundamental group has at most polynomial growth. On the other hand, if S has negative curvature, then its fundamental group has exponential growth in the sense that $n(\lambda)$ grows exponentially, where $n(\lambda)$ is (essentially) the number of different "words" of length λ which can be made in the fundamental group.

3

Homology and Cohomology

This chapter focuses on the theories of homology and cohomology groups for general surfaces. The content of this chapter belongs to algebraic topology. The basic idea is to convert the topological problems to algebraic problems and use simple linear algebraic methods to solve them.

3.1 Simplicial Homology

The basic approach to studying topological space is to decompose the space into a collection of simplest pieces. The coherent way of gluing them together implies the most important topological information.

3.1.1 Simplicial Complex

All smooth surfaces embedded in the Euclidean space can be triangulated. In engineering fields, most surfaces are approximated by piecewise linear triangular meshes. If the triangulation is very refined, then the triangular mesh is very close to the original smooth surface. Fig. 3.1 illustrates the surface of Michelangelo's David head approximated by a triangular mesh.

Definition 3.1 (Simplex). *Suppose* $k+1$ *points* $\{v_0, v_1, \cdots, v_k\}$ *are in general positions in* $\mathbb{R}^n, n \geqslant k+1$, *the* standard simplex $[v_0, v_1, \cdots, v_k]$ *is the minimal convex set including all of them,*

$$\sigma = [v_0, v_1, \cdots, v_k] = \left\{ x \in \mathbb{R}^n | x = \sum_{i=0}^{k} \lambda_i v_i, \sum_{i=0}^{k} \lambda_i = 1, \lambda_i \geqslant 0 \right\},$$

we call $v_0, v_1, \cdots, v_k$ *as the* vertices *of the simplex* σ. *Suppose* $\tau \subset \sigma$ *is also a simplex, then we say* τ *is a* facet *of* σ.

Simplexes are oriented. Each simplex has two orientations, defined as the following:

Fig. 3.1. A smooth surface is approximated by a piecewise linear triangular mesh.

Definition 3.2 (Orientation of a Simplex). *Suppose $k+1$ points $\{v_0, v_1, \cdots, v_k\}$ are in the general positions in $\mathbb{R}^n, n \geqslant k+1$. The orientation of a simplex $[v_{i_0}, v_{i_1}, \cdots, v_{i_k}]$ is positive, if the permutation $(i_0, i_1, \cdots, i_k)$ differs from $(0, 1, \cdots, k)$ by an even number of two-element swaps; otherwise, the orientation is negative.*

Simplexes can be coherently glued together to form complexes.

Definition 3.3 (Simplicial Complex). *A* simplicial complex *Σ is a union of simplices, such that*

1. *If a simplex σ belongs to K, then all its facets also belongs to Σ.*
2. *If $\sigma_1, \sigma_2 \subset K$, $\sigma_1 \cap \sigma_2 \neq \emptyset$, then the intersection of σ_1 and σ_2 is also a common facet.*

The commonly used triangular meshes in digital geometry processing are simplicial complexes. The vertex, oriented edges, and oriented faces are 0-simplexes, 1-simplexes, and 2-simplexes respectively. Fig. 3.2 demonstrates a smooth surface is approximated by different triangular meshes with different resolutions.

3.1.2 Geometric Approximation Accuracy

The approximation accuracy can be measured by different metrics. For example, we can find the pair of closest points, one is on the original surface, the other is on the mesh, and use the distance between them as the distance between the surface and the mesh.

Definition 3.4 (Hausdorff Distance). *Let $M_1, M_2 \subset \mathbb{R}^3$ be non empty point sets, the* Hausdorff distance *between M_1 and M_2 is defined as*

$$d_H(M_1, M_2) = inf\{\varepsilon > 0 | M_1 \subset U_\varepsilon(M_2), M_2 \subset U_\varepsilon(M_1)\},$$

where $U_\varepsilon(M) = \{x \in \mathbb{R}^3 | \exists y \in M : d(x,y) < \varepsilon\}$.

In real applications, the shading of surfaces is the major interest, which is mainly determined by the normal of the surface. Therefore, the approximation accuracy of normals is more important.

Suppose S is a smooth surface embedded in $\mathbb{R}^3$ and M is a triangular mesh approximating S. Then we can define the *shortest distance map* as follows.

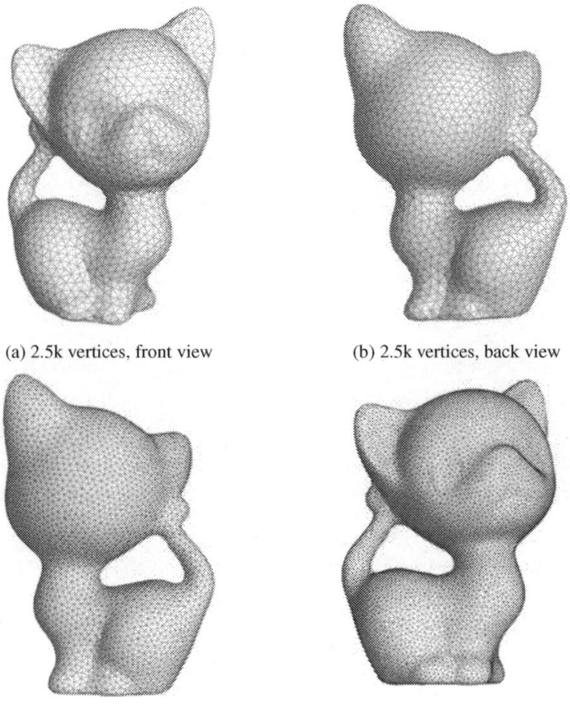

(a) 2.5k vertices, front view (b) 2.5k vertices, back view

(c) 10k vertices, back view (d) 20k vertices, front view

Fig. 3.2. Approximating a smooth surface by triangular meshes with different resolutions.

Definition 3.5 (Shortest Distance Map). *Suppose S and M are two surfaces embedded in $\mathbb{R}^3$. The shortest distance map $g : M \to S$ is defined to map $p \in M$ to its nearest point $g(p)$ on S,*

$$g : M \to S, \quad s.t. \quad d(p, g(p)) = \inf_{q \in S} |q - p|,$$

where $|q - p|$ is the Euclidean distance between points p and q.

It is proven in [44] that if the Hausdorff distance between S and M are small enough, then the shortest distance map is well-defined and is a homeomorphism.

The metric to measure the normal difference between two surfaces is defined as follows.

Definition 3.6 (Normal Distance). *Suppose a smooth surface S and a piecewise linear mesh M are embedded in $\mathbb{R}^3$ with a Hausdorff distance small enough, the normal distance between them is*

$$d_n(S, M) = \max_{p \in S} |\mathbf{n}(p) - \mathbf{n}(g^{-1}(p))|.$$

Furthermore, because surfaces are embedded in the Euclidean space, the lengths of curves on the surfaces can be measured. We represent a curve as a map

$$\gamma : [0, 1] \to S.$$

Then the length of the curve can be computed directly as

$$l(\gamma) := \int_0^1 \left| \frac{d\gamma(t)}{dt} \right| dt.$$

We can define the metric distance between two surfaces as

Definition 3.7 (Metric Distance). *Suppose a smooth surface S and a piecewise linear mesh M are embedded in $\mathbb{R}^3$ with a Hausdorff distance small enough, the metric distance between them is*

$$d_g(S,M) = \sup\left\{ \left| \frac{l(\gamma)}{l(g(\gamma))} - 1 \right| \forall \gamma : [0,1] \to M \right\}.$$

Given any two points p and q on a compact smooth surface S, if the two points are close enough, we can find a unique path connecting them on the surface, such that among all of the paths, this one is with the shortest distance, we call it a *geodesic* from p to q. Given a point $p \in S$ on the surface, and a small positive number $r \in \mathbb{R}^+$, we can define a *geodesic* circle centered at p with radius r, which is the locus of points whose geodesic distance to p equals to r.

Definition 3.8 (Geodesic Delaunay Triangulation). *Suppose S is a smooth surface embedded in $\mathbb{R}^3$, $P \subset S$ is a sample point set on S. T is a triangulation of P. T is called a* Geodesic Delaunay Triangulation, *if all of the edges of T are geodesics and the circum-circle for each geodesic triangle contains no vertices of the triangulation in its interior.*

Fig. 3.3 shows one example of a planar Delaunay triangulation. The circum-circles of two faces are depicted, which contain no vertices in their interiors. If we replace the Euclidean line segments by geodesics, then we can naturally generalize the planar Delaunay triangulation to the geodesic Delaunay triangulation.

Definition 3.9 (Medial Axis). *The medial axis of a surface S embedded in $\mathbb{R}^3$ is the closure of the set of points with more than one nearest neighbor in S.*

Leibon [45] proved that if the sample point set P is dense enough, then the geodesic Delaunay triangulation uniquely exists. The density requirement of P is determined by the principle curvature of the surface and the distance between the surface and its medial axis, the so-called *local feature size*.

A fundamental result for digital geometry processing is that any smooth surface can be approximated by a piece-wise linear triangular meshes as well as required. The detailed construction method can be found in the theoretical work of Dai and Luo et al. [44].

Theorem 3.10 (Geometric Approximation Accuracy). *Given a surface S with C^2 continuity embedded in $\mathbb{R}^3$, for an arbitrary small ε, one can select a dense sample point set P, such that the geodesic Delaunay triangulation T of P uniquely exists. The induced piecewise linear mesh is denoted as M. Then the shortest distance map is a homeomorphism. The Hausdorff, normal, and metric distances between S and M are all less than ε.*

The density of the sample point set is determined by the principal curvature of S, the distance from S to its medial axis, and the injective radius of S.

In the future discussion, we use piecewise linear triangular meshes to approximate arbitrary smooth surfaces in $\mathbb{R}^3$. The computational errors can be accurately estimated based on this theorem.

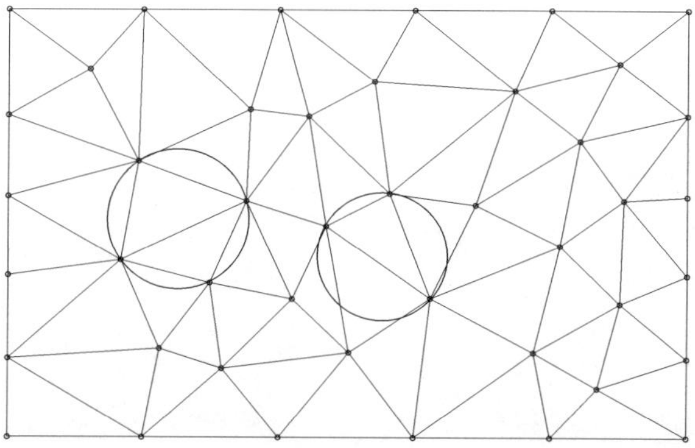

Fig. 3.3. Planar Delaunay triangulation with the empty circum-circle property.

3.1.3 Chain Complex

Given a simplicial complex M, we are interested in their sub-complexes. For example, a curve on the surface can be represented as a set of consecutive oriented edges; an oriented surface patch can be represented as a collection of adjacent oriented faces.

Definition 3.11 (Chain Group). *A* q-chain *is a linear combination of all* q*-simplexes in* M*,*

$$\sum_i \lambda_i \sigma_i, \quad \lambda_i \in \mathbb{Z}.$$

The set $C_q(M)$ *of* q*-chains in* M *forms an abelian group with addition defined by*

$$\sum \alpha_j \sigma_j + \sum \beta_j \sigma_j = \sum (\alpha_j + \beta_j) \sigma_j.$$

The zero element is $\sum 0\sigma_j$ *and the inverse of* $\sum \alpha_j \sigma_j$ *is* $\sum(-\alpha_j)\sigma_j$*.* $C_q(M)$ *is called the* q-dimensional chain group of M.

A curve on the mesh is a 1-chain, and a surface patch is a 2-chain, as shown in Fig. 3.4.

Given a q-chain on the complex, we can compute its boundary, which is a $(q-1)$-chain. For example, suppose a 1-chain connecting two vertices p and q, where p is the source vertex and q is the target vertex. Then its boundary is $q-p$. The boundary of a 2-chain $[v_0, v_1, v_2]$ is the loop formed by three oriented edges $[v_0, v_1]$, $[v_1, v_2]$, and $[v_2, v_0]$. We can formally define the boundary operators acting on the chain spaces.

Definition 3.12 (Boundary Operator). *The* q*-dimensional boundary operator is a homomorphism,*

$$\partial_q : C_q \to C_{q-1},$$

such that

$$\partial_q [v_0, v_1, v_2, \cdots, v_q] = \sum_i (-1)^i [v_0, v_1, \cdots, v_{i-1}, v_{i+1}, \cdots, v_q].$$

and

$$\partial_q \sum_i \alpha_i \sigma_i = \sum_i \alpha_i \partial_q \sigma_i, \quad \alpha_i \in \mathbb{Z}.$$

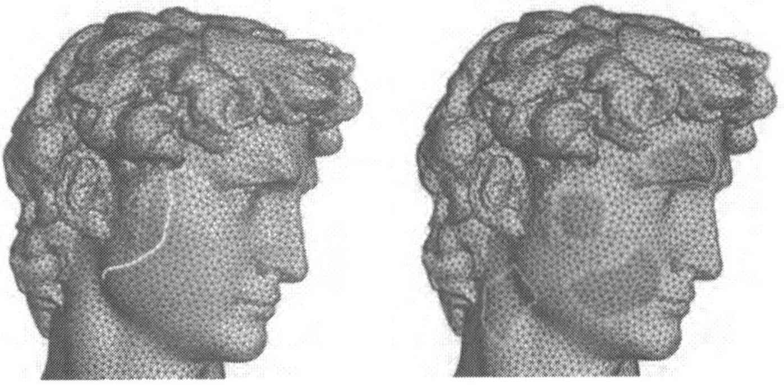

Fig. 3.4. 1-chain and 2-chain on a simplex complex.

It is very intuitive that the boundary operator extracts the boundary of a chain. Fig. 3.5 demonstrates the boundary operator acting on a complex with holes, the results are several loops surrounding the holes.

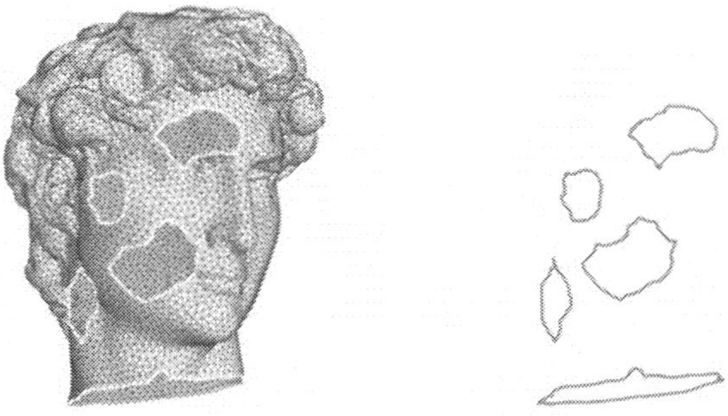

Fig. 3.5. Boundary operator extracts boundaries of a complex.

The central idea of homology is to study the difference between the closed curves and the boundary curves. A curve is *closed* if it has no boundary points, whereas, an open curve has boundary vertices, as shown in Fig. 3.6. On the left image, there are four closed 1-chains, on the right figure, there are several open 1-chains.

Definition 3.13 (Closed Chain Group). *A q-chain $\sigma \in C_q(M)$ is closed if $\partial_q \sigma = 0$. All of the closed q-chains form a subgroup in C_q, which is the kernel of the homomorphism $\partial_q : C_q \to C_{q-1}$, denoted as $Z_q := ker\partial_q$, called the* q-dimensional closed chain group.

Similarly, we can define *boundary* 1-chains which are the boundaries of 2-chains (surface patches) as shown in Fig. 3.7. A boundary chain is also called an *exact chain*.

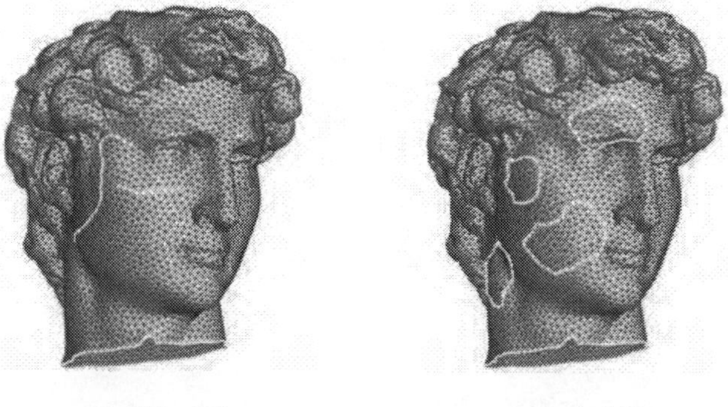

closed 1-chain open 1-chain

Fig. 3.6. Closed 1-chain and open 1-chain.

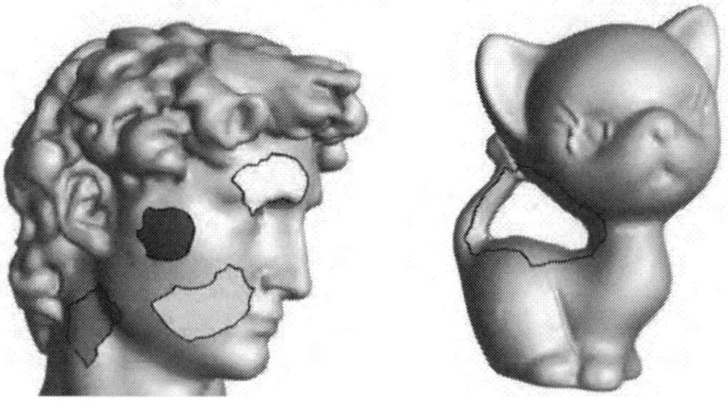

exact 1-chain closed 1-chain

Fig. 3.7. Exact 1-chain and closed 1-chain.

Definition 3.14 (Boundary Chain Group). *A q-chain $\gamma \in C_q(M)$ is exact, if there exists a $(q+1)$-chain σ, such that $\gamma = \partial_{q+1}\sigma$. The exact q-chains form a subgroup in C_q, which is the image of the homomorphism ∂_{q+1}, denoted as $B_q(M) := img\partial_{q+1}$, and called the* q-dimensional boundary chain group.

It is obvious that exact chains are closed. The boundary of a surface patch must be a closed loop, and the boundary of a volume must be a closed surface. Therefore,

$$B_q \subset Z_q \subset C_q.$$

Theorem 3.15 (Boundary of Boundary). *The boundary of a boundary is empty:*

$$\partial_{q-1} \circ \partial_q \equiv 0.$$

Namely, exact chains are closed. In general though, closed chains are not necessarily exact. The second image in Fig. 3.7 demonstrates a closed loop on the kitten surface. We denote the loop as γ and the surface as M. We can slice the surface along γ, then $M-\gamma$ is a topological cylinder, which is a 2-chain. The surface patch $M-\gamma$ has two boundary loops, one is γ, the other one is $-\gamma$. Therefore, the loop γ itself is not the boundary of any 2-chain.

The difference between the closed chains and the exact chains indicates the topology of the surface. The closed and non-exact 1-chain surrounds a tunnel, a handle or a hole of the surface.

Definition 3.16 (Chain Complex). *A chain complex* $C = \{C_q, \partial_q\}$ *is a series of abelian groups* C_q *(q-dimensional chain group) and a series of homomorphisms* $\partial_q : C_q \to C_{q-1}$ *(q-dimensional boundary operators), arranged as a sequence*

$$\cdots \longrightarrow C_{q+1} \xrightarrow{\partial_{q+1}} C_q \xrightarrow{\partial_q} C_{q-1} \xrightarrow{\partial_{q-1}} C_{q-2} \longrightarrow \cdots \tag{3.1}$$

satisfying the condition, for each dimension q,

$$\partial_q \circ \partial_{q+1} = 0.$$

Definition 3.17 (Homology Group). *The q-dimensional homology group* $H_q(C)$ *of a chain complex C is the quotient group,*

$$H_q(C) = \frac{Z_q(C)}{B_q(C)},$$

where $Z_q = ker\partial_q$ *is the q-dimensional closed chain group and* $B_q = img\partial_{q+1}$ *is the q-dimensional boundary chain group. Each element in* H_q *is called a q-*dimensional homologous class. *A closed q-chain* $z_q \in Z_q(C)$ *represents a homologous class, denoted as* $[z_q] \in H_q(C)$.

Usually, we put all of the homology groups of M together, and denote it as

$$H_*(C) := \{H_q(C)\}.$$

Two q-chains z_q^1, z_q^2 are homologous if they bound a (k+1)-chain z_{q+1},

$$z_q^1 - z_q^2 = \partial_{q+1} z_{q+1}.$$

3.1.4 Chain Map and Induced Homomorphism

Suppose we have two topological spaces, M and N, and a map $F : M \to N$. C and D are the chain complexes of M and N, respectively. Then the map F induces a chain map between C and D. Namely, we convert the topological spaces and maps to algebraic chain complexes and chain maps.

Definition 3.18 (Chain Map). *Suppose C and D are chain complexes, a* chain map $f : C \to D$ *is a series of homomorphisms*

$$f = \{f_q : C_q \to D_q\},$$

such that, for each dimension q,

$$\partial_q \circ f_q = f_{q-1} \circ \partial_q,$$

Namely, the following diagram commutes:

$$\begin{array}{ccccccccc}
\cdots \longrightarrow & C_{q+1} & \xrightarrow{\partial_{q+1}} & C_q & \xrightarrow{\partial_q} & C_{q-1} & \xrightarrow{\partial_{q-1}} & C_{q-2} & \longrightarrow \cdots \\
& \downarrow{\scriptstyle f_{q+1}} & & \downarrow{\scriptstyle f_q} & & \downarrow{\scriptstyle f_{q-1}} & & \downarrow{\scriptstyle f_{q-2}} & \\
\cdots \longrightarrow & D_{q+1} & \xrightarrow{\partial_{q+1}} & D_q & \xrightarrow{\partial_q} & D_{q-1} & \xrightarrow{\partial_{q-1}} & D_{q-2} & \longrightarrow \cdots
\end{array}$$

A chain map $f : C \to D$ induces a homomorphism between the homology groups $f_* : H_*(C) \to H_*(D)$,

$$f_*([z_q]) := [f_q(z_q)], \quad \forall z_q \in Z_q(C).$$

The homomorphism $f_q : H_q(C) \to H_q(D)$ maps an abelian group $H_q(C)$ to an abelian group $H_q(D)$. Therefore, f_q can be represented as a matrix, where all entries are integers. By studying the matrix, we can obtain the information of the original topological map $F : M \to N$. For example, if F is a homomorphism, F is invertible, then the matrices of all f_q's must be invertible. Because each matrix is an integer matrix, their determinants must be equal to $+1$ or -1.

3.1.5 Simplicial Map

Suppose M and N are topological spaces represented as simplicial complexes and $f : M \to N$ is a continuous map. Then f can be approximated by a simplicial map. A simplicial map transforms each simplex in M to a simplex in N.

Definition 3.19 (Simplicial Map). *Suppose M and N are simplicial complexes, $f : M \to N$ is a* simplicial map *if for any simplex $\sigma \in M$, $f(\sigma)$ is a simplex in N.*

Start with two finite complexes M and N, and a continuous map f from M to N. Note that f need not be a simplicial map. It could smear the simplexes of M all over N. We are going to slide the image $f(M)$ about, stretching here and shrinking there, but never pushing f across simplex boundaries. If $f(\sigma)$ lies in a certain simplex of N, it is always going to be in that simplex. If the resolutions of M and N are not high enough, we can subdivide them by inserting the centers of each simplex (such subdivision is called a *barycentric subdivision*). In the end we will have a map g, homotopic to f, that is a simplicial map on a barycentric subdivision of N. At a high resolution, g acts on the simplexes of M, squashing f into faces, edges, or vertices, or expanding f to fill the destination simplex in N. In fact, g can be as close to f as possible.

Theorem 3.20 (Simplicial Approximation). *Suppose $f : M \to N$ is a continuous map between two finite complexes embedded in $\mathbb{R}^n$. Then for arbitrary $\varepsilon > 0$, there exists a simplicial map $g : \bar{M} \to \bar{N}$, where $\bar{M}$ and $\bar{N}$ are barycentric subdivisions of M and N (maybe subdivided several times), such that $|f(p) - g(p)| < \varepsilon, \forall p \in M$.*

A simplicial map $f : M \to N$ induces a homomorphism between the chain complexes $f_\# : C(M) \to C(N)$ naturally. Therefore, by studying the chain map $f_\#$, we can obtain the topological information of f.

3.1.6 Chain Homotopy

Suppose both $f, g : M \to N$ are two maps between spaces M and N. f and g induce chain maps $f_\#, g_\# : C(M) \to C(N)$ respectively. We would like to study the relation between f and g by examining their induced chain maps $f_\#$ and $g_\#$.

Definition 3.21 (Homotopy). *Suppose M and N are topological spaces and $f, g : M \to N$ are continuous maps from M to N. A homotopy connecting f and g is a continuous map, $F : M \times [0,1] \to N$, such that*

$$F(\cdot, 0) = f, \quad F(\cdot, 1) = g.$$

All of the continuous maps between M and N can be classified by the homotopy relation. The homotopic relation can be reflected by their chain maps also. Intuitively, f and g are homotopic to each other, if and only if their induced chain maps $f_\#$ and $g_\#$ are chain homotopic to each other.

Definition 3.22 (Chain Homotopy). *Two chain maps $f, g : C \to D$ are* chain homotopic *if there exists a series of homomorphisms $T = \{T_q : C_q \to D_{q+1}\}$,*

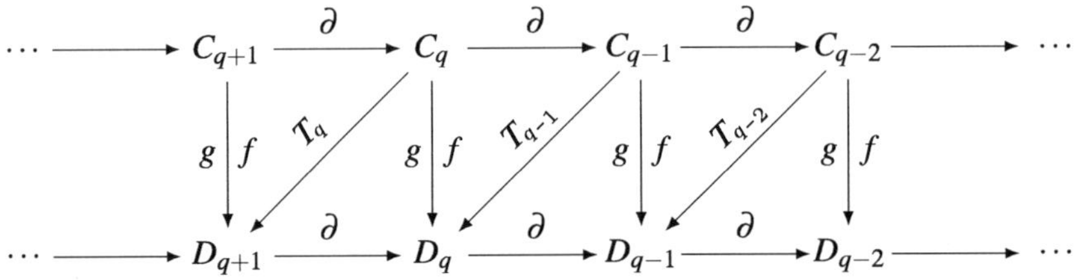

such that, for each dimension q,

$$g_q - f_q = \partial_{q+1} \circ T_q + T_{q-1} \circ \partial_q.$$

T is called a chain homotopy *connecting f, g, denoted as*

$$f \cong g : C \to D.$$

Theorem 3.23. *Suppose two chain maps $f \cong g : C \to D$. Then $f_* = g_* : H_*(C) \to H_*(D)$. Namely, chain homotopy induces the same homomorphism between homology groups.*

The chain homotopic relation is an equivalent relation. All of the chain maps between two finite complexes can be classified by the chain homotopic equivalence.

3.1.7 Homotopy Equivalence

The fundamental task of topology is to classify topological spaces by the homeomorphic equivalence relation.

Definition 3.24 (Homeomorphism). *Suppose M and N are two topological spaces. A continuous map* $f : M \rightarrow N$ *is called a* homeomorphism *between M and N, if f is invertible and* $f^{-1} : N \rightarrow M$ *is also continuous.*

Definition 3.25 (Homeomorphic Equivalent). *Suppose M and N are two topological spaces. They are homeomorphic to each other if there exists a homeomorphism between them.*

The ultimate goal of topology is to find complete invariants, which indicate whether two spaces are homeomorphic to each other. If two spaces are homeomorphic, then they share the same homology groups. However, two spaces with the same homology group are not necessarily homeomorphic to each other. Instead, they are homotopy equivalent. As defined before in 2.5, two topological spaces M, N are *homotopy equivalent* or of the same homotopy type if there exist maps $f : M \rightarrow N$ and $g : N \rightarrow M$, such that $f \circ g$ is homotopic to the identity map id_N and $g \circ f$ is homotopic to the identity map id_M.

The homotopy equivalence relation can be represented by their chain complexes.

Definition 3.26 (Chain Homotopy Equivalence). *Two chain complexes C and D are* chain homotopy equivalent, *if there exist chain maps* $f : C \rightarrow D$ *and* $g : D \rightarrow C$, *such that*

$$g \circ f \cong id_C : C \rightarrow C, \quad f \circ g \cong id_D : D \rightarrow D.$$

f and g are called the chain homotopy equivalences *between C and D, denoted as* $C \cong D$.

Theorem 3.27. *Suppose continuous maps* $f, g : M \rightarrow N$ *between topological spaces M and N are homotopy equivalent. Then their induced chain maps* $f_\#, g_\# : C(M) \rightarrow C(N)$ *are chain homotopy equivalent. The induced homomorphisms* $f_*, g_* : H_*(C(M)) \rightarrow H_*(C(N))$ *are equal.*

Furthermore, the following theorem shows that the homology groups are the homotopy invariants.

Theorem 3.28 (Homology Groups of Homotopy Equivalent Spaces). *If two topological spaces M and N are homotopy equivalent, then their chain complexes* $C(M)$ *and* $C(N)$ *are chain homotopy equivalent, and their homology groups* $H_*(M)$ *and* $H_*(N)$ *are isomorphic.*

If two spaces are homeomorphic to each other, then they must be homotopy equivalent, therefore their homology groups are isomorphic.

3.1.8 Relation Between Homology Group and Homotopy Group

If two loops γ_1 and γ_2 are homotopic on a surface M, they must be homologous. The converse is not true. Suppose γ_1 and γ_2 are homologous. Then

$$\gamma_1 - \gamma_2 = \partial_2 \sigma,$$

where σ is a surface patch. If σ is a cylinder, then γ_1 and γ_2 are homotopic. If σ has handles, then γ_1 and γ_2 are not homotopic. Suppose there are k handles on σ with homotopy basis a_i, b_i on each handle. Then

$$\gamma_1\gamma_2^{-1} \sim a_1b_1a_1^{-1}b_1^{-1}\cdots a_kb_ka_k^{-1}b_k^{-1},$$

In a homology group, $a_ib_ia_i^{-1}b_i^{-1}$ is treated as the identity element. Namely, homology group $H_1(M)$ can be obtained by adding more relations $a_ib_ia_i^{-1}b_i^{-1}$ to the fundamental group $\pi_1(M)$. This is exactly the process to convert a non-abelian group to an abelian group by quotienting its commutator subgroup.

Because $\pi(M)$ is non-abelian, the abelianization is to compute the quotient group:

$$\pi(M)/[\pi_1(M),\pi_1(M)] \cong H_1(M).$$

where the commutator subgroup of $\pi(M)$ is the subgroup generated by all the commutators of elements of $\pi(M)$,

$$[\pi(M),\pi(M)] = \langle g^{-1}h^{-1}gh | g,h \in \pi(M)\rangle.$$

For a genus g oriented surface M,

$$\pi(M) = \langle\{a_1,b_1,a_2,b_2,\cdots,a_g,b_g\}|\{a_1b_1a_1^{-1}b_1^{-1}a_2b_2a_2^{-1}b_2^{-1}\cdots a_gb_ga_g^{-1}b_g^{-1}\}\rangle$$

and

$$H_1(M) = \langle\{a_1,b_1,a_2,b_2,\cdots,a_g,b_g\}|\{a_1b_1a_1^{-1}b_1^{-1},a_2b_2a_2^{-1}b_2^{-1},\cdots,a_gb_ga_g^{-1}b_g^{-1}\}\rangle.$$

3.1.9 Lefschetz Fixed Point

From the above discussion, we can see the methodology of algebraic topology. There are mainly three categories, $\mathbf{C}_1 = \{$topological space, continuous map$\}$, $\mathbf{C}_2 = \{$chain complex, chain map$\}$, and $\mathbf{C}_3 = \{$group, homomorphism$\}$:

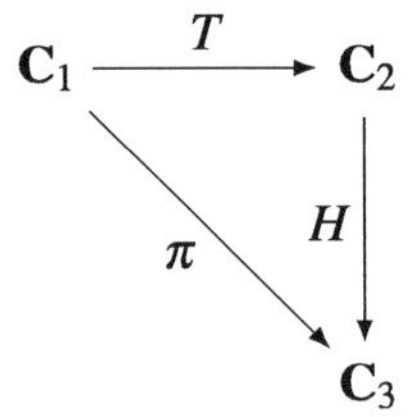

where T is the functor of triangulation, π is the functor of the homotopy group, and H is the functor of the homology group. Therefore, all of the topological problems in category $\mathbf{C}_1$ are converted to combinatorial problems in category $\mathbf{C}_2$, and to algebraic problems in $\mathbf{C}_3$.

In engineering fields, surfaces are obtained by scanning devices, and represented as point clouds. Through meshing and remeshing techniques, they are represented as simplicial complexes. The maps among them are usually approximated by simplicial maps. The computational algorithms are mainly based on the chain complex.

For most topological problems, the computation can be carried out either in the chain complex category or in the homology group category, the results are usually consistent. The following examples compute the fixed point of a map.

Suppose $f : M \to M$ is a continuous map, a point $p \in M$ is called a *fixed point* if $f(p) = p$. Many important existence problems can be formulated as finding fixed points of special maps. We show tow examples here:

- A polynomial f with complex coefficients maps the complex plane $\mathbb{C}$ to itself. A root of f, $f(x) = 0$ is the fixed point of $g(x) = f(x) + x$, $g : \mathbb{C} \to \mathbb{C}$.
- A closed geodesic on a genus zero closed surface can be treated as the fixed point of a map from the circle bundle of the surface to itself.

We can assume f is a simplicial map from a finite simplicial complex to itself. Then f induces homomorphisms $f_*^q : H_q(M,\mathbb{Z}) \to H_q(M,\mathbb{Z})$, which are linear maps.

Definition 3.29 (Lefschetz Number). *The alternative algebraic summation of the traces of the linear maps*

$$L(f) := \sum_{q=0}^{n} (-1)^q trace f_*^q \tag{3.2}$$

is called the Lefschetz number *of the map* f.

Because f is a simplicial map, it induces a chain map. We examine the linear map $f_q : C_q \to C_q$. We use q-simplexes in M as the basis of the linear space C_q, and represent f_q as a matrix. If $trace(f_q)$ is non-zero, there must be a q-simplex σ, such that $f_q(\sigma) = \sigma$, therefore, there exists a fixed point of f. Therefore, if

$$\tilde{L}(f) := \sum_{q=0}^{n} (-1)^q trace(f_q)$$

is nonzero, then the map f must have at leat a fixed point. It is easy to show that $\tilde{L}(f) = L(f)$. Furthermore, we know that $L(f)$ is independent of the choice of the triangulation, and if f is homotopic to g, then $L(f) = L(g)$.

Definition 3.30 (Euler Number). *If f is the identity map $id_M : M \to M$, then $L(id_M)$ is called the* Euler number *of M, denoted as $\chi(M)$,*

$$\chi(M) := L(id_M).$$

Specially, suppose M is a surface represented as a finite complex. Then $\chi(M) = \tilde{L}(id_M) = V + F - E$, where V,E, and F are the number of vertices, edges, and faces, respectively. Later, we will see that the Euler number reflects the topological obstruction and becomes a bridge between topology and global differential geometry.

3.1.10 Mayer-Vietoris Homology Sequence

The Mayer-Vietoris sequence (named after Walther Mayer and Leopold Vietoris) is an exact sequence that often helps one to compute homology groups. It is analogous to the Seifert-van Kampen theorem in homotopy theory.

Homology groups can often be computed directly using the tools of linear algebra (in simplicial homology). However, for large complexes, such computations become cumbersome. It is useful to have tools that allow one to compute homology groups from others that one already knows. The Mayer-Vietoris sequence is one of the most useful tools for this purpose.

Definition 3.31 (Exact Sequence). *Consider a sequence of groups and homomorphisms,*

$$\cdots \longrightarrow C_q \xrightarrow{f_q} C_{q+1} \xrightarrow{f_{q+1}} C_{q+2} \xrightarrow{f_{q+2}} C_{q+3} \longrightarrow \cdots$$

if $ker f_{k+1} = img f_k, \forall k$, then the sequence is called an exact sequence.

Let M be a topological space, $U_1, U_2 \subset M$, such that $M = U_1 \cup U_2$. Then there is an *exact sequence* of homology groups:

$$\cdots \longrightarrow H_q(U_1 \cap U_2) \xrightarrow{i_* \oplus -j_*} H_q(U_1) \oplus H_q(U_2) \xrightarrow{i_* + j_*} H_q(U_1 \cup U_2) \xrightarrow{\partial_*}$$
$$H_{q-1}(U_1 \cap U_2) \longrightarrow \cdots$$

This is called the *Mayer-Vietoris sequence*. Here, i_* is induced by the inclusions:

$$i : U_1 \hookrightarrow M,$$

and j_* is induced by

$$j : U_2 \hookrightarrow M.$$

∂_* is the following map: suppose γ is a chain, $[\gamma] \in H_q(M)$. Then it can be written as the sum of a chain in U_1 and one in U_2,

$$\gamma = \gamma_1 + \gamma_2, \gamma_1 \subset U_1, \gamma_2 \subset U_2.$$

Because $\partial\gamma = 0$ then $\partial\gamma_1 = -\partial\gamma_2$, thus

$$\partial\gamma_1 \subset U_1 \cap U_2,$$

therefore

$$[\partial\gamma_1] \in H_{q-1}(U_1 \cap U_2).$$

Therefore, $\partial_*([\gamma]) = [\partial\gamma_1]$.

3.1.11 Tunnel Loop and Handle Loop

Let M be a connected, closed and orientable surface, embedded in the Euclidean space $\mathbb{R}^3$. We compactify $\mathbb{R}^3$ to the three-dimensional sphere $\mathbb{S}^3$ by adding one infinity point ∞ and the open sets of the infinity

$$U_r = \{(x,y,z) | x^2 + y^2 + z^2 > r^2\}.$$

The surface separates $\mathbb{S}^3$ into two parts, the inside part I and the outside part O. Let

$$\mathscr{I} = I \cup M, \quad \mathscr{O} = O \cup M.$$

M admits an open tubular neighborhood in $\mathbb{S}^3$, denoted as Σ.

Definition 3.32 (Tunnel Loop). *A loop on M is* a tunnel loop *if the homology class carried by it is trivial in $H_1(\mathscr{O})$ and non trivial in $H_1(\mathscr{I})$.*

Definition 3.33 (Handle Loop). *A loop on M is* a handle loop *if the homology class carried by it is trivial in $H_1(\mathscr{I})$ and non trivial in $H_1(\mathscr{O})$.*

The tunnel loop and handle loop on a genus two surface are shown in Fig 3.8. The following theorem shows that the tunnel and handle loops are unique and form a homology basis. The first proof is given by Dey et.al. in [46].

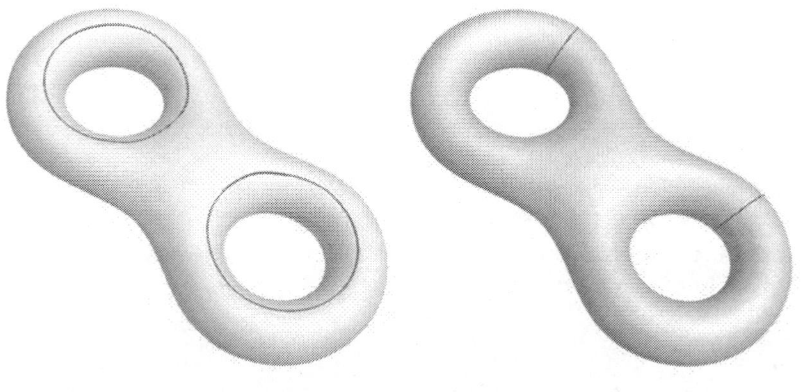

(a) Tunnel loops (b) Handle loops

Fig. 3.8. Tunnel loops and handle loops.

Theorem 3.34 (Handle Loops and Tunnel Loops[46]). *For any connected closed surface $M \hookrightarrow \mathbb{S}^3$ embedded in the three-dimensional sphere $\mathbb{S}^3$ of genus g, there exist g handle loops $\{h_i\}_{i=1}^g$ forming a basis for $H_1(\mathscr{O})$ and g tunnel loops $\{t_i\}_{i=1}^g$ forming a basis for $H_1(\mathscr{I})$. Furthermore, $\{h_i\}_{i=1}^g$ and $\{t_i\}_{i=1}^g$ form a basis for $H_1(M)$.*

Proof. Let $U_1 = \mathscr{I} \cup \Sigma$ and $U_2 = \mathscr{O} \cup \Sigma$. Then $\mathbb{S}^3 = U_1 \cup U_2$ and $\Sigma = U_1 \cap U_2$. We have the following Mayer-Vietoris sequence, which is exact

$$\cdots \longrightarrow H_2(\mathbb{S}^3) \xrightarrow{\alpha} H_1(\Sigma) \xrightarrow{\beta} H_1(\mathscr{I} \cup \Sigma) \oplus H_1(\mathscr{O} \cup \Sigma)$$

$$\xrightarrow{\gamma} H_1(\mathbb{S}^3) \longrightarrow \cdots$$

Because $H_1(\mathbb{S}^3)$ and $H_2(\mathbb{S}^3)$ are trivial, β is an isomorphism induced by the inclusion map. Since, Σ, $\mathscr{I} \cup \Sigma$ and $\mathscr{O} \cup \Sigma$ deformation retract to M, $\mathscr{I}$ and $\mathscr{O}$ respectively, there is an isomorphism from $H_1(M)$ to $H_1(\mathscr{I}) \oplus H_2(\mathscr{O})$, which is induced by inclusion. Therefore,

$$rank(H_1(\mathscr{I})) + rank(H_1(\mathscr{O})) = rank(H_1(M)) = 2g.$$

From Theorem 19 in [47] p.172,

$$rank(H_1(\mathscr{I})) \geqslant g, \quad rank(H_1(\mathscr{O})) \geqslant g,$$

therefore the ranks of $H_1(\mathscr{I})$ and $H_1(\mathscr{O})$ are g. The pre-images of the basis $H_1(\mathscr{I})$ are tunnel loops, the pre-images of the basis of $H_1(\mathscr{O})$ handle loops. □

3.2 Cohomology

Homology is more geometric and easier to be visualized or imagined. Cohomology is more analytic and easier to compute and manipulate. In this section, we introduce simplicial cohomology groups of a topological space. The concepts and construction method can be directly implemented on a computer.

3.2.1 Cohomology Group

Suppose M is a simplicial complex, $C = \{C_q, \partial_q\}$ is the associated simplicial complex,

$$\cdots \longrightarrow C_{q+1} \xrightarrow{\partial_{q+1}} C_q \xrightarrow{\partial_q} C_{q-1} \xrightarrow{\partial_{q-1}} C_{q-2} \longrightarrow \cdots$$

C_q is the q-dimensional chain group, $\partial_q : C_q \to C_{q-1}$ is the q-dimensional boundary operator.

Definition 3.35 (Cochain Group). *Suppose M is a simplicial complex, C_q is the q-chain group. Then a q-*dimensional cochain

$$\sigma^q : C_q \to \mathbb{Z}$$

is a homomorphism between C_q and $\mathbb{Z}$. Suppose σ_1^q and σ_2^q are two q-cochains. Then

$$(\sigma_1^q + \sigma_2^q)(c) = \sigma_1^q(c) + \sigma_2^q(c), \quad \forall c \in C_q.$$

Therefore all of the q-cochains form a group, the zero element is

$$\sigma^q \equiv 0.$$

*which is called the q-*dimensional cochain group *of M.*

We often denote $\sigma(c)$ as $\langle \sigma, c \rangle$, where $\sigma \in C^q, c \in C_q$. Namely, the cochain space is formed by the linear functions on the chain space.

Definition 3.36 (Coboundary Operator). *The coboundary operator $\delta^q : C^q \to C^{q+1}$ is a linear operator, defined as the dual operator of the boundary operator $\partial_{q+1} : C_{q+1} \to C_q$. Suppose $\sigma^q \in C^q$, $z_{q+1} \in C_{q+1}$. Then*

$$\langle \delta^q \sigma^q, z_{q+1} \rangle := \langle \sigma^q, \partial_{q+1} z_{q+1} \rangle.$$

For example, suppose M is a surface represented as a simplicial complex, a 1-cochain $\sigma : C_1 \to \mathbb{Z}$ is a linear function defined on the oriented edges of M. Then $\delta^1\sigma : C_2 \to \mathbb{Z}$ is a 2-cochain, a linear function defined on the oriented faces, such that

$$\delta^1\sigma([v_0, v_1, v_2]) = \sigma(\partial_2[v_0, v_1, v_2]) = \sigma([v_0, v_1]) + \sigma([v_1, v_2]) + \sigma([v_2, v_0]).$$

Assume $\sigma \in C^q$, $z \in C_{q+2}$. Then

$$\langle \delta^{q+1}\delta^q \sigma, z \rangle = \langle \delta^q \sigma, \partial_{q+2} z \rangle = \langle \sigma, \partial_{q+1}\partial_{q+2} z \rangle = 0.$$

Because of $\partial_{q+1} \circ \partial_{q+2} = 0$, we obtain

$$\delta^{q+1} \circ \delta^q = 0.$$

Definition 3.37 (Cochain Complex). *A cochain complex $C = \{C^q, \delta^q\}$ is a series of abelian groups C^q (q-dimensional cochain group) and a series of homomorphisms $\delta^q : C^q \to C^{q+1}$ (q-dimensional coboundary operators), arranged as a sequence*

$$\cdots \longleftarrow C^{q+2} \xleftarrow{\delta^{q+1}} C^{q+1} \xleftarrow{\delta^q} C^q \xleftarrow{\delta^{q-1}} C^{q-1} \longleftarrow \cdots \tag{3.3}$$

satisfying the condition: for each dimension q,

$$\delta^{q+1} \circ \delta^q = 0.$$

The *q-dimensional closed cochain group* is defined as the kernel of the coboundary operator δ^q,

$$Z^q(C) := ker\delta^q,$$

where each element in $Z^q(C)$ is called a *q-dimensional closed cochain*. The *q-dimensional exact cochain group* is the image of the coboundary δ^{q-1},

$$B^q(C) := img\delta^{q-1},$$

where each element is called a *q-dimensional exact cochain*.

Definition 3.38 (Cohomology Group). *The quotient group*

$$H^q(C) := \frac{Z^q(C)}{B^q(C)}$$

is called the q-dimensional cohomology group of C.

3.2.2 Cochain Map

Suppose that we have two topological spaces M and N, and a map $F : M \to N$. C and D are the cochain complexes of M and N, respectively. The map F induces a chain map $f_q : C_q \to D_q$ from C_q to D_q, it also induces a cochain map $f^q : D^q \to C^q$ from D^q to C^q. Note that the order is reversed.

Suppose F is a simplicial map, a q-chain $z_q \in C_q$ is in C_q. Then $f_q(z_q)$ is a q-chain in D_q. Suppose σ^q is a cochain in D^q. Then $f^q(\sigma^q)$ is a cochain in C^q, defined as

$$\langle f^q(\sigma^q), z_q\rangle := \langle \sigma^q, f_q(z_q)\rangle.$$

We have defined chain map and chain homotopy; similarly, we can define cochain map and cochain homotopy in parallel.

Definition 3.39 (Cochain Map). *Suppose C and D are cochain complexes. A* cochain map $f : D \to C$ *is a series of homomorphisms*

$$f = \{f^q : D^q \to C^q\},$$

such that, for each dimension q,

$$\delta^q \circ f^q = f^{q+1} \circ \delta^q,$$

namely, the following diagram is commutative:

$$\begin{array}{ccccccccc}
\cdots \longleftarrow & C^{q+2} & \xleftarrow{\delta^{q+1}} & C^{q+1} & \xleftarrow{\delta^q} & C^q & \xleftarrow{\delta^{q-1}} & C^{q-1} & \longleftarrow \cdots \\
& f^{q+2}\uparrow & & f^{q+1}\uparrow & & f^q\uparrow & & f^{q-1}\uparrow & \\
\cdots \longleftarrow & D^{q+2} & \xleftarrow{\delta^{q+1}} & D^{q+1} & \xleftarrow{\delta^q} & D^q & \xleftarrow{\delta^{q-1}} & D^{q-1} & \longleftarrow \cdots
\end{array}$$

A cochain map $f : D \to C$ induces a homomorphism between the cohomology groups, $f^* : H^*(D) \to H^*(C)$,

$$f^*([z^q]) := [f^q(z^q)], \quad \forall z^q \in Z^q(D).$$

3.2.3 Cochain Homotopy

Suppose both $f,g : M \to N$ are two maps between spaces M and N. f and g induce cochain maps $f^{\#}, g^{\#} : C(N) \to C(M)$ respectively. If f and g are homotopic, then their induced cochain maps are cochain homotopic to each other.

Definition 3.40 (Cochain Homotopy). *Two cochain maps $f,g : C \to D$ are* cochain homotopic, *if there exists a series of homomorphisms $T = \{T^q : C^q \to D^{q-1}\}$,*

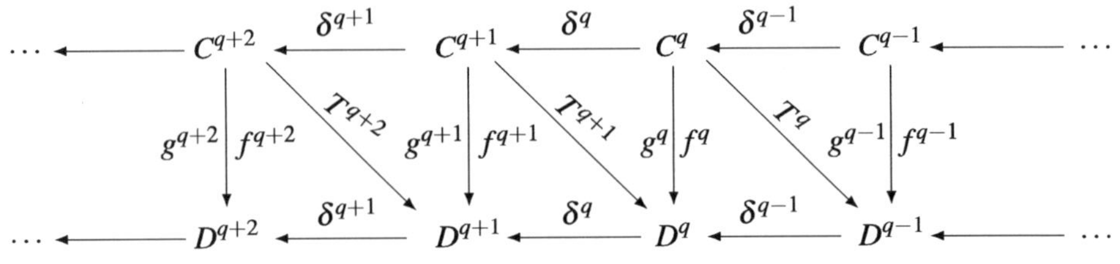

such that, for each dimension q,

$$g^q - f^q = \delta^{q-1} \circ T^q + T^{q+1} \circ \delta^q.$$

T is called a cochain homotopy *connecting f and g, denoted as*

$$f \cong g : C \to D.$$

Theorem 3.41. *Suppose two cochain maps $f \cong g : C \to D$. Then $f^* = g^* : H^*(D) \to H^*(C)$. Namely, cochain homotopy induces the same homomorphism between cohomology groups.*

The cochain homotopy relation is an equivalent relation. All of the cochain maps between two finite complexes can be classified by the cochain homotopy equivalence.

Definition 3.42 (Cochain Homotopy Equivalence). *Two cochain complexes C and D are* cochain homotopy equivalent *if there exist cochain maps $f : C \to D$ and $g : D \to C$, such that*

$$g \circ f \cong id_C : C \to C, \quad f \circ g \cong id_D : D \to D.$$

f and g are called the cochain homotopy between C and D, denoted as $C \cong D$.

Theorem 3.43. *Suppose continuous maps $f,g : M \to N$ between topological spaces M and N are homotopic equivalent. Then their induced cochain maps $f^{\#}, g^{\#} : C(N) \to C(M)$ are cochain homotopic. The induced homomorphisms $f^*, g^* : H^*(C(N)) \to H^*(C(M))$ are equal.*

If two topological spaces M and N are homotopy equivalent, then their cochain complexes $C(M)$ and $C(N)$ are cochain homotopy equivalent, and their cohomology groups $H^*(M)$ and $H^*(N)$ are isomorphic.

Problems

3.1. Non-Orientable Surface-Klein Bootle
(a) Triangulate the Klein bottle.
(b) Compute the homology group of the Klein bottle.

3.2. Three-Dimensional Sphere and Solid Torus
Consider a three-dimensional sphere $\mathbb{S}^3$ with an inside solid torus removed, denoted as K.
(a) Compute the homology groups of K.
(b) Prove K is a solid torus.

3.3. Embedding of Non-Orientable Surfaces
Let N_q represent the surface obtained in the following way. First we remove q disks from a sphere, then glue the holes by q Möbius bands. Prove that if one more disk is removed

from N_q, then the result surface with a boundary can be embedded in $\mathbb{R}^3$ without self-intersection.

3.4. Double Covering on Non-Orientable Surfaces
(a) Show that any non-orientable surface has a double covering, which is orientable.
(b) Prove that the double covering of N_q is of genus $q-1$.

3.5. Degree of Gauss Map
Suppose S is a closed smooth surface embedded in $\mathbb{R}^3$, the Gauss map $g : S \to \mathbb{S}^2, p \to \mathbf{n}(p)$ maps each point p to its normal $\mathbf{n}(p)$. What is the degree of the Gauss map?

3.6. Fixed Point
Suppose $f : S \to S$ is a automorphism of S. $(\tilde{S}, \pi)$ is the universal covering space of S. Then we can lift f to $\tilde{f} : \tilde{S} \to \tilde{S}$, such that

$$\pi \circ \tilde{f} = f \circ \pi.$$

Show that if for all such $\tilde{f}$ there is no fixed point, then there is no fixed point of f either.

3.7. Poincaré Theorem
Suppose S is a closed surface with genus $g \neq 1$. show that any smooth vector field on S has singularities, otherwise an automorphism can be constructed homotopic to the identity, which has no fixed point.

3.8. Vector Field
Suppose S is a genus g closed surface embedded in $\mathbb{R}^3$. Show that a vector field can be designed such that there is only one singularity.

3.9. Circle Bundle
Let S be a smooth surface embedded in $\mathbb{R}^3$. All of the unit tangent vectors form a manifold, called the circle bundle of S.
(a) Let S be the unit sphere. Find a cellular decomposition of its circle bundle.
(b) Compute the homology groups of the circle bundle of S.
(c) Let S be a general surface with genus g. Compute its homology groups.

3.10. Direct Product
Let K be a simplicial complex and $K \times K$ be the direct product of K.
(a) Construct a cellular decomposition of $K \times K$.
(b) Compute the homology of $K \times K$.
(c) Prove that

$$\chi(K \times K) = \chi(K) \cdot \chi(K).$$

4

Exterior Differential Calculus

4.1 Smooth Manifold

Definition 4.1 (Manifold). *A* manifold *is a topological space M covered by a set of open sets $\{U_\alpha\}$. A homeomorphism $\phi_\alpha : U_\alpha \to \mathbb{R}^n$ maps U_α to the Euclidean space $\mathbb{R}^n$. (U_α, ϕ_α) is called a* coordinate chart *of M, the set of all charts $\{(U_\alpha, \phi_\alpha)\}$ form the* atlas *of M. Suppose $U_\alpha \cap U_\beta \neq \varnothing$. Then*

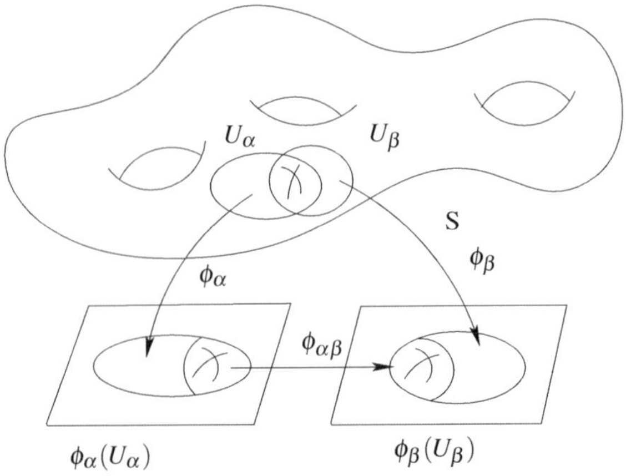

Fig. 4.1. Manifold

$$\phi_{\alpha\beta} = \phi_\beta \circ \phi_\alpha^{-1} : \phi_\alpha(U_\alpha \cap U_\beta) \to \phi_\beta(U_\alpha \cap U_\beta)$$

is a transition map.

Transition maps satisfy the *cocycle condition*. Suppose $U_\alpha \cap U_\beta \cap U_\gamma \neq \emptyset$. Then

$$\phi_{\beta\gamma} \circ \phi_{\alpha\beta} = \phi_{\alpha\gamma}.$$

$$\begin{array}{ccc} \phi_\alpha(U_\alpha \cap U_\beta \cap U_\gamma) & \xrightarrow{\phi_{\alpha\beta}} & \phi_\beta(U_\alpha \cap U_\beta \cap U_\gamma) \\ & \searrow^{\phi_{\alpha\gamma}} & \downarrow \phi_{\beta\gamma} \\ & & \phi_\gamma(U_\alpha \cap U_\beta \cap U_\gamma) \end{array}$$

If all transition functions $\phi_\alpha \circ \phi_\beta^{-1} \in C^\infty(\mathbb{R}^n)$ are smooth, then the manifold is a *differential manifold*, or a *smooth manifold*. The atlas is a *differential atlas*. The maximal differential atlas is called a *differential structure*.

Suppose M is a smooth manifold embedded in $\mathbb{R}^n$, a curve on the manifold is a map

$$\gamma : (-\varepsilon, \varepsilon) \to M.$$

Definition 4.2 (Tangent Vector). *A vector* $\mathbf{v} \in \mathbb{R}^{n+1}$ *is said to be* tangent *to* $M \subset \mathbb{R}^{n+1}$ *at a point* $p \in M$, *if there exists a smooth curve* γ *with* $\gamma(0) = p$ *and* $\gamma'(0) = \mathbf{v}$.

The set T_pM represents all of the vectors tangent to the manifold $M \subset \mathbb{R}^{n+1}$ at the point p. Then T_pM is a vector space. A tangent vector can be abstractly defined without embedding as follows.

Definition 4.3. *A tangent vector ξ at the point p is an association to every coordinate chart $(x^1, x^2, \cdots, x^n)$ at p an n-tuple $(\xi^1, \xi^2, \cdots, \xi^n)$ of real numbers, such that if $(\tilde{\xi}^1, \tilde{\xi}^2, \cdots, \tilde{\xi}^n)$ is associated with another coordinate system $(\tilde{x}^1, \tilde{x}^2, \cdots, \tilde{x}^n)$, then it satisfies the transition rule*

$$\tilde{\xi}^i = \sum_{j=1}^{n} \frac{\partial \tilde{x}^i}{\partial x^j}(p)\xi^j.$$

Namely, tangent vectors are represented by elements of $\mathbb{R}^n$ for each coordinate chart transforming via differentials of a coordinate change at the reference point.

A *smooth vector field* ξ assigns a tangent vector for each point of M, it has a local representation

$$\xi(x^1, x^2, \cdots, x^n) = \sum_{i=1}^{n} \xi_i(x^1, x^2, \cdots, x^n)\frac{\partial}{\partial x_i}.$$

$\{\frac{\partial}{\partial x_i}\}$ represents the vector fields of the velocities of iso-parametric curves on M. They form a basis of all vector fields.

Suppose $\phi : M \to N$ is a differentiable map from M to N, $\gamma : (-\varepsilon, \varepsilon) \to M$ is a curve, $\gamma(0) = p$, $\gamma'(0) = \mathbf{v} \in T_pM$. Then $\phi \circ \gamma$ is a curve on N, $\phi \circ \gamma(0) = \phi(p)$. We define the tangent vector

$$\phi_*(\mathbf{v}) = (\phi \circ \gamma)'(0) \in T_{\phi(p)}N,$$

as the *push-forward* tangent vector of $\mathbf{v}$ induced by ϕ.

4.2 Differential Forms

Definition 4.4 (Differential One-Form). *The tangent space T_pM is an n-dimensional vector space, its dual space T_p^*M is called the* cotangent space *of M at p. Suppose $\omega \in T_p^*M$. Then $\omega : T_pM \to \mathbb{R}$ is a linear function defined on T_pM, ω is called a* differential 1-form *at p.*

Similarly, we can define the differential 1-form field, it has the local representation

$$\omega(x^1, x^2, \cdots, x^n) = \sum_{i=1}^{n} \omega_i(x^1, x^2, \cdots, x^n)dx_i,$$

where $\{dx_i\}$ are the dual differential 1-forms to $\{\frac{\partial}{\partial x_j}\}$, such that

$$dx_i(\frac{\partial}{\partial x_j}) = \delta_{ij}, \quad i, j = 1, 2, \cdots, n,$$

where

$$\delta_{ij} = \begin{cases} 1, & i = j, \\ 0, & i \neq j. \end{cases}$$

Definition 4.5 (Tensor). *A* tensor *Θ of type (m, n) on a manifold M is a correspondence that associates to each point $p \in M$ a multi-linear map*

$$\Theta_p : T_pM \times T_pM \cdots \times T_pM \times T_p^*M \times \cdots \times T_p^*M \to \mathbb{R},$$

where the tangent space T_pM appears m times and cotangent space T_p^*M appears n times.

Then we can generalize 1-form to m-form, An exterior m-form is a tensor ω of type $(m,0)$ which is skew symmetric in its arguments, namely

$$\omega_p(\xi_{\sigma(1)},\cdots,\xi_{\sigma(m)}) = (-1)^\sigma \omega_p(\xi_1,\cdots,\xi_m)$$

for any tangent vectors $\xi_1,\cdots,\xi_m \in T_pM$ and any permutation $\sigma \in S_m$. The local representation of ω in $(x^1,\cdots,x^m)$ is

$$\omega = \omega_{i_1 i_2 \cdots i_m} dx^{i_1} \wedge \cdots \wedge dx^{i_m} = \omega_I dx^I,$$

where $1 \leqslant i_1 < i_2 < \cdots < i_m \leqslant n$ and $I = (i_1,\cdots,i_m)$, $\omega_I = \omega_{i_1\cdots i_m}$, $dx^I = dx^{i_1} \wedge \cdots \wedge dx^{i_m}$. ω_I is a function of the reference point. ω is said to be differentiable if each ω_I is differentiable.

Definition 4.6 (Wedge Product). *The* wedge product *of an m_1-form ω_1 and an m_2-form ω_2 is an (m_1+m_2)-form which is defined in local coordinates by*

$$(\omega_{I_1} dx^{I_1}) \wedge (\omega_{I_2} dx^{I_2}) = \omega_{I_1}\omega_{I_2} dx^{I_1} \wedge dx^{I_2}.$$

A coordinate free representation of wedge product is

$$(\omega_1 \wedge \omega_2)(\xi_1,\cdots,\xi_{m_1+m_2}) = \sum_{\sigma \in S_{m_1+m_2}} \frac{(-1)^\sigma}{m_1!m_2!} \omega_1(\xi_{\sigma(1)},\cdots,\xi_{\sigma(m_1)}) \cdot \omega_2(\xi_{\sigma(m_1+1)},\cdots,\xi_{\sigma(m_1+m_2)}).$$

Suppose ω is an m-form on N, and $\phi : M \to N$ is a differentiable map from M to N. Then the *pullback* $\phi^*\omega$ is an m-form on M defined by

$$(\phi^*\omega)_p(\xi_1,\cdots,\xi_m) = \omega_{\phi(p)}(\phi_*\xi_1,\cdots,\phi_*\xi_m), \quad p \in M,$$

for $\xi_1,\xi_2,\cdots,\xi_m \in T_pM$, where $\phi_*\xi_j \in T_{\phi(p)}M$ is the pushforward of $\xi_j \in T_pM$.

4.3 Integration

In this section, we define the integration of an n-form on an orientable n-manifold M. If a manifold M is orientable, then the Jacobian of all chart transitions are positive.

Supposing that $M = U \subset \mathbb{R}^n$ is an open set, and

$$\omega = f(x) dx^1 \wedge dx^2 \wedge \cdots \wedge dx^n,$$

we set

$$\int_U \omega = \int_U f(x) dx^1 dx^2 \cdots dx^n.$$

Now suppose $U \subset M$ is an open set of M, a chart $\phi : U \to \Omega \subset \mathbb{R}^n$ is defined. Then

$$\int_U \omega = \int_\Omega (\phi^{-1})^* \omega.$$

This integration is independent of the choice of charts. Let $\psi : U \to \psi(U)$ be an orientation preserving diffeomorphism from one local coordinate system $(x_1, x_2, \cdots, x_n)$ to the other $(u_1, u_2, \cdots, u_n)$. Then

$$\int_{\psi(U)} f(x)dx^1 \cdots dx^n = \int_U f(x(u)) \det\left(\frac{\partial x^i}{\partial u^j}\right) du^1 \cdots du^n.$$

Now, consider a covering of M by coordinate charts (U_α, ϕ_α) and choose a partition of unity $\{f_i\}$, $i \in I$, such that $f_i(p) \geqslant 0$,

$$\sum_i f_i(p) \equiv 1, \quad \forall p \in M,$$

for each point p, it is covered by the supports of some f_i's, the number of such f_i's is finite. For each f_i, its support is contained in a U_α.

Then $\omega_i = f_i \omega$ is an n-form on M with compact support in some U_α, we can set the integration as

$$\int_M \omega = \sum_i \int_M \omega_i.$$

4.4 Exterior Derivative and Stokes Theorem

Suppose $f : M \to \mathbb{R}$ is a differentiable function. Then the *exterior derivative* of f is a 1-form,

$$df = \sum_i \frac{\partial f}{\partial x_i} dx^i.$$

The *exterior derivative* of an m-form on M is an $(m+1)$-form on M defined in local coordinates by

$$d\omega = d(\omega_I dx^I) = (d\omega_I) \wedge dx^I,$$

where $d\omega_I$ is the differential of the function ω_I.

Stokes theorem relates the integral of ω over the boundary of a manifold to the integral of $d\omega$ over the manifold itself.

Theorem 4.7 (Stokes Theorem). *Let M be an n-manifold with the boundary ∂M and ω be a differentiable $(n-1)$-form with compact support on M. Then*

$$\int_{\partial M} \omega = \int_M d\omega. \tag{4.1}$$

4.5 De Rham Cohomology Group

Let M be a differentiable manifold, $\Omega^n(M)$ represent all the n-forms on M, d be the exterior derivative. Then we get the *de Rham complex*

$$\cdots \xrightarrow{d^{q-3}} \Omega^{q-2} \xrightarrow{d^{q-2}} \Omega^{q-1} \xrightarrow{d^{q-1}} \Omega^q \xrightarrow{d^q} \Omega^{q+1} \xrightarrow{d^{q+1}}$$

$$\Omega^{q+2} \xrightarrow{d^{q+2}} \cdots$$

where $\Omega^0(M)$ is the space of smooth functions on M, and $\Omega^1(M)$ is the space of 1-forms. The exterior differentiation operator

$$d^m : \Omega^m(M) \to \Omega^{m+1}(M)$$

is a linear operator with the property

$$d^m \circ d^{m-1} \equiv 0.$$

Suppose $\omega \in \Omega^m(M)$ is an m-form. If there exists an $(m-1)$-form $\eta \in \Omega^{m-1}(M)$, such that $\omega = d^{m-1}\eta$, then ω is called an *exact form*. If $d^m\omega = 0$, then ω is called a *closed form*. Exact forms are closed.

Definition 4.8 (de Rham Cohomology Group). *Suppose M is a differential manifold. The m-th de Rham cohomology is defined as*

$$H^m_{dR}(M) = \frac{ker d^m}{Img d^{m-1}}$$

Suppose ω is a closed 1-form, γ_1, γ_2 are closed homologous 1-chains, namely, there exists a 2-chain σ, such that

$$\partial\sigma = \gamma_1 - \gamma_2.$$

Then

$$\int_{\gamma_1} \omega - \int_{\gamma_2} \omega = \int_{\partial\sigma} \omega = \int_D d\omega = 0,$$

therefore $f_\omega(\gamma) = \int_\gamma \omega$ only depends on the homologous class of γ. Therefore

$$f_\omega : H_1(M, \mathbb{R}) \to \mathbb{R}$$

is a linear function defined on $H_1(M, \mathbb{R})$.

On the other hand, suppose ω_1, ω_2 are two cohomologous closed 1-forms. Then there exists a 0-form $\eta \in \Omega^0(M)$, such that

$$\omega_1 - \omega_2 = d\eta.$$

Let γ be a closed 1-chain, we obtain

$$\int_\gamma \omega_1 - \int_\gamma \omega_2 = \int_\gamma d\eta = \int_{\partial\gamma} \eta = 0,$$

therefore $f_{\omega_1} = f_{\omega_2}$.

From above discussion, we see that $H^m_{dR}(M)$ is the dual space of $H_m(M, \mathbb{R})$. Because $H^m(M, \mathbb{R})$ is also the dual space of $H_m(M, \mathbb{R})$, we get the following result.

Theorem 4.9. *The de Rham cohomology group $H^m_{dR}(M)$ is isomorphic to the cohomology group $H^m(M, \mathbb{R})$,*

$$H^m_{dR}(M) \cong H^m(M, \mathbb{R}).$$

4.6 Harmonic Forms

A Riemannian metric on M is an inner product on each T_pM. If the manifold is embedded in $\mathbb{R}^{n+1}$, then it has an induced Euclidean metric.

Suppose M is equipped with a Riemannian metric. Then we can locally find oriented orthonormal basis of vector fields, and choose parameterization, such that

$$\left\{\frac{\partial}{\partial x_1}, \frac{\partial}{\partial x_2}, \cdots, \frac{\partial}{\partial x_n}\right\}$$

form an oriented orthonormal basis. Let

$$\{dx_1, dx_2, \cdots, dx_n\}$$

be the dual 1-form basis.

Definition 4.10 (Hodge Star Operator). *The* Hodge star *operator $*$ is a linear map $*$: $\Omega^k(M) \to \Omega^{n-k}(M)$, defined as*

$$^*(dx_1 \wedge dx_2 \wedge \cdots \wedge dx_k) = dx_{k+1} \wedge dx_{k+2} \wedge \cdots \wedge dx_n.$$

Let $\sigma = (i_1, i_2, \cdots, i_n)$ be a permutation of $(1, 2, \cdots, n)$. Then the Hodge star operator

$$^*(dx_{i_1} \wedge dx_{i_2} \wedge \cdots \wedge dx_{i_k}) = (-1)^\sigma dx_{i_{k+1}} \wedge dx_{i_{k+2}} \wedge \cdots \wedge dx_{i_n}.$$

Suppose η, ζ are both k-forms on M. Then the Hodge star operator induces an L^2 norm

$$(\eta, \zeta) = \int_M \eta \wedge {}^*\zeta.$$

Therefore $\Omega^k(M)$ is a Hilbert space.

Definition 4.11 (Codifferential Operator). *The codifferential operator δ*

$$\delta : \Omega^k(M) \to \Omega^{k-1}(M)$$

is defined as

$$\delta = (-1)^{k+1+k(n-k)} * d *,$$

where d is the exterior derivative.

The codifferential is the adjoint of the exterior derivative, in that

$$(\delta\zeta, \eta) = (\zeta, d\eta).$$

The most important operator is the Laplace operator.

Definition 4.12 (Laplace Operator). *The Laplace operator $\Delta : \Omega^k(M) \to \Omega^k(M)$,*

$$\Delta = \delta d + d\delta.$$

Proposition 4.13. *The Laplace operator is symmetric*

$$(\Delta\zeta, \eta) = (\zeta, \Delta\eta)$$

and non-negative

$$(\Delta\eta, \eta) \geqslant 0.$$

Proof.

$$(\Delta\zeta, \eta) = ((d\delta + \delta d)\zeta, \eta) = (d\zeta, d\eta) + (\delta\zeta, \delta\eta).$$

□

4.7 Hodge Theorem

Definition 4.14 (Harmonic Form). *Suppose $\omega \in \Omega^k(M)$. Then ω is called a k-harmonic form if*

$$\Delta\omega = 0.$$

Definition 4.15 (Harmonic Form Group). *All of the k-harmonic forms form a group, denoted as $H^k_\Delta(M,\mathbb{R})$.*

Let $\omega \in \Omega^k(M)$. We say ω is *closed* if $d\omega = 0$ and *co-closed* if $d*\omega = 0$.

Lemma 4.16. *A harmonic form is both closed and co-closed.*

Proof. Suppose ω is a harmonic form. Then

$$0 = (\Delta\omega,\omega) = (d\omega,d\omega) + (\delta\omega,\delta\omega) = |d\omega|^2 + |\delta\omega|^2.$$

Therefore, $d\omega = 0$, $\delta\omega = 0$. Furthermore, $\delta\omega = *d*\omega$, therefore $d*\omega = 0$. □

Let E be the exact forms $\{d\eta|\eta \in \Omega^{k-1}(M)\}$, E^* be the co-exact forms $\{^*d\eta|\eta \in \Omega^{k-1}(M)\}$. The complement spaces of E and E^* are

$$\begin{aligned} E^\perp &= \{\omega \in \Omega^k(M)|(\omega,\ d\eta) = 0, \forall\eta \in \Omega^{k-1}(M)\}, \\ E^{*\perp} &= \{\omega \in \Omega^k(M)|(\omega,\ ^*d\eta) = 0, \forall\eta \in \Omega^{k-1}(M)\}. \end{aligned}$$

Proposition 4.17. *The complement space of exact forms is the space of co-closed forms. The complement space of co-exact forms is the space of closed forms.*

1. $\omega \in E^\perp$ *if and only if* ω *is co-closed.*
2. $\omega \in E^{*\perp}$ *if and only if* ω *is closed.*

Proof. By definitions,

1. If $\omega \in E^\perp$, then $\forall\eta \in \Omega^{k-1}(M)$, $(\omega,d\eta) = 0$, hence $(\delta\omega,\eta) = 0$, $\delta\omega = 0$, namely ω is co-closed.
2. If $\omega \in E^{*\perp}$, then $\forall\eta \in \Omega^{k-1}(M)$,

$$0 = (\omega,^*d\eta) = -(\omega,\delta*\eta) = (d\omega,*\eta),$$

therefore $d\omega = 0$, namely ω is closed. □

Theorem 4.18 (Hodge Decomposition Theorem).

$$\Omega^k(M) = E \oplus E^* \oplus H_\Delta.$$

Proof. Exact forms are closed, therefore $E \subset E^{*\perp}$; co-exact forms are co-closed, therefore $E^* \subset E$. So E and E^* are orthogonal to each other.

$$E \perp E^*.$$

Let a form ω be in $E^\perp \cap E^{*\perp}$. Then ω is both closed and co-closed, therefore ω is harmonic.

$$E^\perp \cap E^{*\perp} = H_\Delta.$$

We obtain

$$\Omega^k(M) = E \oplus E^* \oplus (E \oplus E^*)^\perp = E \oplus E^* \oplus (E^\perp \cap E^{*\perp}) = E \oplus E^* \oplus H.$$ □

Each closed form ω is orthogonal to E^*, therefore $\omega \in E \oplus H$ can be decomposed to

$$\omega = \alpha + \beta, \quad \alpha \in E, \; \beta \in H_\Delta.$$

We define the projection to the harmonic form space H as

$$h : \omega \to \beta.$$

Suppose ω_1 and ω_2 are two closed forms,

$$\omega_1 = \alpha_1 + \beta_1, \quad \omega_2 = \alpha_2 + \beta_2.$$

Then

$$\omega_1 - \omega_2 = (\alpha_1 - \alpha_2) + (\beta_1 - \beta_2) \in E \oplus H_\Delta.$$

If ω_1 and ω_2 are cohomologous, then $\omega_1 - \omega_2 \in E$ is an exact form, and the harmonic component $\beta_1 - \beta_2$ is 0. Therefore the projection h is defined on the cohomologous class, namely, h is a map from H_{dR} to H_Δ. On the other hand, $H_\Delta \subset H_{dR}$, the inclusion map is the inverse of h. So, we have proven that the de Rham cohomology group and the harmonic form cohomology group are isomorphic.

Theorem 4.19 (Hodge Theorem). *Suppose ω is a closed form, its harmonic component is $h(\omega)$. Then the map*

$$h : H^k_{dR}(M, \mathbb{R}) \to H^k_\Delta(M, \mathbb{R})$$

is an isomorphism.

In other words, each cohomologous class has a unique harmonic form as its representative.

Problems

4.1. Differential forms on the unit sphere $\mathbb{S}^2$,
(a) Construct an atlas of $\mathbb{S}^2$, using the stereographic projection with the projection center at the south pole and the north pole.
(b) Construct a closed differential 1-form.
(c) Show the 1-form is also exact.
(d) Locate the zero points of the 1-form.

4.2. Suppose S is an annulus on the plane.
(a) Construct a closed 1-form, which is also exact.
(b) Construct a closed 1-form, which is not exact.
(c) Suppose ω_1 and ω_2 are two closed 1-forms, but not exact. Is $\omega_1 - \omega_2$ closed? Exact?

4.3. Suppose γ is a closed curve on the surface, ω is a closed 1-form.
(a) If $\tilde{\gamma}$ is another closed curve homotopic to γ, show that

$$\int_\gamma \omega = \int_{\tilde{\gamma}} \omega.$$

(b) If $\tilde{\omega}$ is another closed 1-form, ω and $\tilde{\omega}$ are cohomologous, show that

$$\int_\gamma \omega = \int_\gamma \tilde{\omega}.$$

(c) Generalize the above propositions to higher dimensional forms.

4.4. Consider vector fields on the plane

$$\mathbf{v}(x,y) = f(x,y)\mathbf{e}_1 + g(x,y)\mathbf{e}_2.$$

Let $\langle , \rangle$ denote the inner product on the plane. Then for any vector field $\mathbf{w}$, we can define a 1-form $\omega_\mathbf{v}$,

$$\omega_\mathbf{v}(\mathbf{w}) = \langle \mathbf{v}, \mathbf{w} \rangle.$$

(a) Show that if $\omega_\mathbf{v}$ is a closed 1-form, i. e., $d\omega_\mathbf{v} = 0$, then

$$\mathrm{curl}\mathbf{v} = 0.$$

(b) Show that if $\omega_\mathbf{v}$ is a co-closed 1-form, i. e., $\delta\omega_\mathbf{v} = 0$, then

$$\mathrm{div}\mathbf{v} = 0.$$

(c) Show that if $\omega_\mathbf{v}$ is harmonic, then $\mathbf{v}$ is curl free and divergence free.
(d) Generalize the above results to 1-forms in $\mathbb{R}^3$.

4.5. Suppose $p = (x,y,z)$ is a point in $\mathbb{R}^3 - (0,0,0)$. Then let $r = \sqrt{x^2+y^2+z^2}$, define a 1-form

$$\omega = \frac{1}{r}(dx + dy + dz).$$

(a) Show that ω is a closed 1-form.
(b) Show that ω is an exact 1-form.
(c) What is the 0-form whose exterior differential equals to ω?

4.6. Let the lattice $\Gamma = \{m\mathbf{a} + n\mathbf{b} | m,n \in \mathbb{Z}\}$, $\mathbf{a},\mathbf{b}$ are two vectors in $\mathbb{R}^2$. The quotient space $T = \mathbb{R}^2/\Gamma$ is a torus.
(a) Construct two closed 1-forms on T, which are not exact and not cohomologous.
(b) Construct two harmonic 1-forms on T, which are not cohomologous.

5

Differential Geometry of Surfaces

5.1 Curve Theory

Definition 5.1 (Regular Curve). *A curve* $\gamma:(0,1)\to\mathbb{R}^3$ *is called a* regular curve, *if every component is a* C^∞ *function and for all* $t\in(0,1)$, $|\frac{d\gamma}{dt}|>0$.

Considering a spacial regular curve

$$\gamma(t) = (x(t), y(t), z(t)), \quad t \in (0,1)$$

and computing the derivative, we get the velocity vector of the curve:

$$\frac{d\gamma}{dt} = \left(\frac{dx(t)}{dt}, \frac{dy(t)}{dt}, \frac{dz(t)}{dt}\right),$$

called the *tangent vector* of γ.

The length of the tangent vector is measured using Euclidean metric as

$$\left|\frac{d\gamma}{dt}\right| = \sqrt{\left(\frac{dx(t)}{dt}\right)^2 + \left(\frac{dy(t)}{dt}\right)^2 + \left(\frac{dz(t)}{dt}\right)^2} > 0.$$

Let $[a,b] \subset (0,1)$. Then the *arc length* of the curve segment $\gamma(t)$ $(a \leqslant t \leqslant b)$ is

$$\int_a^b \left|\frac{d\gamma}{dt}\right| dt.$$

Fixing the starting time a and replacing b by t, we get the arc length from time a to time t as

$$s(t) = \int_a^t \left|\frac{d\gamma}{d\tau}\right| d\tau.$$

Because $\frac{ds}{dt}(t) = |\frac{d\gamma}{dt}| > 0$, therefore, s is a monotonous increasing function of t, and t can be formulated as a function of s, $t = t(s)$. Therefore, using the arc length as the parameter, we get

$$\gamma(s) = (x(s), y(s), z(s)).$$

Denote

$$\dot{\gamma}(s) = \frac{d\gamma}{ds}.$$

It is obvious that $|\dot{\gamma}(s)| = 1$.

Let

$$\mathbf{t}(s) = \dot{\gamma}(s).$$

Differentiating $\langle \mathbf{t}, \mathbf{t} \rangle = 1$, we get $\langle \dot{\mathbf{t}}, \mathbf{t} \rangle = 0$, therefore $\dot{\mathbf{t}}$ is orthogonal to the tangent vector, we call it one of the *normal vectors* of the curve.

Definition 5.2 (Curve Curvature). *The* curvature vector *of the curve* $\gamma(s)$ *at* s *is*

$$\ddot{\gamma}(s) = (\ddot{x}(s), \ddot{y}(s), \ddot{z}(s)).$$

The curvature is

$$\kappa(s) = |\ddot{\gamma}(s)| = \sqrt{\ddot{x}^2 + \ddot{y}^2 + \ddot{z}^2}.$$

We define the *principal normal vector* of the curve as

$$\mathbf{n}(s) = \frac{1}{\kappa(s)} \dot{\mathbf{t}}(s),$$

and the *binormal vector* as

$$\mathbf{b}(s) = \mathbf{t}(s) \wedge \mathbf{n}(s).$$

Therefore, there is an orthonormal frame moving along the curve:

$$\{\gamma(s); \mathbf{t}(s), \mathbf{n}(s), \mathbf{b}(s)\},$$

which is called the *Frenet frame* of the curve.

We take the derivative of the Frenet frame with respect to the arc length parameter. Because $\mathbf{t}, \mathbf{n}, \mathbf{b}$ are unit vectors, $\langle \dot{\mathbf{t}}, \mathbf{t} \rangle = \langle \dot{\mathbf{n}}, \mathbf{n} \rangle = \langle \dot{\mathbf{b}}, \mathbf{b} \rangle = 0$. Let $\tau = \langle \dot{\mathbf{n}}, \mathbf{b} \rangle$. Then

$$\dot{\mathbf{n}} = -\kappa \mathbf{t} + \tau \mathbf{b},$$

τ is called the *torsion* of the curve. Because $\langle \mathbf{t}, \mathbf{b} \rangle = \langle \mathbf{n}, \mathbf{b} \rangle = 0$, $\langle \dot{\mathbf{t}}, \mathbf{b} \rangle = -\langle \mathbf{t}, \dot{\mathbf{b}} \rangle$ and $\langle \dot{\mathbf{n}}, \mathbf{b} \rangle = -\langle \mathbf{n}, \dot{\mathbf{b}} \rangle$. We get the following Frenet frame motion equation:

$$\frac{d}{ds} \begin{pmatrix} \mathbf{t} \\ \mathbf{n} \\ \mathbf{b} \end{pmatrix} = \begin{pmatrix} 0 & \kappa & 0 \\ -\kappa & 0 & \tau \\ 0 & -\tau & 0 \end{pmatrix} \begin{pmatrix} \mathbf{t} \\ \mathbf{n} \\ \mathbf{b} \end{pmatrix}$$

If τ is zero everywhere, then the curve is a planar curve; furthermore, if κ is zero, then the curve becomes a straight line.

The curvature and the torsion are the invariants of the curve under rigid motion in the Euclidean space.

Theorem 5.3 (Uniqueness of Curves). *Suppose* $\gamma_1(s)$ *and* $\gamma_2(s)$ *are two arc length parametric curves in* $\mathbb{R}^3$, *defined on the same parameter interval* $(0,1)$. *If* $\kappa_1(s) = \kappa_2(s) > 0, \tau_1(s) = \tau_2(s), \forall s \in (0,1)$, *then there exists a rigid motion in* $\mathbb{R}^3$, *mapping* $\gamma_2(s)$ *to* $\gamma_1(s)$.

Theorem 5.4 (Existence of Curves). *Let $\kappa(s), \tau(s)$ be two smooth functions defined on the interval $(0,1)$, furthermore $\kappa(s) > 0$. There exists an arc length parametric curve $\gamma(s)$, $s \in (0,1)$, in $\mathbb{R}^3$, whose arc length, curvature and torsion are s, κ and τ respectively.*

Both theorems can be proved by solving the Frenet frame motion equation using the ordinary differential equation method directly.

5.2 Local Theory of Surfaces

In this section, we focus on the local differential geometry of surfaces.

5.2.1 Regular Surface

Suppose a map from a planar domain $D = \{(u,v)\}$ to $\mathbb{R}^3$,

$$\mathbf{r}(u,v) = (x(u,v), y(u,v), z(u,v)),$$

such that all component functions are C^∞, and the vectors $\mathbf{r}_u = \left(\frac{\partial x}{\partial u}, \frac{\partial y}{\partial u}, \frac{\partial z}{\partial u}\right)$ and $\mathbf{r}_v = \left(\frac{\partial x}{\partial v}, \frac{\partial y}{\partial v}, \frac{\partial z}{\partial v}\right)$ are linearly independent, namely,

$$\mathbf{r}_u \wedge \mathbf{r}_v \neq 0.$$

Then we call $\mathbf{r}$ a *regular surface patch* in $\mathbb{R}^3$, and (u,v) are called the coordinate parameters of the surface.

$$\mathbf{n}(u,v) = \frac{\mathbf{r}_u \wedge \mathbf{r}_v}{|\mathbf{r}_u \wedge \mathbf{r}_v|},$$

is called the *normal* of the surface.

The same surface may have a different parametric representations. Consider a surface patch

$$\mathbf{r} : D \to \mathbb{R}^3,$$

and a parameter transformation

$$\sigma : (\bar{u}, \bar{v}) \in \bar{D} \to (u,v) \in D,$$

σ is a diffeomorphism, the determinant of the Jacobian matrix

$$\frac{\partial(u,v)}{\partial(\bar{u},\bar{v})} = \begin{vmatrix} \frac{\partial u(\bar{u},\bar{v})}{\partial \bar{u}} & \frac{\partial v(\bar{u},\bar{v})}{\partial \bar{u}} \\ \frac{\partial u(\bar{u},\bar{v})}{\partial \bar{v}} & \frac{\partial v(\bar{u},\bar{v})}{\partial \bar{v}} \end{vmatrix} \neq 0.$$

Then the surface has another parametric representation

$$\mathbf{r}(\bar{u},\bar{v}) = \mathbf{r} \circ \sigma(\bar{u},\bar{v}) = \mathbf{r}(u(\bar{u},\bar{v}), v(\bar{u},\bar{v})) : \bar{D} \to \mathbb{R}^3.$$

Suppose $(u(t), v(t))$ is a curve in the domain D. Then $\mathbf{r}(t) = \mathbf{r}(u(t), v(t))$ is a curve on the surface. The velocity vector of the curve

$$\left.\frac{d\mathbf{r}(t)}{dt}\right|_{t=0} = \mathbf{r}_u \left.\frac{du}{dt}\right|_{t=0} + \mathbf{r}_v \left.\frac{dv}{dt}\right|_{t=0}$$

is a *tangent vector* of the surface at $\mathbf{r}(u(0),v(0))$. All of the tangent vectors at a point $p \in S$ form a linear space, which is called the *tangent space* of S at p, and denoted as T_pS. From the definition, we can see that the tangent space is orthogonal to the normal $\mathbf{n}(u(0),v(0))$. The position $\mathbf{r}$ of p, the tangent vectors $\mathbf{r}_u, \mathbf{r}_v$ and the normal vector $\mathbf{n}$ together form a local frame of the surface S, $\{\mathbf{r}; \mathbf{r}_u, \mathbf{r}_v, \mathbf{n}\}$. It is a subtle question that whether the local frame can be continuously extended globally to cover the whole surface. The answer to the question depends on the topology of the surface.

We use the following symbols to represent the base tangent vectors:

$$\frac{\partial}{\partial u} := \mathbf{r}_u, \quad \frac{\partial}{\partial v} := \mathbf{r}_v.$$

If we change the parameters, the basis transformation of the tangent space can be calculated in a direct way:

$$\mathbf{r}_{\bar{u}} = \mathbf{r}_u \frac{\partial u}{\partial \bar{u}} + \mathbf{r}_v \frac{\partial v}{\partial \bar{u}},$$
$$\mathbf{r}_{\bar{v}} = \mathbf{r}_u \frac{\partial u}{\partial \bar{v}} + \mathbf{r}_v \frac{\partial v}{\partial \bar{v}}.$$

Namely

$$\begin{pmatrix} \frac{\partial}{\partial \bar{u}} \\ \frac{\partial}{\partial \bar{v}} \end{pmatrix} = \begin{pmatrix} \frac{\partial u}{\partial \bar{u}} & \frac{\partial v}{\partial \bar{u}} \\ \frac{\partial u}{\partial \bar{v}} & \frac{\partial v}{\partial \bar{v}} \end{pmatrix} \begin{pmatrix} \frac{\partial}{\partial u} \\ \frac{\partial}{\partial v} \end{pmatrix}.$$

A *smooth vector field* on the surface S can be represented as

$$f(u,v)\frac{\partial}{\partial u} + g(u,v)\frac{\partial}{\partial v},$$

where $f(u,v)$ and $g(u,v)$ are smooth functions.

5.2.2 First Fundamental Form

Suppose S is a regular surface in $\mathbb{R}^3$, its parametric representation is $\mathbf{r}(u,v)$. A tangent vector is represented as

$$d\mathbf{r} = \mathbf{r}_u du + \mathbf{r}_v dv,$$

where (du, dv) can be treated as the coordinates of $d\mathbf{r}$ on the tangent plane. The length of the tangent vector is

$$\langle d\mathbf{r}, d\mathbf{r} \rangle = du^2 \langle \mathbf{r}_u, \mathbf{r}_u \rangle + 2dudv \langle \mathbf{r}_u, \mathbf{r}_v \rangle + dv^2 \langle \mathbf{r}_v, \mathbf{r}_v \rangle.$$

We denote

$$E = \langle \mathbf{r}_u, \mathbf{r}_u \rangle, \quad F = \langle \mathbf{r}_u, \mathbf{r}_v \rangle, \quad G = \langle \mathbf{r}_v, \mathbf{r}_v \rangle.$$

Then the length of a tangent vector can be represented as

$$ds^2 = \begin{pmatrix} du & dv \end{pmatrix} \begin{pmatrix} E(u,v) & F(u,v) \\ F(u,v) & G(u,v) \end{pmatrix} \begin{pmatrix} du \\ dv \end{pmatrix},$$

where ds^2 is called the *first fundamental form*, which defines the inner product in the tangent space of the surface.

Suppose $(\bar{u},\bar{v})$ are other coordinate parameters. Then the first fundamental form has different representations

$$ds^2 = Edu^2 + 2Fdudv + Gdv^2 = \bar{E}d\bar{u}^2 + 2\bar{F}d\bar{u}d\bar{v} + \bar{G}d\bar{v}^2.$$

Note that

$$\begin{aligned} du &= \tfrac{\partial u}{\partial \bar{u}}d\bar{u} + \tfrac{\partial u}{\partial \bar{v}}d\bar{v}, \\ dv &= \tfrac{\partial v}{\partial \bar{u}}d\bar{u} + \tfrac{\partial v}{\partial \bar{v}}d\bar{v}. \end{aligned}$$

Let

$$J = \begin{pmatrix} \frac{\partial u}{\partial \bar{u}} & \frac{\partial v}{\partial \bar{u}} \\ \frac{\partial u}{\partial \bar{v}} & \frac{\partial v}{\partial \bar{v}} \end{pmatrix}.$$

Then

$$\begin{pmatrix} \bar{E} & \bar{F} \\ \bar{F} & \bar{G} \end{pmatrix} = J \begin{pmatrix} E & F \\ F & G \end{pmatrix} J^T.$$

We use a symbol $\mathbf{g}$ to represent the first fundamental form, which defines an inner product in the tangent space, denoted as $\langle , \rangle_{\mathbf{g}}$. Suppose that we have two tangent vectors in T_pS, $\mathbf{v} = \mathbf{r}_u du + \mathbf{r}_v dv$ and $\mathbf{w} = \mathbf{r}_u \delta u + \mathbf{r}_v \delta v$. The angle between them can be calculated as

$$\theta = \cos^{-1} \frac{\langle \mathbf{v},\mathbf{w}\rangle_{\mathbf{g}}}{\sqrt{\langle \mathbf{v},\mathbf{v}\rangle_{\mathbf{g}}}\sqrt{\langle \mathbf{w},\mathbf{w}\rangle_{\mathbf{g}}}}.$$

5.2.3 Second Fundamental Form

The first fundamental form $\langle d\mathbf{r}, d\mathbf{r}\rangle$ defines the lengths on the surface. The second fundamental form defines how the surface is embedded in the Euclidean space. The normal vector $\mathbf{n}$ indicates the relation between the surface and the outer space. We know the tangent vector has the form

$$d\mathbf{r} = \mathbf{r}_u du + \mathbf{r}_v dv.$$

Considering the parametric surface $\mathbf{n}(u,v)$, whose tangent vector is

$$d\mathbf{n} = \mathbf{n}_u du + \mathbf{n}_v dv,$$

we define the *second fundamental form* as

$$II = -\langle d\mathbf{r}, d\mathbf{n}\rangle,$$

where $\langle , \rangle$ is the Euclidean inner product.

The normal is perpendicular to the tangent plane, therefore

$$\langle \mathbf{r}_u, \mathbf{n}\rangle = 0, \quad \langle \mathbf{r}_v, \mathbf{n}\rangle = 0.$$

Taking partial derivatives of the above equation, we get

$$\begin{aligned} &\langle \mathbf{r}_{uu}, \mathbf{n}\rangle + \langle \mathbf{r}_u, \mathbf{n}_u\rangle = 0, \quad &\langle \mathbf{r}_{uv}, \mathbf{n}\rangle + \langle \mathbf{r}_u, \mathbf{n}_v\rangle = 0, \\ &\langle \mathbf{r}_{vu}, \mathbf{n}\rangle + \langle \mathbf{r}_v, \mathbf{n}_u\rangle = 0, \quad &\langle \mathbf{r}_{vv}, \mathbf{n}\rangle + \langle \mathbf{r}_v, \mathbf{n}_v\rangle = 0. \end{aligned}$$

Let

$$\begin{aligned} L &= \langle \mathbf{r}_{uu}, n \rangle = -\langle \mathbf{r}_u, \mathbf{n}_u \rangle, \\ M &= \langle \mathbf{r}_{uv}, n \rangle = -\langle \mathbf{r}_u, \mathbf{n}_v \rangle = -\langle \mathbf{r}_v, \mathbf{n}_u \rangle, \\ N &= \langle \mathbf{r}_{vv}, n \rangle = -\langle \mathbf{r}_v, \mathbf{n}_v \rangle. \end{aligned}$$

Then the second fundamental form is

$$II = \begin{pmatrix} du & dv \end{pmatrix} \begin{pmatrix} L(u,v) & M(u,v) \\ M(u,v) & N(u,v) \end{pmatrix} \begin{pmatrix} du \\ dv \end{pmatrix}$$

If we change to another parameter $(\bar{u}, \bar{v})$, similarly, we can show that

$$\begin{pmatrix} \bar{L} & \bar{M} \\ \bar{M} & \bar{N} \end{pmatrix} = J \begin{pmatrix} L & M \\ M & N \end{pmatrix} J^T.$$

5.2.4 Weingarten Transformation

The so-called *Gauss map* maps a point on the surface to its normal vector on the unit sphere

$$G : S \to \mathbb{S}^2, \quad \mathbf{r}(u,v) \to \mathbf{n}(u,v).$$

We can compute the *derivative map* of the Gauss map, namely,

$$\mathscr{W} : T_pS \to T_{n(p)}\mathbb{S}^2, \quad d\mathbf{r} \to d\mathbf{n},$$

which is called the *Weingarten map*. We calculate the Weingarten map using the first and the second fundamental forms as follows.

Because the tangent space T_pS and $T_{n(p)}\mathbb{S}^2$ are both orthogonal to the normal $\mathbf{n}(p)$, therefore, they are parallel to each other in the Euclidean space $\mathbb{R}^3$. We can treat them as the same plane. Then the Weingarten map is a linear map from a plane to itself.

We explicitly compute the Weingarten transformation coefficients matrix

$$\begin{pmatrix} a & b \\ c & d \end{pmatrix}$$

using the basis $\{\mathbf{r}_u, \mathbf{r}_v\}$,

$$\begin{aligned} \mathscr{W}(\mathbf{r}_u) &= -\mathbf{n}_u = a\mathbf{r}_u + b\mathbf{r}_v, \\ \mathscr{W}(\mathbf{r}_v) &= -\mathbf{n}_v = c\mathbf{r}_u + d\mathbf{r}_v. \end{aligned}$$

Then

$$-\begin{pmatrix} \mathbf{n}_u \\ \mathbf{n}_v \end{pmatrix} \begin{pmatrix} \mathbf{r}_u & \mathbf{r}_v \end{pmatrix} = \begin{pmatrix} a & b \\ c & d \end{pmatrix} \begin{pmatrix} \mathbf{r}_u \\ \mathbf{r}_v \end{pmatrix} \begin{pmatrix} \mathbf{r}_u & \mathbf{r}_v \end{pmatrix}.$$

Therefore the coefficients matrix of the Weingarten transformation is

$$\begin{pmatrix} a & b \\ c & d \end{pmatrix} = \begin{pmatrix} L & M \\ M & N \end{pmatrix} \begin{pmatrix} E & F \\ F & G \end{pmatrix}^{-1} = \frac{1}{EG - F^2} \begin{pmatrix} LG - MF & ME - LF \\ MG - NF & NE - MF \end{pmatrix}.$$

It is obvious that the Weingarten map is symmetric, namely $\langle \mathscr{W}(\mathbf{v}), \mathbf{w} \rangle = \langle \mathbf{v}, \mathscr{W}(\mathbf{w}) \rangle$. The eigen values k_1, k_2 are called the *principal curvatures* of the surface. The eigen directions are called the *principal directions*. $H = \frac{1}{2}(k_1 + k_2)$ is called the *mean curvature* of the surface, $K = k_1 k_2$ is called the *Gaussian curvature*.

From the Weingarten matrix, we know that the principle curvatures are the roots of the following equation:

$$k^2 - \frac{LG - 2MF + NE}{EG - F^2} k + \frac{LN - M^2}{EG - F^2} = 0.$$

From the relations between the roots and the coefficients of a quadratic equation, we obtain

$$H = \frac{1}{2} \frac{LG - 2MF + NE}{EG - F^2}, \tag{5.1}$$

$$K = \frac{LN - M^2}{EG - F^2}. \tag{5.2}$$

K is the Jacobian of the Gauss map, therefore, it is the *area ratio* between a surface region and its Gauss image. A surface area element is $\mathbf{r}_u \wedge \mathbf{r}_v$, the corresponding area element on the Gauss sphere is $\mathbf{n}_u \wedge \mathbf{n}_v$, therefore,

$$K = \frac{\mathbf{n}_u \wedge \mathbf{n}_v}{\mathbf{r}_u \wedge \mathbf{r}_v}.$$

Definition 5.5 (Normal Curvature). *Suppose* $\mathbf{v} \in T_pS$ *is a tangent vector, the normal* $\mathbf{n}$ *and* $\mathbf{v}$ *determine a sectional plane intersecting the surface at a sectional curve* γ. *The curvature of* γ *at* p *is called the* normal curvature *of the surface along* $\mathbf{v}$.

Because the curve is planar, its normal coincides with the normal of the surface, therefore,

$$k_n(\mathbf{v}) = \left\langle \frac{d^2\mathbf{r}}{ds^2}, \mathbf{n} \right\rangle = \frac{\langle d^2\mathbf{r}, \mathbf{n} \rangle}{ds^2} = \frac{II(\mathbf{v},\mathbf{v})}{I(\mathbf{v},\mathbf{v})}.$$

Suppose $\mathbf{v}$ is a unit vector. Then the normal curvature can be computed as

$$k_n(\mathbf{v}) = \langle \mathbf{n}, d\mathbf{v} \rangle = -\langle d\mathbf{n}, \mathbf{v} \rangle = \langle \mathcal{W}(\mathbf{v}), \mathbf{v} \rangle.$$

Choose $\mathbf{e}_1$ and $\mathbf{e}_2$ as being the unit vectors along the eigen directions of the Weingarten map. Then $\mathbf{e}_1 \perp \mathbf{e}_2$. Suppose that $\mathbf{v} \in T_pS$ is a unit vector, and assume

$$\mathbf{v} = \cos\theta \mathbf{e}_1 + \sin\theta \mathbf{e}_2,$$

where θ is the angle between $\mathbf{v}$ and $\mathbf{e}_1$. Then the normal curvature along $\mathbf{v}$ is

$$k_n(\mathbf{v}) = \langle \mathcal{W}(\mathbf{v}), \mathbf{v} \rangle = k_1 \cos^2\theta + k_2 \sin^2\theta.$$

5.3 Orthonormal Movable Frame

We introduce the *orthonormal movable frame* method to construct the motion equation to study the geometry of a surface.

Suppose a regular surface S is embedded in $\mathbb{R}^3$, a parametric representation is $\mathbf{r}(u,v)$. Select two vector fields $\mathbf{e}_1, \mathbf{e}_2$, such that

$$\langle \mathbf{e}_1, \mathbf{e}_1 \rangle = \langle \mathbf{e}_2, \mathbf{e}_2 \rangle = 1, \quad \langle \mathbf{e}_1, \mathbf{e}_2 \rangle = 0.$$

Furthermore, $\mathbf{e}_1$,$\mathbf{e}_2$ are smooth with respect to (u,v).

Let $\mathbf{e}_3 = \mathbf{e}_1 \wedge \mathbf{e}_2$ be the unit normal field of the surface. Then

$$\{\mathbf{r}; \mathbf{e}_1, \mathbf{e}_2, \mathbf{e}_3\}$$

form the *orthonormal frame field* of the surface.

The tangent vector can be represented as the linear combination of the movable frame:

$$d\mathbf{r} = \omega_1 \mathbf{e}_1 + \omega_2 \mathbf{e}_2,$$

where ω_1 and ω_2 are two differential 1-forms, $\omega_1(\mathbf{v}) = \langle \mathbf{e}_1, \mathbf{v} \rangle$ and $\omega_2(\mathbf{v}) = \langle \mathbf{e}_2, \mathbf{v} \rangle$,

The first fundamental form of the surface is represented as

$$I = \langle d\mathbf{r}, d\mathbf{r} \rangle = \omega_1 \omega_1 + \omega_2 \omega_2.$$

Differentiate $\mathbf{e}_i$, and represent $d\mathbf{e}_i$ in the frame $\{\mathbf{e}_1, \mathbf{e}_2, \mathbf{e}_3\}$. Then

$$d\mathbf{e}_i = \omega_{i1} \mathbf{e}_1 + \omega_{i2} \mathbf{e}_2 + \omega_{i3} \mathbf{e}_3, \quad i \in \{1,2,3\},$$

where the differential 1-form

$$\omega_{ij} = \langle d\mathbf{e}_i, \mathbf{e}_j \rangle, \quad i,j \in \{1,2,3\}.$$

Because $\langle \mathbf{e}_i, \mathbf{e}_j \rangle = \delta_i^j$,

$$\langle d\mathbf{e}_i, \mathbf{e}_j \rangle + \langle \mathbf{e}_i, d\mathbf{e}_j \rangle = 0,$$

therefore

$$\omega_{ij} + \omega_{ji} = 0, \quad \omega_{ii} = 0.$$

The second fundamental form can be represented as

$$II = -\langle d\mathbf{r}, d\mathbf{e}_3 \rangle = -\omega_1 \omega_{31} - \omega_2 \omega_{32} = \omega_1 \omega_{13} + \omega_2 \omega_{23}.$$

Therefore, we get the *motion equation* of the orthonormal frame of the surface:

$$d\mathbf{r} = \omega_1 \mathbf{e}_1 + \omega_2 \mathbf{e}_2,$$

$$\begin{pmatrix} d\mathbf{e}_1 \\ d\mathbf{e}_2 \\ d\mathbf{e}_3 \end{pmatrix} = \begin{pmatrix} 0 & \omega_{12} & \omega_{13} \\ -\omega_{12} & 0 & \omega_{23} \\ -\omega_{13} & -\omega_{23} & 0 \end{pmatrix} \begin{pmatrix} \mathbf{e}_1 \\ \mathbf{e}_2 \\ \mathbf{e}_3 \end{pmatrix} \tag{5.3}$$

and the first and the second fundamental forms are

$$I = \omega_1 \omega_1 + \omega_2 \omega_2, \tag{5.4}$$

$$II = \omega_1 \omega_{13} + \omega_2 \omega_{23}. \tag{5.5}$$

Suppose that we choose another orthonormal frame field $\{\mathbf{r}; \tilde{\mathbf{e}}_1, \tilde{\mathbf{e}}_2, \tilde{\mathbf{e}}_3\}$, $\tilde{\mathbf{e}}_3 = \mathbf{e}_3$, the angle between $\mathbf{e}_1$ and $\tilde{\mathbf{e}}_1$ is $\theta(u,v)$. Then

$$\begin{cases} \tilde{\mathbf{e}}_1 = \cos\theta\mathbf{e}_1 + \sin\theta\mathbf{e}_2, \\ \tilde{\mathbf{e}}_2 = -\sin\theta\mathbf{e}_1 + \cos\theta\mathbf{e}_2. \end{cases}$$

Assuming that the coefficients of the motion equation of the frame $\{\mathbf{r};\tilde{\mathbf{e}}_1,\tilde{\mathbf{e}}_2,\tilde{\mathbf{e}}_3\}$ are $\{\tilde{\omega}_{ij} : i,j = 1,2,3\}$, by direct computation, we obtain

$$\begin{pmatrix} \tilde{\omega}_1 \\ \tilde{\omega}_2 \end{pmatrix} = \begin{pmatrix} \cos\theta & \sin\theta \\ -\sin\theta & \cos\theta \end{pmatrix} \begin{pmatrix} \omega_1 \\ \omega_2 \end{pmatrix},$$

$$\begin{pmatrix} \tilde{\omega}_{31} \\ \tilde{\omega}_{32} \end{pmatrix} = \begin{pmatrix} \cos\theta & \sin\theta \\ -\sin\theta & \cos\theta \end{pmatrix} \begin{pmatrix} \omega_{31} \\ \omega_{32} \end{pmatrix},$$

therefore,

$$\begin{aligned} \tilde{\omega}_1\tilde{\omega}_1 + \tilde{\omega}_2\tilde{\omega}_2 &= \omega_1\omega_1 + \omega_2\omega_2, \\ \tilde{\omega}_1\tilde{\omega}_{31} + \tilde{\omega}_2\tilde{\omega}_{32} &= \omega_1\omega_{31} + \omega_2\omega_{32}. \end{aligned}$$

We have proved the following theorem.

Theorem 5.6. *The first fundamental form and the second fundamental form are independent of the choice of the orthonormal frames.*

ω_{13} and ω_{23} can be represented as the linear combinations of ω_1 and ω_2. Therefore, let

$$\begin{pmatrix} \omega_{13} \\ \omega_{23} \end{pmatrix} = \begin{pmatrix} h_{11} & h_{12} \\ h_{21} & h_{22} \end{pmatrix} \begin{pmatrix} \omega_1 \\ \omega_2 \end{pmatrix}$$

Then the matrix

$$H = \begin{pmatrix} h_{11} & h_{12} \\ h_{21} & h_{22} \end{pmatrix}$$

represents the Weingarten mapping

$$d\mathbf{r} = \omega_1\mathbf{e}_1 + \omega_2\mathbf{e}_2 \to -d\mathbf{e}_3 = \omega_{31}\mathbf{e}_1 + \omega_{32}\mathbf{e}_2.$$

Therefore, H is symmetric. The eigen values of H, k_1, k_2, represent the principal curvature of the surface, the determinant $\det H = h_{11}h_{22} - h_{12}h_{21}$ is the Gaussian curvature, the trace $\frac{1}{2}\mathrm{tr}H = \frac{1}{2}(h_{11} + h_{22})$ is the mean curvature.

5.3.1 Structure Equation

Suppose $\{\mathbf{r};\mathbf{e}_1,\mathbf{e}_2,\mathbf{e}_3\}$ is an orthonormal frame of a surface S. Then the motion equations are

$$\begin{aligned} d\mathbf{r} &= \omega_1\mathbf{e}_1 + \omega_2\mathbf{e}_2, \\ d\mathbf{e}_i &= \Sigma_{j=1}^{3}\,\omega_{ij}\mathbf{e}_j, \quad \omega_{ij} + \omega_{ji} = 0, \quad i = 1,2,3. \end{aligned}$$

From $d^2\mathbf{r} = 0$, we get

$$\omega_1 \wedge \omega_{13} + \omega_2 \wedge \omega_{23} = 0,$$

namely,

$$(h_{12} - h_{21})\omega_1 \wedge \omega_2 = 0,$$

therefore the matrix H is symmetric.

Furthermore, we get

$$\begin{cases} d\omega_1 = \omega_2 \wedge \omega_{21}, \\ d\omega_2 = \omega_1 \wedge \omega_{12}. \end{cases} \tag{5.6}$$

From $d^2\mathbf{e}_i = 0,\ i = 1,2$, we obtain the following equations:

$$d\omega_{12} = \omega_{13} \wedge \omega_{32}, \tag{5.7}$$

and further

$$\begin{cases} d\omega_{13} = \omega_{12} \wedge \omega_{23} \\ d\omega_{23} = \omega_{21} \wedge \omega_{13}. \end{cases} \tag{5.8}$$

Equations (5.6) — (5.8) are together called the *orthonormal frame structure equations*, they are the necessary and sufficient conditions for the integrability of the orthonormal frame motion equations.

Equation (5.8) is called the *Codazzi equation*. Equation (5.7) is called the *Gauss equation*. $\omega_{13} \wedge \omega_{23}$ is the area element on the Gauss sphere, $\omega_{13} = h_{11}\omega_1 + h_{12}\omega_2$, $\omega_{23} = h_{21}\omega_1 + h_{22}\omega_2$, therefore,

$$d\omega_{12} = -(h_{11}h_{22} - h_{12}^2)\omega_1 \wedge \omega_2 = -K\omega_1 \wedge \omega_2.$$

The Gauss equation shows that although the original definition of Gaussian curvature involves the normal vector $\mathbf{e}_3$, namely, the embedding of the surface in $\mathbb{R}^3$, in fact, Gaussian curvature can be computed only using the tangent vectors $\mathbf{e}_1, \mathbf{e}_2$ and their derivatives, $\omega_{12} = \langle d\mathbf{e}_1, \mathbf{e}_2 \rangle$. This means that the Gaussian curvature is solely determined by the Riemannian metric and independent of the embedding.

Theorem 5.7 (Gauss's Theorema Egregium). *The Gaussian curvature is intrinsic, solely determined by the first fundamental form.*

In summary, the *motion equation* for the frames are

$$\begin{aligned} d\mathbf{r} &= \omega_1\mathbf{e}_1 + \omega_2\mathbf{e}_2, \\ d\mathbf{e}_1 &= \omega_{12}\mathbf{e}_2 + \omega_{13}\mathbf{e}_3, \\ d\mathbf{e}_2 &= \omega_{21}\mathbf{e}_1 + \omega_{23}\mathbf{e}_3, \\ d\mathbf{e}_3 &= \omega_{31}\mathbf{e}_1 + \omega_{32}\mathbf{e}_2, \end{aligned}$$

where $\omega_{ij} + \omega_{ji} = 0$. The structure equation for the surface are

$$\begin{aligned} d\omega_1 &= \omega_{12} \wedge \omega_2, \\ d\omega_2 &= \omega_{21} \wedge \omega_1, \\ d\omega_{12} &= \omega_{13} \wedge \omega_{32}, \\ d\omega_{13} &= \omega_{12} \wedge \omega_{23} \\ d\omega_{23} &= \omega_{21} \wedge \omega_{13}. \end{aligned}$$

The coefficients for the motion equation are differential 1-forms

$$\{\omega_1, \omega_2, \omega_{12}, \omega_{13}, \omega_{23}\},$$

which are independent of the choice of the parameters, but dependent of the choice of the orthonormal frames.

Let $\{\mathbf{r};\bar{\mathbf{e}}_1,\bar{\mathbf{e}}_2,\bar{\mathbf{e}}_3\}$ be another orthonormal frame field of the surface, $\{\bar{\omega}_1,\bar{\omega}_2,\bar{\omega}_{12},\bar{\omega}_{13},\bar{\omega}_{23}\}$ be the corresponding differential 1-forms. Assume $\bar{\mathbf{e}}_1$ is obtained by rotating $\mathbf{e}_1$ about $\mathbf{e}_3$ by an angle θ. Then

$$\begin{aligned}\bar{\mathbf{e}}_1 &= \cos\theta\mathbf{e}_1+\sin\theta\mathbf{e}_2,\\ \bar{\mathbf{e}}_2 &= -\sin\theta\mathbf{e}_1+\cos\theta\mathbf{e}_2,\end{aligned}$$

and

$$\begin{aligned}\bar{\omega}_1 &= \cos\theta\,\omega_1+\sin\theta\,\omega_2,\\ \bar{\omega}_2 &= -\sin\theta\,\omega_1+\cos\theta\,\omega_2,\end{aligned}$$

$$\begin{aligned}\bar{\omega}_{13} &= \cos\theta\,\omega_{13}+\sin\theta\,\omega_{23},\\ \bar{\omega}_{23} &= -\sin\theta\,\omega_{13}+\cos\theta\,\omega_{23}.\end{aligned}$$

Furthermore,

$$\bar{\omega}_{12}=\omega_{12}+d\theta. \tag{5.9}$$

Therefore, the following are geometric quantities, independent of the choice of the orthonormal frames.

1. *First fundamental form*
$$I=\omega_1\omega_1+\omega_2\omega_2.$$
2. *Area element of the surface*
$$dA=\omega_1\wedge\omega_2=\bar{\omega}_1\wedge\bar{\omega}_2.$$
3. *Second fundamental form*
$$II=\omega_1\omega_{13}+\omega_2\omega_{23}=\bar{\omega}_1\bar{\omega}_{13}+\bar{\omega}_2\bar{\omega}_{23}.$$

Furthermore, the *third fundamental form*

$$III=\langle d\mathbf{e}_3,d\mathbf{e}_3\rangle=\omega_{13}\omega_{13}+\omega_{23}\omega_{23}=\bar{\omega}_{13}\bar{\omega}_{13}+\bar{\omega}_{23}\bar{\omega}_{23};$$

the area element induced by the

$$\omega_{13}\wedge\omega_{23}=\bar{\omega}_{13}\wedge\bar{\omega}_{23}.$$

Suppose $\sigma:S\to\tilde{S}$ is an isometric map between two surfaces. Then we can select orthonormal frames $\{\tilde{\mathbf{r}};\tilde{\mathbf{e}}_1,\tilde{\mathbf{e}}_2,\tilde{\mathbf{e}}_3\}$ for $\tilde{S}$, such that

$$\omega_1=\tilde{\omega}_1,\quad \omega_2=\tilde{\omega}_2.$$

$\sigma:S\to\tilde{S}$ is a conformal map between two surfaces if and only if there exists a positive function λ, such that at the corresponding points,

$$\tilde{I}=\lambda^2 I.$$

Theorem 5.8 (Isothermal Coordinates [48]). *For any point on the surface, there exists a neighborhood, which can be conformally mapped to a planar domain, namely, we can find (u,v) coordinates for a neighborhood of the surface, the first fundamental form is*

$$I=\lambda^2(u,v)(du^2+dv^2),\quad \lambda(u,v)>0.$$

The following theorem is fundamental for the existence and the uniqueness of surfaces.

Theorem 5.9 (Existence of Surfaces). *If I and II can be found, so that I is positive definite, the Gauss equation and the Codazzi equation hold, then there exists a surface uniquely determined up to a Euclidean rigid motion for which I and II are the first and second fundamental forms.*

5.4 Covariant Differentiation

In this section, we discuss the intrinsic geometry of the surface.

From the previous discussion, we know that $\{\omega_1, \omega_2, \omega_{12}\}$ satisfy

$$d\omega_1 = \omega_{12} \wedge \omega_2, \quad d\omega_2 = -\omega_{12} \wedge \omega_1,$$

furthermore, $d\omega_{12} = -K\omega_1 \wedge \omega_2$.

We use isothermal coordinates. Then

$$ds^2 = \lambda^2(du^2 + dv^2) = \omega_1\omega_1 + \omega_2\omega_2.$$

Let

$$\omega_1 = \lambda du, \quad \omega_2 = \lambda dv.$$

Then

$$\begin{aligned} d\omega_1 &= -\lambda_v du \wedge dv = -\tfrac{\lambda_v}{\lambda} du \wedge \omega_2, \\ d\omega_2 &= \lambda_u du \wedge dv = -\tfrac{\lambda_u}{\lambda} dv \wedge \omega_1, \end{aligned}$$

therefore,

$$\begin{aligned} \omega_{12} &= -\tfrac{\lambda_v}{\lambda} du + \tfrac{\lambda_u}{\lambda} dv, \\ d\omega_{12} &= -\tfrac{1}{\lambda^2}\left(\tfrac{\partial^2}{\partial u^2} + \tfrac{\partial^2}{\partial v^2}\right) \ln\lambda. \end{aligned}$$

In the motion equation of the orthonormal frame,

$$\begin{aligned} d\mathbf{e}_1 &= \omega_{12}\mathbf{e}_2 + \omega_{13}\mathbf{e}_3, \\ d\mathbf{e}_2 &= \omega_{21}\mathbf{e}_1 + \omega_{23}\mathbf{e}_3. \end{aligned}$$

The differentiation of the frame has a tangential component and a normal component. We define the tangential component as the *covariant differentiation*:

$$D\mathbf{e}_1 = \omega_{12}\mathbf{e}_2, \quad D\mathbf{e}_2 = -\omega_{12}\mathbf{e}_1.$$

Definition 5.10 (Covariant Differentiation). *Covariant differentiation is the generalization of directional derivative with the following properties. Suppose* $\mathbf{v}, \mathbf{w}$ *are tangent vector fields,* f *is a function defined on the surface. Then*

1. $D(\mathbf{v} + \mathbf{w}) = D\mathbf{v} + D\mathbf{w}$.
2. $D(f\mathbf{v}) = df\mathbf{v} + fD\mathbf{v}$.
3. $D\langle \mathbf{v}, \mathbf{w}\rangle = \langle D\mathbf{v}, \mathbf{w}\rangle + \langle \mathbf{v}, D\mathbf{w}\rangle$.

Now we can define the concept of *parallel transportation* on the surface.

Definition 5.11 (Parallel Transportation). *Let S be a surface embedded in $\mathbb{R}^3$, $\mathbf{r}(u,v)$ be its parametric representation, $p,q \in S$, a curve $\gamma(t) = (u(t),v(t))$ connect p and q. Suppose $\mathbf{v}(t)$ is a vector field along the curve. Then we say $\mathbf{v}(t)$ is* parallel along the curve *γ, if*

$$\frac{D\mathbf{v}}{dt} = 0.$$

Let $\mathbf{v}(t) = f_1\mathbf{e}_1 + f_2\mathbf{e}_2$. Then

$$\frac{D\mathbf{v}}{dt} = \left(\frac{df_1}{dt} + f_2\frac{\omega_{21}}{dt}\right)\mathbf{e}_1 + \left(\frac{df_2}{dt} + f_1\frac{\omega_{12}}{dt}\right)\mathbf{e}_2,$$

where $\frac{\omega_{12}}{dt}$ means $\omega_{12}(\mathbf{w})$, $\mathbf{w} = \frac{d\gamma}{dt}$ is the velocity vector of the curve.

Therefore, a parallel vector field satisfies the following ordinary equation

$$\begin{cases} \frac{df_1}{dt} + f_2\frac{\omega_{21}}{dt} = 0, \\ \frac{df_2}{dt} + f_1\frac{\omega_{12}}{dt} = 0. \end{cases}$$

Therefore, given any tangent vector on T_pS, we can find a unique vector field parallel along the curve, and get a unique tangent vector at T_qS. Namely, we build a one to one map between T_pS and T_qS to connect the two different tangent spaces. That is the reason why we call ω_{12} the *connection*.

Suppose two vector fields $\mathbf{v}(t)$ and $\mathbf{w}(t)$ are parallel along the curve. Then

$$\frac{d}{dt}\langle \mathbf{v},\mathbf{w}\rangle = \left\langle \frac{D\mathbf{v}}{dt},\mathbf{w}\right\rangle + \left\langle \mathbf{v},\frac{D\mathbf{w}}{dt}\right\rangle = 0.$$

Namely, parallel transportation preserves the length and the angle.

5.4.1 Geodesic Curvature

Let S be a surface in $\mathbb{R}^3$, $\mathbf{r} = \mathbf{r}(u^1,u^2)$ be the parametric representation, $\gamma(s) = \mathbf{r}(u^1(s), u^2(s))$ be a curve parameterized using the arc length. Choose an orthonormal frame field $\{\mathbf{e}_1,\mathbf{e}_2,\mathbf{e}_3\}$ along the curve such that $\mathbf{e}_1 = \frac{d\mathbf{r}}{ds}$, $\mathbf{e}_3$ is the normal. Then

Definition 5.12 (Geodesic Curvature). $k_g = \langle \frac{D\mathbf{e}_1}{ds},\mathbf{e}_2\rangle$ *is called the* geodesic curvature *of the curve,* $\mathbf{k}_g = \frac{D\mathbf{e}_1}{ds} = k_g\mathbf{e}_2$ *is called the* geodesic curvature vector.

Given a curve on the surface, then its curvature vector is

$$\frac{d^2\gamma}{ds^2} = k_g\mathbf{e}_2 + k_n\mathbf{e}_3.$$

Intuitively, the geodesic curvature vector is the tangential component of the curvature vector, and the normal curvature vector is the normal component of the curvature vector, therefore

$$k^2 = k_g^2 + k_n^2.$$

k_n depends on both the first fundamental form and the second fundamental form, therefore it is extrinsic. k_g is independent of the embedding of the surface in $\mathbb{R}^3$, therefore it is intrinsic.

Definition 5.13 (Geodesic). *A curve on the surface is called a* geodesic, *if its geodesic curvature is zero.*

A geodesic can be formulated as a curve whose velocity vector field is parallel along itself. A geodesic also can be described as the curve, whose principal normal is parallel to the normal of the surface, namely, the tangential projection of the curvature vector is always zero everywhere along the curve. A geodesic can be computed by solving an ordinary differential equation.

Let $\mathbf{e}_1$ be the unit velocity vector field of a geodesic $\gamma : (0,1) \to M$, $\mathbf{e}_2$ be the unit tangent vector field orthogonal to $\mathbf{e}_1$, $f : [0,1] \to \mathbb{R}$, $f(0) = f(1) = 0$. We choose a family of curves on the surface:

$$F(s,t) : (-\varepsilon, \varepsilon) \times (0,1) \to M,$$

such that $F(0,t) = \gamma(t)$, furthermore,

$$F(s,0) = p, \quad F(s,1) = q, \quad \frac{\partial F(s,t)}{\partial s} = f(t)\mathbf{e}_2(t).$$

Then we define the length of the curve $\gamma_s = F(s,\cdot)$, and $\{\gamma_s\}$ form a variation of the curve γ_0,

$$L(s) = \int_0^1 \left| \frac{\partial \gamma_s(t)}{dt} \right| dt.$$

Then

$$\frac{L(s)}{ds} = -\int_0^1 f k_g(s) ds.$$

If γ is the shortest curve connecting the two points, then $L(s)$ reaches its critical point at 0, $k_g = 0$, and γ is a geodesic.

Theorem 5.14. *Suppose a curve γ is the shortest curve connecting p and q on the surface. Then γ is a geodesic.*

If we consider the second derivative $\frac{d^2L(\lambda)}{d\lambda^2}$, it is determined by the Gaussian curvature. If $K < 0$ along the geodesic, then the second derivative is positive, meaning the geodesic reaches the minimum; if $K > 0$, the second derivative is positive, the geodesic reaches the maximum, and therefore it is unstable.

If the surface has a handle, then there must be a shortest loop with a non-trivial homotopy class, and the geodesic is stable. Suppose the Gaussian curvature is positive everywhere. Then any closed geodesic is unstable, therefore, the surface must have zero genus.

5.5 Gauss-Bonnet Theorem

Suppose D is a topological disk on the surface, ∂D is a closed loop, denoted as $\gamma(s)$, where s is the arc length parameter. Suppose the angle between $\dot{\gamma}(s)$ and $\mathbf{e}_1(s)$ is $\alpha(s)$, i. e.,

$$\dot{\gamma} = \cos\alpha \mathbf{e}_1 + \sin\alpha \mathbf{e}_2.$$

Then

$$\mathbf{e}_3 \wedge \dot{\gamma} = -\sin\alpha\mathbf{e}_1 + \cos\alpha\mathbf{e}_2.$$

The geodesic curvature of γ is

$$k_g = \left\langle \frac{D\dot{\gamma}}{ds}, \mathbf{e}_3 \wedge \dot{\gamma} \right\rangle = \frac{d\alpha + \omega_{12}}{ds}.$$

Then

$$\int_D K\omega_1 \wedge \omega_2 = -\int_D d\omega_{12} = -\int_{\partial D} \omega_{12},$$

therefore

$$\int_D KdA + \int_{\partial D} k_g ds = \int_{\partial D} d\alpha.$$

Because $\alpha(0) = \alpha(l)$, l is the length of $\partial D = \gamma$, therefore $\int_{\partial D} d\alpha = 2k\pi$. We can shrink D to form a small circle on an isothermal coordinates, which preserve angles, therefore we get $k = 1$.

If γ is not smooth with outer angles α_i, we can use a smooth curve to approximate it as close as possible, therefore, we have proven the following Gauss-Bonnet theorem.

Theorem 5.15 (Gauss-Bonnet Theorem). *Suppose D is a simply connected region on the surface S, ∂D is a piecewise smooth closed curve, and assume α_i's are the outer angles of the vertices of ∂D. Then*

$$\int_D KdA + \int_{\partial D} k_g ds + \sum \alpha_i = 2\pi.$$

5.6 Index Theorem of Tangent Vector Field

Suppose $\mathbf{v}$ is a tangent vector field on a closed surface S. Then we can construct a family map $f_t : S \to S$, in the following way. Suppose $p \in S$. Then we find a path $\gamma : [0,t] \to S$, such that

$$\gamma(0) = p, \quad \frac{d\gamma}{dt}(t) = \mathbf{v}(\gamma(t)).$$

Then the map is defined as

$$f_t : S \to S, \gamma(0) \to \gamma(t).$$

If $\mathbf{v}$ is non zero everywhere, and t is small enough, then $\gamma(0) \neq \gamma(t)$ for all point $p \in S$. Therefore, f_t has no fixed point and is homotopic to the identity map. According to Lefschetz fixed point theorem, $\chi(S)$ must be zero. Therefore, we show that

Proposition 5.16. *For a closed surface with non zero Euler number, any tangent vector field must have zero points.*

Suppose $\mathbf{v}$ is a tangent field on the surface, the singularities are isolated, $P = \{p_1, p_2, \cdots, p_n\}$. Let $\mathbf{e}_1 = \frac{\mathbf{v}}{|\mathbf{v}|}$ defined on $S - P$, $\mathbf{e}_3$ to be the normal, $\mathbf{e}_2 = \mathbf{e}_3 \wedge \mathbf{e}_1$. Then we have the connection ω_{12}. Suppose p_i is an isolated singularity, γ_ε is a geodesic circle with radius ε.

Definition 5.17 (Index of Singularity). *The index of the singularity is*

$$I(\mathbf{v}, p_i) = \frac{1}{2\pi} \lim_{\varepsilon \to 0} \int_{\gamma_\varepsilon} \omega_{12}.$$

We choose isothermal coordinates of a neighborhood of p_i,

$$I = e^{2\lambda}(du^2 + dv^2),$$

$\tilde{\omega}_{12} = -\lambda_v du + \lambda_u dv$, the angle between $\mathbf{v}$ and the iso-parametric curve $v = const$ is θ. Then

$$\omega_{12} = d\theta + \tilde{\omega}_{12},$$

and

$$I(\mathbf{v}, p_i) = \frac{1}{2\pi} \int_{\gamma_\varepsilon} d\theta + \tilde{\omega}_{12} = \frac{1}{2\pi} \int_{\gamma_\varepsilon} d\theta,$$

therefore $I(\mathbf{v}, p_i)$ must be an integer. The singularity indices are shown in Fig. 5.1.

Now, we want to show that the total singularity index is independent of the choice of the tangent vector field. Suppose we choose a neighborhood D_i of p_i, which is small enough, such that

$$2\pi \sum_i I(\mathbf{v}, p_i) = \sum_i \int_{\partial D_i} \omega_{12} = -\int_{S-\cup D_i} d\omega_{12} = \int_{S-\cup D_i} K dA.$$

Letting the radius of D_i go to zero, we get

$$2\pi \sum_i I(\mathbf{v}, p_i) = \int_S K dA.$$

The right-hand side is independent of the choice of the vector field $\mathbf{v}$, therefore this shows that

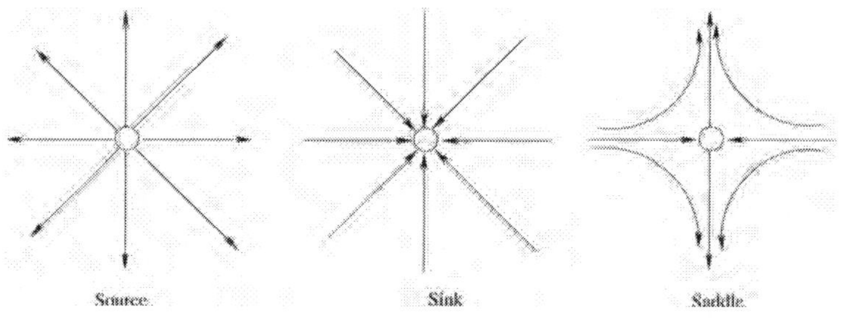

Fig. 5.1. Index of singularities of a vector field.

Lemma 5.18. *The total index of singularities of a tangent vector field is independent of the choice of the tangent vector field.*

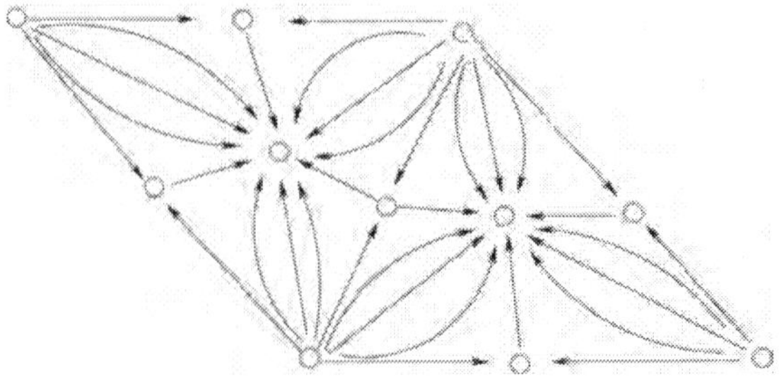

Fig. 5.2. Special vector field on a surface.

Now we construct a special vector field as shown in Fig. 5.2. First we triangulate the surface, then we define a smooth vector field, such that the vertices, the centers of the edges, and the centers of the faces are singularities. The index of vertex singularity, edge singularity, and face singularity are $+1, -1, +1$ respectively. Therefore the total index equals to $\chi(S)$. Therefore, the total index

$$\sum_i I(\mathbf{v}, p_i) = V + F - E = \chi(S).$$

For surfaces with boundaries, we can use the double covering technique to prove the same result. Therefore, we get the following theorem.

Theorem 5.19 (Poincaré-Hopf Index Theorem). *For any smooth vector field on a surface,*

$$2\pi \sum_i I(\mathbf{v}, p_i) = \frac{1}{2\pi} \int_S K dA = \chi(S),$$

where $\chi(S)$ is the Euler number.

This also shows the Gauss-Bonnet theorem for surfaces with arbitrary topologies.

Theorem 5.20 (Gauss-Bonnet Theorem). *The total curvature on a surface is a topological invariant:*

$$\int_S K dA + \int_{\partial S} k_g dA = 2\pi \chi(S).$$

where $\chi(S)$ is the Euler number.

5.7 Minimal Surface

The classical Plateau problem is as follows: given a Jordan curve γ, whether there exists a surface S, such that $\partial S = \gamma$, and the area of S is the minimal among all such surfaces. Such kind of surface is called the *minimal surface*.

Suppose S is a surface in $\mathbb{R}^3$, $\mathbf{e}_3$ is the normal field of the surface, D is a compact region; assume f is a function on S, and f is zero outside D. A family of surfaces

$$\mathbf{r}^t = \mathbf{r} + t\mathbf{e}_3, \quad t \in (-\varepsilon, \varepsilon),$$

is called the *normal variation of the surface*. Assume the variation of surface S smoothly depends on the parameter t, such that $S_0 = S$. Choosing isothermal coordinates of the surface (u,v), the area element of the surface is $\mathbf{r}_u \wedge \mathbf{r}_v dudv$,

$$\mathbf{r}^t_u = \mathbf{r}_u + t\mathbf{n}_u, \quad \mathbf{r}^t_v = \mathbf{r}_v + t\mathbf{n}_v,$$

the area element of $\mathbf{r}^t$ is

$$\begin{aligned}\mathbf{r}^t_u \wedge \mathbf{r}^t_v dudv = & \mathbf{r}_u \wedge \mathbf{r}_v dudv \\ & + t(\mathbf{r}_u \wedge \mathbf{n}_v + \mathbf{n}_u \wedge \mathbf{r}_v) dudv \\ & + t^2(\mathbf{n}_u \wedge \mathbf{n}_v) dudv.\end{aligned}$$

Using the Weingarten map $\mathbf{n}_u = \mathscr{W}(\mathbf{r}_u)$, $\mathbf{n}_v = \mathscr{W}(\mathbf{r}_v)$, we get

$$\frac{d}{dt}(\mathbf{r}^t_u \wedge \mathbf{r}^t_v dudv) = 2H\mathbf{r}_u \wedge \mathbf{r}_v dudv,$$

and

$$\frac{d^2}{dt^2}(\mathbf{r}^t_u \wedge \mathbf{r}^t_v dudv) = 2K\mathbf{r}_u \wedge \mathbf{r}_v.$$

Therefore, we get

Proposition 5.21. *If S is a surface, ∂S is a Jordan curve, S reaches the area minimum in the normal variation, then $H = 0$.*

Furthermore, the sign of the Gauss curvature K affects the stability of the minimal surface.

Definition 5.22 (Minimal Surface). *A surface in $\mathbb{R}^3$ is called a* minimal surface*, if its mean curvature is zero everywhere.*

We choose isothermal coordinates, $ds^2 = \lambda^2(du^2 + dv^2)$, and define the *Laplace operator*

$$\Delta_S = \frac{1}{\lambda^2}\left(\frac{\partial^2}{\partial u^2} + \frac{\partial^2}{\partial v^2}\right).$$

Then

Proposition 5.23.

$$\Delta_S \mathbf{r} = 2H\mathbf{n}.$$

Proof. Because (u,v) are isothermal coordinates, we have

$$\langle \mathbf{r}_u, \mathbf{r}_u \rangle = \langle \mathbf{r}_v, \mathbf{r}_v \rangle.$$

Differentiating the above equation, we get

$$\langle \mathbf{r}_{uu}, \mathbf{r}_u \rangle = \langle \mathbf{r}_{uv}, \mathbf{r}_v \rangle. \tag{5.10}$$

Furthermore,

$$\langle \mathbf{r}_u, \mathbf{r}_v \rangle = 0,$$

by differentiation we get

$$\langle \mathbf{r}_{vv}, \mathbf{r}_u \rangle = -\langle \mathbf{r}_{uv}, \mathbf{r}_v \rangle. \tag{5.11}$$

By adding Equations (5.10) and (5.11), we get

$$\langle \mathbf{r}_{uu} + \mathbf{r}_{vv}, \mathbf{r}_u \rangle = 0.$$

Similarly, we can show that

$$\langle \mathbf{r}_{uu} + \mathbf{r}_{vv}, \mathbf{r}_v \rangle = 0,$$

therefore $\mathbf{r}_{uu} + \mathbf{r}_{vv}$ is along the normal direction. E and G in the first fundamental form equal λ^2, F equals zero;

$$\langle \Delta_S \mathbf{r}, \mathbf{n} \rangle = \frac{1}{\lambda^2}(\langle \mathbf{r}_{uu}, \mathbf{n} \rangle + \langle \mathbf{r}_{vv}, \mathbf{n} \rangle) = \frac{LG - 2MF + NE}{EG - F^2} = 2H. \qquad \square$$

Definition 5.24 (Mean Curvature Flow). *Let* $\mathbf{r}(u,v)$ *is a regular surface. The mean curvature flow is defined as*

$$\frac{d\mathbf{r}}{dt} = -\Delta \mathbf{r}.$$

The mean curvature flow is commonly used to compute minimal surface with fixed boundaries. If the surface is a convex closed surface, mean curvature flow will deform the surface to a sphere and shrinks to a point. Mean curvature flow of general surfaces may produce singularities.

5.7.1 Weierstrass Representation

Suppose $\mathbf{r} = (x_1, x_2, x_3)$ is the position vector of a minimal surface. Then $\Delta_S \mathbf{r} = 0$. Choose isothermal coordinates (u,v),

$$ds^2 = \lambda^2 \left(du^2 + dv^2\right).$$

Then

$$\Delta_S \mathbf{r} = 2H\mathbf{n} = 0,$$

therefore

$$\frac{\partial^2 x_k}{\partial u^2} + \frac{\partial^2 x_k}{\partial v^2} = 0,$$

then x_k are harmonic functions with respect to the isothermal coordinates. Let

$$\phi_k = \frac{\partial}{\partial z} x_k = \left(\frac{\partial}{\partial u} - i\frac{\partial}{\partial v}\right) x_k, \quad k = 1,2,3.$$

Then ϕ_k are the holomorphic functions of $z = u + iv$.

$$\frac{\partial}{\partial \bar{z}} \phi_k = \frac{\partial}{\partial \bar{z}} \frac{\partial}{\partial z} x_k = \Delta x_k = 0.$$

Therefore $\phi_k dz$'s are holomorphic 1-forms,

$$x_k = \mathrm{Re}\frac{1}{2}\int \phi_k dz.$$

Let

$$f = \phi_1 - i\phi_2, \quad g = \frac{\phi_3}{\phi_1 - i\phi_2}.$$

then we can get

$$\begin{aligned}\phi_1 &= \tfrac{1}{2}f(1-g^2),\\ \phi_2 &= \tfrac{i}{2}f(1+g^2),\\ \phi_3 &= fg.\end{aligned}$$

Then

Definition 5.25 (Weierstrass Representation). *Given complex functions f and g, the* Weierstrass representation of a minimal surface *is*

$$\begin{aligned}x_1 &= \mathrm{Re}\tfrac{1}{2}\int f(1-g^2)dz,\\ x_2 &= \mathrm{Re}\tfrac{i}{2}\int f(1+g^2)dz,\\ x_3 &= \mathrm{Re}\int fgdz.\end{aligned}$$

The first fundamental form of the surface is

$$ds^2 = \frac{1}{4}|f|^2(1+|g|^2)^2|dz|^2.$$

The Gaussian curvature is

$$K = -\left\{\frac{4|g'|}{|f|(1+|g|^2)^2}\right\}^2.$$

g has a special geometric meaning, it is the stereo-graphic projection of the Gauss map image of the normal field.

The following list some common minimal surfaces in $\mathbb{R}^3$,

1. (f,g) are $(1,0)$, the minimal surface is a plane, defined on the entire complex plane $\mathbb{C}$.
2. (f,g) are $(1,z)$, the minimal surface is an Enneper surface, defined on the entire $\mathbb{C}$, which is immersed in $\mathbb{R}^3$.
3. (f,g) are $(1,\frac{1}{z})$, the minimal surface is a Helicoid,

   $$\mathbf{r}(x,y) = \left(x, y, \arctan^{-1}\frac{y}{x}\right),$$

   defined on $\mathbb{C}\backslash\{0\}$, embedded in $\mathbb{R}^3$.
4. (f,g) are $(i,\frac{1}{z})$, the rotational minimal surface is a Catenoid,

   $$x^2 + y^2 = \cosh^2 z.$$

5. (f,g) are $(\frac{1}{1-z^4}, z)$, defined on $\mathbb{C}\backslash\{1\}$, the minimal surface is the Scherk surface

   $$(x, y, \ln\cos x - \ln\cos y),$$

   embedded in $\mathbb{R}^3$.

5.7.2 Costa Minimal Surface

In 1980, the young Brazilian mathematician Costa constructed a genus one minimal surface using a computer: the *Costa minimal surface*. Hoffman and Meeks proved that Costa minimal surface is embedded in $\mathbb{R}^3$. It is an excellent example to show how computer graphics can help research in pure mathematics.

First, we consider a torus,

$$\Sigma = \mathbb{C}/\Gamma, \quad \Gamma = \{m + ni | m, n \in \Gamma\},$$

$\pi : \mathbb{C} \to \Sigma$ is the projection. Then the Weierstrass function defined on Σ with period Γ is

$$\beta(z) = \frac{1}{z^2} + \sum_{\substack{(m,n)\in\mathbb{Z}^2 \\ (m,n)\neq(0,0)}} \left(\frac{1}{(z - m - ni)^2} - \frac{1}{(m + ni)^2} \right).$$

Fig. 5.3. The Costa minimal surface.

Then we define the function

$$\bar{\beta}(z) : \Sigma \to \mathbb{C} \cup \{\infty\}, \quad \bar{\beta} \circ \pi = \beta.$$

The zero points of β' are $z = \{0, \frac{1}{2}, \frac{i}{2}\}$. Let

$$g(z) = \frac{A}{\bar{\beta}'}, \quad f(z) = \bar{\beta}(z),$$

defined on $\Sigma \backslash \{\pi(0), \pi(\frac{1}{2}), \pi(\frac{i}{2})\}$. Then the Weierstrass representation gives us the Costa surface (Fig 5.3, Fig 5.4).

Problems

5.1. Spherical Curve
Prove the following facts:

Fig. 5.4. The Costa minimal surface is a genus one surface with three boundaries.

(a) Suppose γ is a spatial curve, its curvature $k(s)$ and torsion $\tau(s)$ are always not equal to zero. Then the necessary and sufficient condition for γ to be on a sphere is

$$\frac{d}{ds}\left[\frac{1}{\tau(s)}\frac{d}{ds}\frac{1}{k(s)}\right]+\frac{\tau(s)}{k(s)}=0.$$

(b) Suppose γ is a spatial curve, its curvature $k(s)$, derivative of the curvature $\frac{dk(s)}{ds}$ and torsion $\tau(s)$ are always not equal to zero. Then the necessary and sufficient condition for γ to be on a sphere is

$$\frac{1}{k^2(s)}+\frac{1}{\tau^2(s)}\left(\frac{d}{ds}\frac{1}{k(s)}\right)^2=C,$$

where C is a positive constant.

(c) Suppose γ is a spatial curve. Then the necessary and sufficient condition for γ to be on a sphere is, there exits constant A, B, A and B are not zeros at the same time, such that

$$k(s)\left[A\cos\int_0^s\tau(t)dt+B\sin\int_0^s\tau(t)dt\right]=1.$$

5.2. Envelope

The envelope of a one parameter family of surfaces $F(x,y,z,\lambda)=0$ (where λ is the parameter) can be derived by eliminating λ from the following equation group:

$$\begin{cases} F(x,y,z,\lambda) &= 0,\\ F_\lambda(x,y,z,\lambda) &= 0.\end{cases}$$

Given a one parameter family of regular parametric surfaces S_λ:

$$\begin{cases} x = x(u,v,\lambda),\\ y = y(u,v,\lambda),\\ z = z(u,v,\lambda).\end{cases}$$

Prove that its envelope can be derived by eliminating λ from the following equation group:

$$\begin{cases} x = x(u,v,\lambda), \\ y = y(u,v,\lambda), \\ z = z(u,v,\lambda), \\ 0 = \frac{\partial(x,y,z)}{\partial(u,v,\lambda)}. \end{cases}$$

5.3. Delaunay Surface Delaunay triangulation is well known. The following surface was also discovered by Delaunay. A half ellipse $\frac{x^2}{a^2} + \frac{(y-b)^2}{b^2} = 1$, $x > 0$, a,b are positive constants, rolls along the x-axis without sliding. The locus of one of the foci of the ellipse is a curve γ. Rotate γ about the x-axis and a rotational surface S is obtained. Then S is the Delaunay surface. Compute the mean curvature of the Delaunay surface.

5.4. Minimal Surface Suppose $\mathbf{r}(u,v) = (x(u,v), y(u,v), z(u,v))$ is a regular surface, (u,v) are isothermal coordinates. Prove that $\mathbf{r}$ is a minimal surface, if and only if the coordinate functions $x(u,v), y(u,v), z(u,v)$ satisfy

$$\frac{\partial^2 f}{\partial u^2} + \frac{\partial^2 f}{\partial v^2} = 0.$$

5.5. Chebyshev Net The net formed by the iso-parametric curves is called a *Chebyshev net*, if all the quadrilaterals formed by the iso-parametric curves have the following property: the edge lengths of opposite sides are equal.
(a) Show that if the iso-parametric curves form a Chebyshev net, then the first fundamental form of the surface is

$$ds^2 = du^2 + 2\cos\omega du dv + dv^2,$$

where ω is the intersection angle between iso-parametric curves.
(b) Show that if the Gaussian curvature of S is a negative constant, then the asymptotic curves form a Chebyshev net.
(c) Show that if S has no flat point, the asymptotic curves form a Chebyshev net, then it has negative constant Gaussian curvature.

5.6. Developable Surface Prove that if there are two families of geodesics with constant intersection angles, then the surface is developable.

5.7. Parallel Surface Given a surface $S : \mathbf{r}(u,v)$, let its *parallel surface* S_λ be

$$\mathbf{r}(u,v) + \lambda \mathbf{n}(u,v),$$

where λ is a constant, such that

$$1 - 2\lambda H + \lambda^2 K > 0.$$

Prove the following propositions.
(a) All points of S are umbilical points, if and only if all points of S_λ are umbilical points.
(b) S is developable, if and only if S_λ is developable.
(c) S_λ is a minimal surface, if and only if the Gaussian curvature and the mean curvature of S satisfy

$$H = \lambda K.$$

(d) S is a minimal surface without any flat point, if and only if the principal curvatures of S_λ satisfy

$$\frac{1}{k_1}+\frac{1}{k_2}=-2\lambda.$$

(e) Suppose S has no umbilical point. Then the principal curvature lines of S correspond to the principal curvature lines of S_λ.

5.8. Isometric Surfaces Suppose two surfaces S and $\tilde{S}$ are isometric. Prove that if the isometry maps the two families of asymptotic curves on S to the asymptotic curves on $\tilde{S}$, then S and $\tilde{S}$ differ by a rigid motion in $\mathbb{R}^3$.

5.9. Gauss-Bonnet Theorem
(a) Suppose K is negative everywhere on a surface. Prove that each homotopy class has a unique closed geodesic.
(b) Suppose K is positive everywhere. Prove that two simple closed geodesics must intersect each other.

5.10. Minkowski Theorem Suppose S is a topological sphere, K is the Gaussian curvature, and $\mathbf{n}$ is the normal. The function $K(\mathbf{n})$ is defined on the unit sphere. Show that $K(\mathbf{n})$ determines the surface unique up to translations.

6

Riemann Surface

This chapter covers the basic theories of compact Riemann surface.

6.1 Riemann Surface

Definition 6.1 (Harmonic Function). *Suppose $u : D \to \mathbb{R}$ is a real valued function defined on a domain $D \subset \mathbb{C}$. If $u \in C^2(D)$, and for any $z \in D$, $z = x + iy$, we have*

$$\Delta u(z) = \frac{\partial^2 u(z)}{\partial x^2} + \frac{\partial^2 u(z)}{\partial y^2} = 0,$$

then we call $u(z)$ a harmonic function *on D, where $\Delta = \frac{\partial^2}{\partial x^2} + \frac{\partial^2}{\partial y^2}$ is the Laplace operator.*

For convenience, we denote

$$\begin{aligned} dz &= dx + idy, \\ d\bar{z} &= dx - idy, \\ \frac{\partial}{\partial z} &= \frac{1}{2}\left(\frac{\partial}{\partial x} - i\frac{\partial}{\partial y}\right), \\ \frac{\partial}{\partial \bar{z}} &= \frac{1}{2}\left(\frac{\partial}{\partial x} + i\frac{\partial}{\partial y}\right). \end{aligned}$$

Then $\langle dz, \frac{\partial}{\partial z}\rangle = 1$ and $\langle d\bar{z}, \frac{\partial}{\partial \bar{z}}\rangle = 1$. The Laplace operator is

$$\Delta = 4\frac{\partial^2}{\partial z \partial \bar{z}}.$$

Definition 6.2 (Holomorphic Function). *A function $f : \mathbb{C} \to \mathbb{C}, (x,y) \to (u,v)$ is* holomorphic, *if it satisfies the following* Cauchy-Riemann equation:

$$\begin{cases} \frac{\partial u}{\partial x} = \frac{\partial v}{\partial y}, \\ \frac{\partial u}{\partial y} = -\frac{\partial v}{\partial x}. \end{cases} \tag{6.1}$$

If $f = u + iv$ is holomorphic, then

$$\frac{\partial f}{\partial \bar{z}} = 0.$$

Furthermore, it is easy to verify that both u and v are harmonic. We say that u and v are conjugate if $u + iv$ satisfies the Cauchy-Riemann equation. If a holomorphic function $f : D \to \mathbb{C}$ is bijective and f^{-1} is holomorphic, then f is called to be *biholomorphic*, or a *conformal mapping*.

Definition 6.3 (Riemann Surface). *A* Riemann Surface *represents a two-dimensional manifold M with an atlas $\{(U_\alpha, z_\alpha)\}$, such that $\{U_\alpha\}$ is an open covering, $M \subset \bigcup U_\alpha$; $z_\alpha : U_\alpha \to \mathbb{C}$ is a homeomorphism from an open set $U_\alpha \subset M$ to an open set on $\mathbb{C}$, $z_\alpha(U_\alpha)$. If $U_\alpha \cap U_\beta \neq \emptyset$, then*

$$z_\beta \circ z_\alpha^{-1} : z_\alpha(U_\alpha \cap U_\beta) \to z_\beta(U_\alpha \cap U_\beta)$$

is biholomorphic, or a conformal mapping (Fig. 6.1).

The atlas $\{(U_\alpha, z_\alpha)\}$ is called the *complex atlas* or *conformal atlas* of the Riemann surface. Given two conformal atlases $\{(U_\alpha, z_\alpha)\}$ and $\{(V_\beta, \tau_\beta)\}$, if their union is still a conformal atlas, then we say $\{(U_\alpha, z_\alpha)\}$ and $\{(V_\beta, \tau_\beta)\}$ are equivalent. Each equivalence class of conformal atlases is called a *conformal structure* or *complex structure*.

Suppose M has a Riemannian metric $\mathbf{g}$. Then we require the conformal structure to be *compatible* with the Riemannian metric, namely, on each local chart of $\{(U_\alpha, z_\alpha)\}$,

$$\mathbf{g} = e^{2\lambda(z_\alpha)} dz_\alpha d\bar{z}_\alpha,$$

namely, the local parameters are isothermal coordinates.

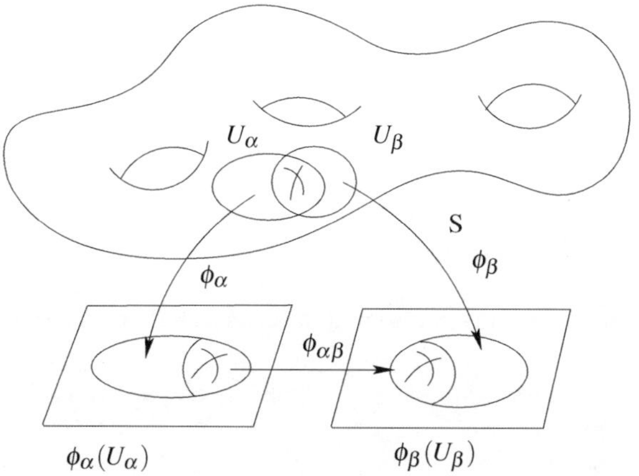

Fig. 6.1.

Proposition 6.4. *Given a metric surface with a differential atlas $\{(U_\alpha, z_\alpha)\}$, if all local coordinates are isothermal coordinates, then $\{(U_\alpha, z_\alpha)\}$ is a conformal structure.*

According to the existence theorem of the isothermal coordinates for arbitrary metric surface, we proved that

Theorem 6.5. *Any metric surface has a conformal structure compatible with its Riemannian metric, therefore any metric surface is a Riemann surface.*

Definition 6.6 (Conformal Mapping). *Suppose M and $\tilde{M}$ are two Riemann surfaces. A mapping $f : M \to \tilde{M}$ is called a* conformal mapping *(holomorphic mapping), if $\forall p \in M$, $\tilde{p} = f(p) \in \tilde{M}$, for any local parameter charts (U, ϕ) and $(\tilde{U}, \tilde{\phi})$, $z = \phi(p)$, $\tilde{z} = \tilde{\phi}(\tilde{z})$,*

$$\begin{array}{ccc} M & \xrightarrow{f} & \tilde{M} \\ \phi\downarrow & & \downarrow\tilde{\phi} \\ z & \xrightarrow{\tilde{\phi}\circ f\circ\phi^{-1}} & \tilde{z} \end{array}$$

under local parameters

$$\tilde{z} = \tilde{\phi} \circ f \circ \phi^{-1}(z)$$

is holomorphic in U.

Definition 6.7 (Conformal Equivalence). *Suppose M and $\tilde{M}$ are two Riemann surfaces. If a mapping $f : M \to \tilde{M}$ is holomorphic, one to one, and onto and f^{-1} is also holomorphic, then M and $\tilde{M}$ are* conformally equivalent.

A holomorphic mapping $f : M \to \mathbb{C}$ is called a *holomorphic function* defined on the Riemann surface M; a holomorphic mapping $f : M \to \bar{\mathbb{C}} = \mathbb{C} \cup \{\infty\}$ is called a *meromorphic function*.

Suppose $f : \mathbb{C} \to \mathbb{C}$ is a holomorphic function, $w = f(z)$. Then

$$dw = \frac{\partial f(z)}{\partial z} dz + \frac{\partial f(z)}{\partial \bar{z}} d\bar{z},$$

$\frac{\partial f(z)}{\partial \bar{z}}$ is zero, therefore

$$dw d\bar{w} = \left| \frac{\partial f(z)}{\partial z} \right|^2 dz d\bar{z}.$$

This shows the pull back metric $dwd\bar{w}$ induced by f is conformal to the original metric $dzd\bar{z}$. Therefore, a holomorphic function between planar domains preserves angles. Similarly, holomorphic mappings between Riemann surfaces with metrics also preserve angles (Fig. 6.2).

Fig. 6.2. Holomorphic mappings preserve angles. The human face surface is mapped onto the plane by a conformal mapping. The checker board texture is mapped back onto the surface. All the corner angles of the checkers are preserved.

It is not obvious whether any two topologically equivalent surfaces are conformally equivalent, namely, can be mapped to each other via a conformal mapping. The following Fig. 6.3 demonstrates three surfaces, which are homeomorphic to one another, but any two of them are not conformally equivalent. The surfaces in the first row are conformally mapped to the planar domains in the second row, which are two-holed annuli. The centers and the radii of the holes on the plane are the conformal invariants. By comparing these invariants, we can verify that the surfaces are not conformally equivalent.

The holomorphic automorphisms of a domain D form a group with the composition as the product rule, denoted as $Aut(D)$ and called the *holomorphic automorphism group* of D. The holomorphic automorphism group of the unit disk Δ is

$$\text{Aut}(\Delta) = \{z \to e^{i\theta} \frac{z - z_0}{1 - \bar{z}_0 z}, \theta \in [0, 2\pi), z_0 \in \Delta\}.$$

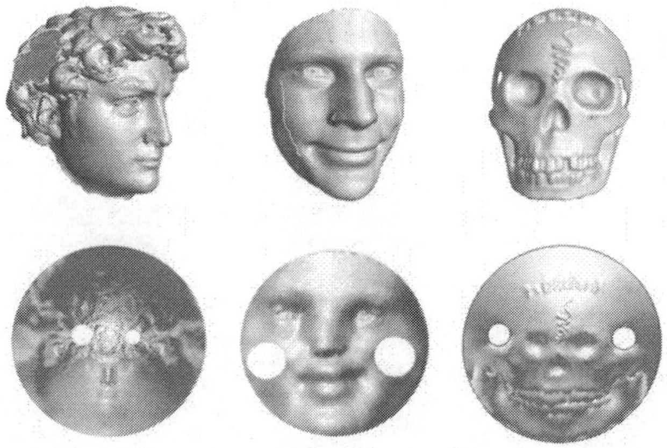

Fig. 6.3. Surfaces which are topologically equivalent but not conformally equivalent.

All such transformations are called the *Möbius transformation* (Fig. 6.4).

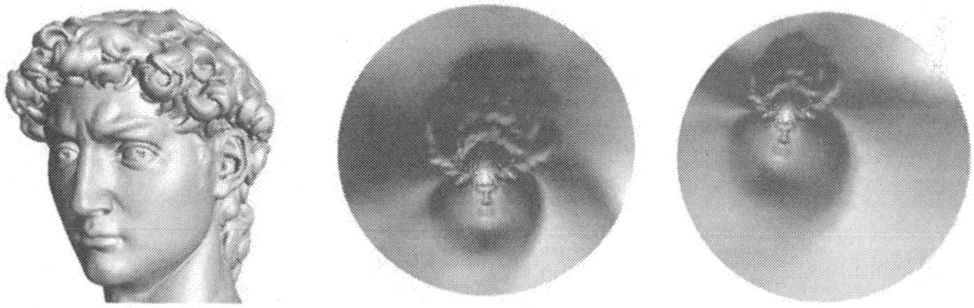

Fig. 6.4. Möbius transformations on the unit disk.

The holomorphic automorphism group of the complex plane is

$$\mathrm{Aut}(\mathbb{C}) = \{z \to az + b : a, b \in \mathbb{C}, a \neq 0\}.$$

The holomorphic automorphism group of the sphere is

$$\mathrm{Aut}(\bar{\mathbb{C}}) = \{z \to \frac{az+b}{cz+d}, ad - bc = 1, a, b, c, d \in \mathbb{C}\}.$$

The map is also called a Möbius transform (Fig. 6.5). The coefficient matrix of a Möbius transform is

$$\begin{pmatrix} a\, b \\ c\, d \end{pmatrix}$$

It is easy to verify that the coefficient matrix of the composition of two Möbius transformations equals to the product of the corresponding coefficient matrices. On the other hand,

Fig. 6.5. Möbius transformation on the unit sphere.

$$\begin{pmatrix} a & b \\ c & d \end{pmatrix}, \quad \begin{pmatrix} -a & -b \\ -c & -d \end{pmatrix}$$

correspond to the same element in $\mathrm{Aut}(\bar{\mathbb{C}})$. Therefore $\mathrm{Aut}(\bar{\mathbb{C}})$ is isomorphic to $SL(2,\mathbb{C})/\{\pm I\}$, namely,

$$\mathrm{Aut}(\mathbb{C}) \cong PSL(2,\mathbb{C}).$$

The upper half complex plane $\mathbb{H} = \{z \in \mathbb{C} : \mathrm{Im} z > 0\}$ is conformally equivalent to the disk Δ. The conformal automorphism group of the upper half complex plane is

$$\mathrm{Aut}(\mathbb{H}) \cong PSL(2,\mathbb{R}).$$

6.2 Riemann Mapping Theorem

Suppose $f_n : D \to \mathbb{C}$ is a family of holomorphic functions defined on the domain D. We say $\{f_n\}$ are *locally uniformly convergent* to f in D, if for all point $z_0 \in D$, there exists a disk $\Delta_\gamma(z_0)$ centered at z_0 with radius γ, such that $\{f_n\}$ uniformly converges to f on the closure of the disk $\Delta_\gamma(z_0)$.

Let $\mathscr{F} = \{f_a : D \to \mathbb{C} | a \in A\}$ be a family of holomorphic functions on D. We call $\mathscr{F}$ a *normal family* if any sequence in $\mathscr{F}$ includes a subsequence which locally uniformly converge in D.

Theorem 6.8 (Montel). *Suppose $\mathscr{F} = \{f_a : D \to \mathbb{C} | a \in A\}$ is a family of holomorphic functions defined on the domain D. If for any compact subset $E \subset D$, $\{f_a\}$ are uniformly bounded on E, then $\mathscr{F}$ is a normal family.*

Theorem 6.9 (Riemann Mapping). *Suppose $D \subset \mathbb{C}$ is a simply connected domain on the complex plane, the boundary ∂D has more than one point, $z_0 \in D$ is an arbitrary interior point. Then there exists a unique holomorphic mapping $\phi : D \to \Delta$ from D to the unit disk Δ, such that $\phi(z_0) = 0$ and $\phi'(z_0) > 0$.*

Proof. Suppose $\mathscr{F} = \{f : D \to \Delta\}$ is the set of all holomorphic univalent functions, which maps D into Δ. We can show that $\mathscr{F}$ is a nonempty normal family. We set

$$\alpha = \sup\{|f'(z_0)|, f \in \mathscr{F}\},$$

and construct a sequence $\{f_n\} \subset \mathscr{F}$,

$$|f_n'(z_0)| > \alpha - \frac{1}{n}.$$

Then select the convergent subsequence, the limit is the desired ϕ. □

In general, a topological mapping or even a differential mapping cannot be extended to the boundary. A conformal mapping can be extended to the boundary, as long as the boundary is a Jordan curve.

6.2.1 Conformal Module

A quadrilateral $Q := \{\Omega : z_1, z_2, z_3, z_4\}$ consists of a Jordan domain Ω and four points z_1, z_2, z_3, z_4 in counter-clockwise order on $\partial\Omega$. According to the Riemann mapping theorem, Q can be conformally mapped onto the upper half plane $\phi : Q \to \mathbb{H}$, and z_1, z_2, z_3, z_4 are mapped to four points on the real axis, $\phi(z_1), \phi(z_2), \phi(z_3), \phi(z_4)$. We use a linear rational $\psi : \mathbb{H} \to \mathbb{H}$ to map them to

$$-\frac{1}{k}, -1, 1, \frac{1}{k},$$

where $k < 1$ is determined by the cross ratio of the four points. Then we use the Christoffel-Schwarz formula

$$w = \int_{-\frac{1}{k}}^{z} \frac{d\zeta}{\sqrt{(1-\zeta^2)(1-k^2\zeta^2)}}$$

to map $\mathbb{H}$ to a rectangle, and the four points are mapped to $0, a, a+ib, bi$, where

$$\begin{aligned} a &= \int_{-\frac{1}{k}}^{-1} \frac{dx}{\sqrt{(1-x^2)(1-k^2x)}}, \\ b &= \int_{-1}^{1} \frac{dx}{\sqrt{(1-x^2)(1-k^2x)}}. \end{aligned}$$

Then we define the *conformal module* of the quadrilateral Q as

$$Mod(Q) = \frac{a}{b},$$

which is a conformal invariant of Q.

Suppose $f : D \to G \subset \mathbb{C}$ is a topological map preserving the orientation. If for any topological quadrilateral Q, the closure of Q is contained in D, $\bar{Q} \subset D$, we have

$$Mod(Q) = Mod(f(Q)),$$

then f is a conformal mapping. (We did not specify the requirements for the differentiability of f.) Hence, module preservation is an intrinsic property of conformal mappings.

6.2.2 Quasi-Conformal Mapping

Quasi-conformal mapping is a generalization of conformal mapping. Conformal mapping preserves modules and is the solution to the Cauchy-Riemann equation; quasi-conformal mapping keeps the change of modules bounded and is the solution to the Beltrami equation.

Definition 6.10 (Quasi-Conformal Mapping). *Suppose $f : D \to G$ is a homeomorphism between domains D and G in $\bar{\mathbb{C}}$. If there exists a constant $K \geqslant 1$, such that for any topological quadrilateral Q in D, $\bar{Q} \subset D$,*

$$Mod(f(Q)) \leqslant KMod(Q),$$

then f is a K-quasiconformal mapping.

The existence of the constant K is essential, but the value of K is not so important. This means that the change between the modules cannot be infinitely large or small.

Theorem 6.11. *Suppose $f : D \to G \subset \mathbb{C}$ is a diffeomorphism. Then the necessary and sufficient condition for f to be a K-quasiconformal mapping is that f is the solution to the following equation:*

$$\frac{\partial f}{\partial \bar{z}} = \mu(z)\frac{\partial f}{\partial z}, \quad |\mu(z)| \leqslant k, \tag{6.2}$$

where

$$k = \frac{K-1}{K+1}.$$

The equation $\frac{\partial f}{\partial \bar{z}} = \mu(z)\frac{\partial f}{\partial z}$ is called the *Beltrami equation*. If $\mu(z) = 0$, then the Beltrami equation becomes the Cauchy-Riemann equation (6.1). $\mu(z)$ is called the *complex dilatation* of Beltrami coefficients of f, denoted as μ_f,

$$K[f] = \frac{1+||\mu_f||_\infty}{1-||\mu_f||_\infty}$$

is called the *maximal dilatation*. The following fundamental theorem for quasi-conformal mappings is a generalization of the Riemann mapping theorem.

Theorem 6.12 (Riemann Mapping theorem for Quasi-Conformal Mapping). *Suppose D is a simply connected domain in $\mathbb{C}$, ∂D has more than one point, $\mu(z)$ is a measurable function on D, $||\mu(z)||_\infty < 1$. Then there exists a quasiconformal mapping $g : D \to \Delta$ which maps D to the unit disk and its complex dilatation is equal to μ.*

6.2.3 Holomorphic Mappings

Holomorphic functions on a compact Riemann surface must be a constant, because of the maximal module property of holomorphic functions.

A meromorphic function on a Riemann surface is a holomorphic mapping between the surface and the sphere. Then the number of poles must equal to the number of zeros. The following discussion explains this fact.

Suppose $f : M \to N$ is a holomorphic mapping between two closed Riemann surfaces M and N. Then it induces a homomorphism between the homology groups

$$f_* : H_2(M,\mathbb{Z}) \to H_2(N,\mathbb{Z}).$$

Because M and N are closed surfaces,

$$H_2(M,\mathbb{Z}) \cong \mathbb{Z}, \quad H_2(N,\mathbb{Z}) \cong \mathbb{Z},$$

therefore $f_*(z) = mz$, where $m \in \mathbb{Z}$ is a constant integer. m is called the *degree of the map* f.

The degree of the mapping can be defined in an analytical way. Suppose $p_0 \in M$ and $f(p_0) \in N$, ζ and w are local coordinates of M and N respectively. Then the local representation of f can be written as

$$w = \zeta^n.$$

n is independent of the choice of the local coordinates, hence n is a topological property of the mapping f itself. n is called the *ramification index* of f at p_0. If $n > 1$, then p_0 is called a *branch point*. We denote the ramification index of f at p_0 as $n_f(p_0)$.

Theorem 6.13. *Suppose $f : M \to N$ is a non-trivial, holomorphic mapping between two compact Riemann surfaces. Then the degree of the map m is an integer, such that any point on N has m pre-images in M (counting the multiplicity),*

$$\sum_{p \in f^{-1}(q)} n_f(p) = m, \quad \forall q \in N.$$

From this result, it is obvious that the number of zeros and poles of a meromorphic function on a compact Riemann surface are equal. This is summarized in the following Riemann-Hurwitz theorem.

Theorem 6.14 (Riemann-Hurwitz). *$f : M \to N$ is a holomorphic mapping between two compact Riemann surfaces. Suppose the genus of M is $g(M)$, the genus of N is $g(N)$, the degree of the map is m. Then*

$$g(M) = m(g(N) - 1) + 1 + \frac{B}{2},$$

where

$$B = \sum_{p \in M} (n_f(p) - 1).$$

Proof. The proof is straightforward. The branch points of a holomorphic mapping are isolated. Suppose the branch point set is $D = \{p \in M | n_f(p) > 1\}$. Because M and N are compact, D and $f(D)$ are finite. We define a triangulation on N, such that all vertices are $f(p), p \in D$, then we pull back the triangulation to M. Each triangle on N has m pre-images on M. Suppose on N there are V vertices, E edges, and F faces. Then on M, there are $mV - B$ vertices, mE edges, and mF faces,

$$V - E + F = 2 - 2g(N), \quad m(V - E + F) - B = 2 - 2g(M). \qquad \square$$

Suppose $f : M \to M$ is a non-trivial holomorphic mapping from M to itself. Then we can get $m = 1$ and $B = 0$, therefore, it must be a conformal automorphism.

6.3 Holomorphic One-Forms

We have defined harmonic functions and harmonic forms in the setting of Riemannian manifolds. Now, we study harmonic forms on Riemann surfaces.

Let M be a Riemann surface with a conformal atlas $\{(U_\alpha, \phi_\alpha)\}$, and let $z_\alpha = x + iy$ be a local complex coordinate. We get

$$dz = dx + idy, \quad d\bar{z} = dx - idy.$$

Suppose $\omega = fdx + gdy$ is a real differential 1-form. Then it has a complex form

$$\omega = udz + vd\bar{z},$$

where $u = \frac{f-ig}{2}, v = \frac{f+ig}{2}$.

Definition 6.15 (Complex Differential 1-form). *Let M be a Riemann surface with conformal atlas $\{(U_\alpha, \phi_\alpha)\}$. For each local coordinate $z_\alpha = \phi_\alpha(p)$, two smooth complex functions u_α and v_α are assigned, such that*

$$u_\alpha dz_\alpha + v_\alpha d\bar{z}_\alpha$$

is invariant under coordinate changes, then it is a complex differential 1-form *on M.*

Namely, if $U_\alpha \cap U_\beta \neq \varnothing$, then

$$u_\alpha dz_\alpha + v_\alpha d\bar{z}_\alpha = u_\beta dz_\beta + v_\beta d\bar{z}_\beta,$$

where $z_\alpha = \phi_\alpha \circ \phi_\beta^{-1}(z_\beta)$. Similarly, we can define complex differential 2-forms,

$$g_\alpha dz_\alpha \wedge d\bar{z}_\alpha,$$

which is invariant under coordinate change,

$$g_\alpha dz_\alpha \wedge d\bar{z}_\alpha = g_\beta dz_\beta \wedge d\bar{z}_\beta$$

in $U_\alpha \cap U_\beta$.

We can define operator $\partial, \bar{\partial}$ on 1-forms. Suppose $\omega = udz + vd\bar{z}$. Then

$$\partial\omega = \frac{\partial v}{\partial z} dz \wedge d\bar{z}, \quad \bar{\partial}\omega = \frac{\partial u}{\partial \bar{z}} d\bar{z} \wedge dz.$$

It is easy to verify that the exterior differential operator

$$d = \partial + \bar{\partial},$$

from $d^2 = 0$ we can obtain

$$\partial\bar{\partial} + \bar{\partial}\partial = 0.$$

Let

$$\Delta = 2i\partial\bar{\partial}.$$

Then for any function f (0-form)

$$\Delta f = \left(\frac{\partial^2 f}{\partial x^2} + \frac{\partial^2 f}{\partial y^2}\right) dx \wedge dy.$$

A complex function $f : M \to \mathbb{C}$ is harmonic if $\partial\bar{\partial} f = 0$. A differential 1-form is harmonic if, locally, it is the derivative of a harmonic function, $\omega = df$.

Definition 6.16 (Holomorphic 1-form). *Suppose M is a Riemann surface with a conformal atlas and ω is a differential 1-form. If on every local coordinate, ω has the representation*

$$\omega = udz,$$

where u is a holomorphic function, then ω is called a holomorphic 1-form.

Proposition 6.17. *A complex differential 1-form ω is a holomorphic 1-form on a Riemann surface M, if and only if*

$$\omega = h + i^* h,$$

where h is a real harmonic differential 1-form, * is the Hodge star operator.

Proof. Suppose f is a harmonic function. Then $\Delta f = 2i\partial\bar{\partial} f = 0$, because of

$$(\partial\bar{\partial} + \bar{\partial}\partial) f = 0,$$

$\bar{\partial}\partial f = 0$, namely ∂f is a holomorphic 1-form.

If h is harmonic, $h = df$ for some harmonic function f, then

$$\omega = h + i^* h = df + i^* df = \partial f,$$

therefore ω is holomorphic.

If ω is holomorphic, then ω and $\bar{\omega}$ are both harmonic, $h = \frac{1}{2}(\omega - \bar{\omega})$ and $^*h = -\frac{i}{2}(\omega + \bar{\omega})$ are harmonic, and $h + i^* h = \omega$. □

We use $\mathscr{H}(M)$ to represent the complex linear space formed by all the holomorphic 1-forms on a compact Riemann surface M.

Theorem 6.18. *Assume M is a compact Riemann surface of genus g. Then the dimension of the complex linear space $\mathscr{H}(M)$ is g.*

Proof. Let H denote the space of all real harmonic 1-forms, whose complex dimension is $2g$. Let $\mathscr{H} = \{h + i^* h | h \in H\}$ and $\tilde{\mathscr{H}} = \{h - i^* h | h \in H\}$. Then any harmonic 1-form $h \in H$ can be written as

$$h = \phi + \tilde{\phi}, \quad \phi \in \mathscr{H}, \quad \tilde{\phi} \in \tilde{\mathscr{H}}.$$

Let $h \in \mathscr{H} \cap \tilde{\mathscr{H}}$,

$$h = h_1 + i^* h_1 = h_2 - i^* h_2.$$

Then $h_1 - h_2 + i^*(h_1 + h_2) = 0$. Holomorphic form

$$(h_1+h_2)+i^*(h_1+h_2)=(h_1+h_2)+(h_2-h_1)=2h_2,$$

therefore, h_2 is holomorphic, $h_2=i^*h_2$, therefore,

$$h=h_2-i^*h_2=h_2-h_2=0,$$

$\mathscr{H}\cap\tilde{\mathscr{H}}=\{0\}$. So $H=\mathscr{H}\oplus\tilde{\mathscr{H}}$.

Consider the map

$$\Phi:\mathscr{H}\to\tilde{\mathscr{H}},\quad h+i^*h\to h-i^*h,$$

which is an isomorphism. Therefore,

$$\dim_{\mathbb{C}}\mathscr{H}=\frac{1}{2}\dim_{\mathbb{R}}\mathscr{H}=\frac{1}{4}\dim_{\mathbb{R}}H=g.$$ □

6.4 Period Matrix

The conformal invariants of a Riemann surface can be represented as a period matrix.

Characteristic Differentials

First we introduce the concept of characteristic differential form of a closed smooth curve γ, denoted as η_γ. Characteristic form has the following property: for any closed 1-form ω,

$$\int_M \eta_\gamma\wedge\omega=\int_\gamma\omega.$$

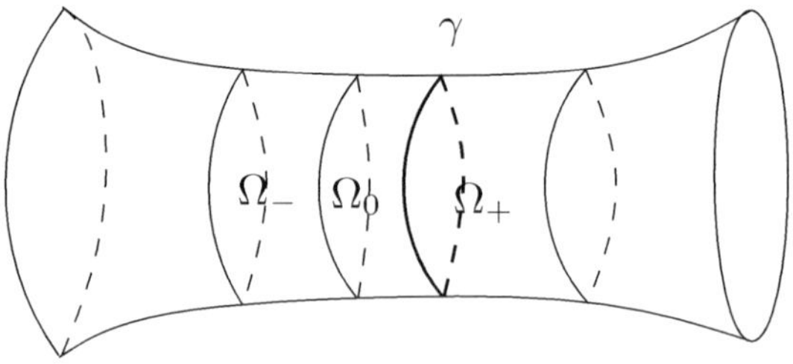

Fig. 6.6. Characteristic form construction.

As shown in Fig. 6.6, we choose a small neighborhood Ω of γ, γ separates Ω into the left and right parts, denoted as Ω_+ and Ω_-. Choose a real subset $\Omega_0\subset\Omega_-$, $\gamma=\partial\Omega_0\cap\partial\Omega_-$. Define a real valued smooth function $f\in C^\infty(M\backslash\gamma)$, such that

$$f(p)=\begin{cases}1, & p\in\Omega_0\cup\gamma,\\ 0, & p\notin\Omega_-\cup\gamma.\end{cases}$$

Define

$$\eta_\gamma(p)=\begin{cases}df, & p\in\Omega_-,\\ 0, & p\notin\Omega_-.\end{cases}$$

Let ω be a closed 1-form,

$$\int_M \eta_\gamma \wedge \omega = \int_{\Omega_-} \eta_\gamma \wedge \omega = \int_{\Omega_-} d(f\omega) = \int_{\partial\Omega_-} f\omega = \int_\gamma \omega.$$

Suppose $\{a_1, b_1, a_2, b_2, \cdots, a_g, b_g\}$ is a set of canonical homotopy group basis,

$$a_i \times b_j = \delta_{ij}, \quad a_i \times a_j = 0, \quad b_i \times b_j = 0, \quad 1 \leqslant i, j \leqslant g.$$

where $a_i \times b_j$ represents the algebraic intersection number between a_i and b_j. Let ω be a closed differential 1-form. Then

$$\int_{a_1} \omega, \int_{a_2} \omega, \cdots, \int_{a_g} \omega$$

is called the *a-period* of ω, similarly,

$$\int_{b_1} \omega, \int_{b_2} \omega, \cdots, \int_{b_g} \omega,$$

is called the *b-period* of ω.

Let

$$\alpha_j = \eta_{b_j}, \quad \beta_j = -\eta_{a_j}, \quad j = 1, 2, \cdots, g.$$

Then

$$\int_{a_j} \alpha_k = \int_{b_j} \beta_k = \delta_{jk},$$

and

$$\int_{a_j} \beta_k = \int_{b_j} \alpha_k = 0.$$

Lemma 6.19. *Let ω and $\tilde{\omega}$ be smooth closed differential 1-forms. Then*

$$\int_M \omega \wedge \tilde{\omega} = \sum_{j=1}^{g} (A_j \tilde{B}_j - B_j \tilde{A}_j),$$

where $(A_1, A_2, \cdots, A_g)$ is the a-period of ω, $(B_1, B_2, \cdots, B_g)$ is the b-period of ω; $(\tilde{A}_1, \tilde{A}_2, \cdots, \tilde{A}_g)$ is the a-period of $\tilde{\omega}$, $(\tilde{B}_1, \tilde{B}_2, \cdots, \tilde{B}_g)$ is the b-period of $\tilde{\omega}$.

Proof.

$$\begin{aligned} \omega &= \Sigma_{j=1}^{g} (A_j \alpha_j + B_j \beta_j) + df, \\ \tilde{\omega} &= \Sigma_{j=1}^{g} (\tilde{A}_j \alpha_j + \tilde{B}_j \beta_j) + d\tilde{f}. \end{aligned}$$

From Stokes theorem,

$$\int_M \omega \wedge \tilde{\omega} = \int_M \sum_{j=1}^{g} (A_j \alpha_j + B_j \beta_j) \wedge \sum_{j=1}^{g} (\tilde{A}_j \alpha_j + \tilde{B}_j \beta_j).$$

Because

$$\int_M \alpha_j \wedge \beta_k = \int_{b_j} \beta_k = \int_{a_k} \alpha_j = \delta_{jk},$$

we obtain

$$\int_M \omega \wedge \tilde{\omega} = \sum_{j=1}^{g} (A_j \tilde{B}_j - B_j \tilde{A}_j).$$

□

There exists a unique basis $\phi_1, \phi_2, \cdots, \phi_g$ of $\mathscr{H}(M)$, whose *a-period matrix*

$$A = (a_{ij})_{g\times g} = (\int_{a_i} \phi_j)_{g\times g}$$

equals to the identity matrix. Then the basis is called a set of *canonical basis* of $\mathscr{H}(M)$. Then the *b-period matrix* of the basis

$$B = (b_{ij})_{g\times g} = (\int_{b_i} \phi_j)_{g\times g}$$

must be symmetric, and its imaginary part ImB must be positive definite.

Theorem 6.20. *Suppose M is a compact Riemann surface of genus g, and $\{a_1, b_1, a_2, b_2,$ $\cdots,$*
$a_g,\ b_g\}$ is a canonical homotopy group basis. A canonical basis of the complex linear space of all holomorphic differential 1-forms $\mathscr{H}(M)$ is given by $(\varphi_1, \varphi_2, \cdots, \varphi_g)$, namely, its a-period matrix is an identity matrix,

$$\int_{a_j} \varphi_k = \delta_{jk}, \quad j,k = 1,2,\cdots,g$$

Then its b-period matrix

$$B = (\int_{b_j} \varphi_k)_{g\times g}$$

is symmetric, and the imaginary part Im B *is positive definite.*

Proof. First we show B is symmetric

$$\int_M \varphi_j \wedge \varphi_k = \sum_{l-1}^{g} (\delta_{jl} b_{lk} - b_{lj}\delta_{kl}) = -b_{kj} + b_{jk},$$

because φ_j, φ_k are holomorphic, therefore $\varphi_j \wedge \varphi_k = 0$, $b_{kj} = b_{jk}$.

Second, we show B is positive definite. Let $\omega = \Sigma_{i=1}^{g} x_i \varphi_i$, where $x_i \in \mathbb{R}$. Then

$$0 \leqslant \|\omega\|^2 = i\int_M \omega \wedge \bar{\omega} = i\sum_{j,k} x_j x_k \int_M \varphi_j \wedge \bar{\varphi}_k.$$

Using Lemma 6.19,

$$i\int_M \varphi_j \wedge \bar{\varphi}_k = i\sum_{l=1}^{g} (\delta_{jl}\bar{b}_{lk} - b_{lj}\delta_{kl}) = 2\mathrm{Im} b_{jk}.$$

Therefore, ImB is positive definite. □

The period matrix is the most important conformal invariant. It can be used to verify whether two Riemann surfaces are conformally equivalent. When $g > 1$, the b-period matrix has $3g-3$ independent variables. This leads to the complex structure of Teichmüller space.

From the above discussion, we can get the *bilinear relation* among holomorphic 1-forms.

Theorem 6.21. *Suppose M is a compact Riemann surface with genus g, φ and $\tilde{\varphi}$ are holomorphic 1-forms, and $\{a_1,\cdots,a_g;b_1,\cdots,b_g\}$ are canonical homotopy group basis. Then*

$$\sum_{j=1}^{g}\left(\int_{a_j}\varphi\int_{b_j}\tilde{\varphi}_j-\int_{b_j}\varphi\int_{a_j}\tilde{\varphi}_j\right)=\int_M\varphi\wedge\tilde{\varphi}=0.$$

We can generalize the bilinear relation to meromorphic 1-forms.

Definition 6.22 (Meromorphic 1-form). *Suppose M is a Riemann surface, ω is a differential 1-form. If on every local coordinates, ω has the representation*

$$\omega=udz,$$

where u is a meromorphic function, then ω is called a meromorphic 1-form.

The *bilinear relation of meromorphic 1-forms* needs to incorporate with the pole information.

Theorem 6.23. *Suppose M is a compact Riemann surface with genus g, ω is a holomorphic 1-form, $\tilde{\omega}$ is a meromorphic 1-form, with a single pole, whose residue is zero. Choose a local parameter of p_0, such that $z(p_0)=0$. Assume the local representation of ω and $\tilde{\omega}$ are*

$$\begin{aligned}\omega&=(a_0+a_1z+\cdots)dz,\\ \tilde{\omega}&=\left(\tfrac{c_{-m}}{z^m}+\cdots+\tfrac{c_{-2}}{z^2}+c_0+c_1z+\cdots\right)dz\quad(m\geqslant 2).\end{aligned}$$

Then

$$\sum_{j=1}^{g}\left(\int_{a_j}\omega\int_{b_j}\tilde{\omega}-\int_{b_j}\omega\int_{a_j}\tilde{\omega}\right)=2\pi i\sum_{n=2}^{m}\frac{c_{-n}a_{n-2}}{n-1}.$$

This result is the foundation to proving the Riemann-Roch theorem.

6.5 Riemann-Roch Theorem

The introduction of *divisor* is to study the meromorphic functions and differentials, which are mainly determined by their poles and zeros. Divisors describe the distributions of their poles and zeros.

Let M be a compact Riemann surface with genus g. The field formed by all the meromorphic functions on M is denoted as $\mathscr{K}(M)$.

Definition 6.24 (Divisor). *Assume there are n points on the surface $p_1,p_2,\cdots,p_n$, and $\alpha_1,\alpha_2,\cdots,\alpha_n$ are n integers. We call*

$$A=\prod_i p_i^{\alpha_i}$$

a divisor, *and $\sum_i\alpha_i$ is called the* degree of the divisor.

Definition 6.25 (Divisor Group). *The product of two divisors is*

$$\prod_{p\in M} p^{\alpha(p)} \prod_{p\in M} p^{\beta(p)} = \prod_{p\in M} p^{\alpha(p)+\beta(p)}.$$

All divisors form an Abelian group, the so-called divisor group, *denoted as* $Div(M)$.

Let $\mathscr{K}^* = \mathscr{K}\backslash\{0\}$ be non-zero meromorphic functions. Suppose $f \in \mathscr{K}^*$. Its divisor is

$$(f) = \prod_{p\in M} p^{\text{ord}_p(f)},$$

where $\text{ord}_p(f)$ is defined in the following way. On a small neighborhood of p, we choose a conformal chart, γ is a small circle,

$$\text{ord}_p(f) = \frac{1}{2\pi i}\int_\gamma \frac{f'}{f} dz.$$

Definition 6.26 (Primary Divisor). *If a divisor is associated with a meromorphic function, then it is called a* primary divisor. *Primary divisors must have zero degrees. All the primary divisors form the* primary divisor group, *denoted as* $D_0(M)$.

If f_1 and f_2 are meromorphic functions sharing the same divisor, then $\frac{f_1}{f_2}$ is a holomorphic function, namely a constant. Therefore, essentially a meromorphic function is determined by its divisor.

Definition 6.27 (Divisor Class Group). *The quotient group*

$$\frac{Div(M)}{D_0(M)}$$

is called the divisor class group.

Assume ω is a meromorphic differential. Then the associated divisor is

$$(\omega) = \prod_{p\in M} p^{\text{ord}_p(\omega)}.$$

The quotient of two meromorphic differentials ω_1/ω_2 is a meromorphic function, therefore, (ω_1) and (ω_2) are in the same divisor class.

Definition 6.28 (Canonical Divisor). *The divisors of meromorphic differentials are called* canonical divisors, *denoted as* Z.

Given two divisors

$$A = \prod_{p\in M} p^{\alpha(p)}, \quad B = \prod_{p\in M} p^{\beta(p)},$$

we say $A \geqslant B$ if for all points $P \in M$, $\alpha(p) \geqslant \beta(p)$.

Further, we define the following two complex linear spaces:

$$L(D) := \{f \in \mathscr{K}(M) | (f) \geqslant D\},$$

$$\Omega(D) := \{\omega | \textit{meromorphic differentials}, (\omega) \geqslant D\}.$$

Let

$$r(D) = \dim_{\mathbb{C}} L(D), \quad i(D) = \dim_{\mathbb{C}} \Omega(D).$$

Lemma 6.29. *$r(D)$ and $i(D)$ only depend on the divisor class, furthermore,*

$$i(D) = r(D/Z),$$

where Z represents the canonical divisor.

Proof. Suppose $f \in \mathscr{K}(M)$, $(f) = D_1/D_2$. Then

$$g \to f^{-1}g$$

is an isomorphism from $L(D_1)$ to $L(D_2)$, so $r(D_1) = r(D_2)$. Similarly,

$$\omega \to f^{-1}\omega$$

is an isomorphism from $\Omega(D_1)$ to $\Omega(D_2)$, so $i(D_1) = i(D_2)$.

Fix a meromorphic differential ω_0, and choose arbitrary meromorphic differential ω. Then ω/ω_0 is a meromorphic function.

$$\omega \to \omega/\omega_0$$

is an isomorphism from $\Omega(D)$ to $L(D/(\omega_0))$. Therefore,

$$i(D) = r(D/(\omega_0)) = r(D/Z).$$

□

Theorem 6.30 (Riemann-Roch). *Suppose M is a compact Riemann surface, with genus g, D is a divisor on M. Then*

$$r(D^{-1}) = \deg(D) - g + 1 + i(D).$$

The Riemann-Roch theorem plays an important role in the whole Riemann surface theory. We use it to show several important results as follows.

Theorem 6.31. *A genus zero compact Riemann surface is conformally equivalent to $\bar{\mathbb{C}}$.*

Proof. Suppose M is a topological sphere, $g = 0$. Choose $p \in M$, and let $D = p$. Then

$$r(D^{-1}) = 1 - g + 1 + i(D) \geqslant 2.$$

Therefore, there exists a meromorphic function, such that p is its pole with degree one. Therefore, the meromorphic function is the conformal mapping between M and the sphere. □

Corollary 6.32. *M is a Riemann surface with genus $g > 1$. For any point $p \in M$, there exists a holomorphic 1-form ω, such that p is not a zero point of ω.*

Proof. Assume p is the zero point of all holomorphic 1-forms, therefore, $\mathscr{H} \subset \Omega(p)$, $i(p) \geqslant g$. Then

$$r(p^{-1}) \geqslant 1 - g + 1 + g = 2,$$

therefore, there exists a meromorphic function with only one pole p (degree one), namely, M is conformally equivalent to a sphere. Contradiction. □

Corollary 6.33.

$$\deg Z = 2g - 2.$$

Proof. Let $D = Z$, then

$$i(D) = r(D/Z) = r(1) = 1,$$

furthermore,

$$r(1/Z) = i(1) = g,$$

therefore,

$$r(1/Z) = \deg(Z) - g + 1 + i(Z),$$

therefore $\deg(Z) = 2g - 2$. □

Definition 6.34 (Holomorphic Quadratic Differential). *A holomorphic quadratic differential on a Riemann surface M has local presentation*

$$\phi(z)dz^2,$$

where $\phi(z)$ is a holomorphic function. Suppose z_α and z_β are two local coordinates. Then

$$\phi_\alpha(z_\alpha)dz_\alpha^2 = \phi_\beta(z_\beta)dz_\beta^2.$$

Theorem 6.35. *Suppose M is a compact Riemann surface with genus $g > 1$. Then the dimension of the complex linear space $\mathscr{H}^2$ formed by all holomorphic quadratic differentials is* $3g - 3$.

Proof. Choose one holomorphic 1-form ω, $(\omega) = D$. Then $\mathscr{H}^2$ and $L(D^{-2})$ are isomorphic. For any $\phi \in \mathscr{H}^2(M)$,

$$\phi/\omega^2 \in L(D^{-2}).$$

For any $f \in L(D^{-2})$,

$$f\omega^2 \in \mathscr{H}^2(M).$$

Therefore,

$$r(D^{-2}) = \deg D^2 - g + 1 + i(D^2),$$

$\deg D^2 = 2\deg D = 4g - 4$, $i(D^2) = 0$, therefore, $\dim \mathscr{H}^2 = r(D^{-2}) = 3g - 3$. □

6.6 Abel Theorem

In the following discussion, we want to find the necessary and sufficient conditions for a divisor to be a primary divisor, which are formulated by the Abel theorem.

Suppose $\{a_1, b_1, \cdots, a_g, b_g\}$ is a set of canonical basis of the fundamental group of M, $\{\phi_1, \phi_2, \cdots, \phi_g\}$ is the corresponding canonical basis of holomorphic differentials. Namely,

$$\int_{a_i} \phi_j = \delta_{ij}, \quad i, j = 1, 2, \cdots, g.$$

Suppose its period matrix is B. Then $\mathrm{Im}B$ is a positive definite matrix. Let E be the identity matrix. Suppose $2g$ columns of (E, B) are

$$e^1, e^2, \cdots, e^g; \quad b^1, b^2, \cdots, b^g.$$

Define *Lattice group*

$$L(M) = \left\{ \sum_{j=1}^{g} m_j e^j + \sum_{j=1}^{g} n_j b^j | m_j, n_j \in \mathbb{Z} \right\},$$

which is an Abelian group.

Pick a base point $p_0 \in M$, and consider

$$I(p) = \left(\int_{p_0}^{p} \phi_1, \int_{p_0}^{p} \phi_2, \cdots, \int_{p_0}^{p} \phi_g \right)^T, \quad p \in M.$$

Let $D = p_1 p_2 \cdots p_r q_1^{-1} q_2^{-1} \cdots q_s^{-1} \in Div(M)$. We define

$$I(D) = \sum_{j=1}^{r} I(p_j) - \sum_{j=1}^{s} I(q_j).$$

Theorem 6.36 (Abel Theorem). *Suppose M is a compact Riemann surface, with genus $g \geqslant 1$, $D \in Div(M)$. Then the necessary and sufficient conditions for D to be a primary divisor are $deg(D) = 0$ and*

$$I(D) = 0 \ (\mathrm{mod}\ L(M)).$$

6.7 Uniformization

The *uniformization theorem* claims that any Riemann surface can be covered either by $\bar{\mathbb{C}}$, $\mathbb{C}$ or Δ; therefore, any Riemann surface has a natural Riemannian metric with constant Gauss curvature, which is one of the spherical metric, the Euclidean metric, or the hyperbolic metric.

Suppose ω is a meromorphic differential on a Riemann surface. Giving a fixed base point, then the integration of ω along a different path gives a multi-valued meromorphic function. If M is simply connected, then the integration is independent of the choice of the path, therefore, it leads to a univalent meromorphic function.

Given a multi-valued meromorphic function $f : M \to \bar{\mathbb{C}}$ on a Riemann surface M, we want to find a special parameter, such that the function becomes univalent under the parameter. A natural idea is to consider the universal covering space $\pi : \tilde{M} \to M$, and consider the lift of f, $\tilde{f} : \tilde{M} \to \bar{\mathbb{C}}$. Because $\tilde{M}$ is simply connected, therefore if $\tilde{M}$ is conformally equivalent to a domain in $\bar{\mathbb{C}}$, then $\tilde{f}$ is a univalent meromorphic function.

Therefore, the problem is to verify if any simply connected Riemann surface is conformally equivalent to a domain in $\bar{\mathbb{C}}$.

Theorem 6.37 (Poincaré-Klein-Koebe Uniformization theorem). *A simply connected Riemann surface is conformally equivalent to one of the following three canonical Riemann surfaces:*

1. *Extended complex plane* $\bar{\mathbb{C}} = \mathbb{C} \cup \{\infty\}$;

2. *Complex plane* $\mathbb{C}$;
3. *Unit disk* $\Delta = \{z \in \mathbb{C} | |z| < 1\}$.

Proof. For a compact genus zero surface, according to the Riemann-Roch theorem, there exists a meromorphic function with a degree one pole, therefore, the surface is conformally equivalent to $\bar{\mathbb{C}}$.

Suppose M is a non-compact and simply connected Riemann surface. Construct a sequence of domains on M, $\{M_n\}$, such that

$$M_n \subset \bar{M}_n \subset M_{n+1}, \quad n = 1, 2, \cdots, \quad M \subset \bigcup_{n=1}^{\infty} M_n.$$

Furthermore, we require ∂M_n to be a real analytic Jordan curve. Then we can choose a conformal atlas $\{(U_\alpha, \phi_\alpha)\}$for each $\bar{M}_n$ to make it a compact Riemann surface with a boundary. Construct another Riemann surface $\bar{M}'_n$ sharing the same topology with $\bar{M}_n$, but with a conformal atlas $\{(U_\alpha, \bar{\phi}_\alpha)\}$. Glue $\bar{M}_n$ and $\bar{M}'_n$ along their boundary to form a closed Riemann surface M_n^d, which is the double covering of $\bar{M}_n$. The double covering is a compact, genus zero Riemann surface.

We choose $p_0, p_1 \in M_1$. Then $p_0, p_1 \in \bigcap M_n$. Suppose $p'_0, p'_1 \in \bar{M}'_n$ are the corresponding points. Then $p'_0 \in \bar{M}_n^d$. Because the double covering is conformally equivalent to $\bar{\mathbb{C}}$, there exists a conformal mapping

$$f_n : M_n^d \to \bar{\mathbb{C}}, \quad f_n(p_0) = 0, \quad f_n(p_1) = 1, \quad f_n(p'_0) = \infty.$$

Let $\phi_n = f_n|_{M_n}$. Then ϕ_n is a conformal mapping, $\phi_n(M_n) \subset \mathbb{C}$, $\phi_n(p_0) = 0$, and $\phi_n(p_1) = 1$. $\{\phi_n\}$ is a normal family.

$M = \bigcup M_n$, there exists a subsequence $\{\phi_{n_k}\}$ which locally uniformly converge:

$$\lim_{k \to \infty} \phi_{n_k} = \phi.$$

Then ϕ is a conformal mapping, $\phi(p_0) = 0, \phi(p_1) = 1$. If $\phi(M) = \mathbb{C}$, then M is conformally equivalent to $\mathbb{C}$. Otherwise, $\phi(M)$ is a simply connected domain in $\mathbb{C}$, which is conformally equivalent to Δ. □

Suppose M is a Riemann surface, its universal covering space $\tilde{M}$ is conformally equivalent to one of $\bar{\mathbb{C}}$, $\mathbb{C}$, and Δ. Let G be the deck transformation group. Then

$$M \cong \tilde{M}/G.$$

Suppose $g \in G$ is a deck transformation. If g is not the identity, then g has no fixed point. Riemann surfaces can be classified according to $\tilde{M}$ and G in the following way:

1. $\tilde{M} \cong \bar{\mathbb{C}}$, *Elliptic Riemann Surface*

The deck transformation group $G = \{id\}$, therefore $M \cong \bar{\mathbb{C}}$. M is the Riemann sphere. Then M can be equipped with the spherical metric

$$ds^2 = \frac{4dzd\bar{z}}{(1 + z\bar{z})^2},$$

the induced Gauss curvature is $K = +1$.

2. $\tilde{M} \cong \mathbb{C}$, *Parabolic Riemann Surface*

G is formed by the Möbius transformations which preserve the complex plane $\mathbb{C}$,

$$g: z \to az+b, \quad a,b \in \mathbb{C}, a \neq 0,$$

furthermore, if $g \neq id$, then g has no fixed point, therefore $a = 1$,

$$g: z \to z+b, \quad b \in \mathbb{C}.$$

If

$$G = \{z \to z + nw | n \in \mathbb{Z}\},$$

$w \neq 0$, then M is conformally equivalent to $\mathbb{C} \setminus \{0\}$. If

$$G = \{z \to z + nw_1 + mw_2 | m,n \in \mathbb{Z}\},$$

$w_1, w_2 \in \mathbb{C}$, $\mathrm{Im}(w_1/w_2) \neq 0$, then M is conformally equivalent to a torus.

Then the Euclidean metric can be defined on M,

$$ds^2 = dzd\bar{z},$$

which induces constant Gauss curvature $K = 0$.

3. $\tilde{M} \cong \Delta$, *Hyperbolic Riemann Surface*

G is a torsion free Fuchs group, $M \cong \Delta/G$. The Poincaré metric can be defined on M,

$$ds^2 = \frac{4dzd\bar{z}}{(1-z\bar{z})^2},$$

which induces constant Gaussian curvature $K = -1$.

6.8 Hyperbolic Riemann Surface

Definition 6.38 (Fuchs Group). *A* Fuchs group G *is formed by Möbius transformations, satisfying the following properties*:

1. *for all* $g \in G$, $g(\Delta) = \Delta$;
2. *for any compact set* $K \subset \Delta$, *the element* $g \in G$,

$$g(K) \cap K \neq \emptyset$$

is at most finite.

If all transformations in a Fuchs group have no fixed point (except the id), then G is a *torsion free Fuchs group*.

Suppose G is a torsion free Fuchs group. Then a hyperbolic Riemann surface can be represented as

$$M = \Delta/G.$$

We want to study the topological property of M from G.

Definition 6.39 (Conjugate Möbius Transformation). *Two Möbius transformations* α, β *are* conjugate, *if there exists another Möbius transformation* γ*, such that*

$$\beta = \gamma \circ \alpha \circ \gamma^{-1}.$$

If a Möbius transformation has a unique fixed point, it is called *parabolic Möbius transformation*, and conjugate to

$$z \to z + 1,$$

whose fixed point is ∞.

For non-parabolic Möbius transforms, it has two fixed points, we use γ to map the fixed points to $0, \infty$,

$$\gamma \circ \alpha \circ \gamma^{-1} : z \to kz, \quad k \neq 0, 1,$$

$k = \lambda e^{i\theta}, \lambda, \theta \in \mathbb{R}$.

1. $\lambda = 1, \theta \neq 0$, the Möbius transformation is a rotation, called an *elliptic Möbius transformation.*
2. $\lambda \neq 1, \theta = 0$, the Möbius transformation is a scaling, called a *hyperbolic Möbius transformation*. Suppose z_1, z_2 are the two fixed points of α. Then z_1 is called the *attractive fixed point*, if

   $$\lim_{n \to \infty} \alpha^n(z) = z_1, \quad \forall z \notin \{z_1, z_2\},$$

   z_2 is called the *repulsive fixed point*, if

   $$\lim_{n \to \infty} \alpha^{-n}(z) = z_1, \quad \forall z \notin \{z_1, z_2\}.$$

3. $\lambda \neq 1, \theta \neq 0$, the Möbius transformation is called a *loxodromic Möbius transformation.*

If α is a parabolic or a hyperbolic transformation preserving Δ, then its fixed points are on $\partial\Delta$. If α is an elliptic transformation preserving Δ, then one of its fixed points is inside Δ, the other is outside Δ.

Theorem 6.40. *Suppose M is a Riemann surface covered by* Δ*, the covering transformation group is G. Then the nontrivial element in G is either hyperbolic or parabolic. If M is compact, then all non-trivial elements are hyperbolic.*

Proof. Suppose α is a non-trivial element in G, it has no fixed point inside Δ, therefore α is not elliptic. The fixed points must be on $\partial\Delta$, therefore, α is not loxodromic. α must be either parabolic or hyperbolic.

Assume α is parabolic, we use a Möbius transformation β, such that $\beta(\Delta)$ is the upper half plane, the fixed point is ∞,

$$\beta \circ \alpha \circ \beta^{-1} : z \to z + 1.$$

Consider a line segment $\gamma_y = \{x + iy | x \in [0, 1]\}$, the length of γ_y under the Poincaré metric is

$$l(\gamma_y) = \int_0^1 \frac{dx}{y} = \frac{1}{y}.$$

Hence, if $y \to +\infty$, then $l(\gamma_y) \to 0$. That means there is a family of closed curves on M, which are topologically non-trivial and whose lengths are arbitrarily small. Therefore M cannot be compact. □

Consider a compact Riemann surface M with the universal covering space $(\tilde{M}, \pi)$. Suppose $\tilde{M} \cong \Delta$. Choose a base point $p_0 \in M$ and a canonical homology basis through p_0,

$$\{a_1, b_1, a_2, b_2, \cdots, a_g, b_g\}.$$

Suppose $z_0 \in \tilde{M}$ is one of the pre-images of p_0,

$$z_0 \in \pi^{-1}(p_0).$$

Lifting a_i, b_i to the universal covering space through z_0, we get

$$\tilde{a}_1, \tilde{b}_1, \cdots, \tilde{a}_g, \tilde{b}_g.$$

Assume the ending points of $\tilde{a}_i$ and $\tilde{b}_j$ are

$$\xi_1, \eta_1, \cdots, \xi_g, \eta_g.$$

Then

$$\pi(\xi_i) = p_0, \quad \pi(\eta_j) = p_0.$$

There are $2g$ elements in G,

$$\alpha_1, \beta_1, \cdots, \alpha_g, \beta_g,$$

such that

$$\alpha_i(z_0) = \xi_i, \quad \beta_i(z_0) = \eta_i, \quad i = 1, 2, \cdots, g.$$

Because

$$a_1 b_1 a_1^{-1} b_1^{-1} a_2 b_2 a_2^{-1} b_2^{-1} \cdots a_g b_g a_g^{-1} b_g^{-1}$$

is homotopic to $[p_0]$, therefore,

$$\alpha_1 \circ \beta_1 \circ \alpha_1^{-1} \circ \beta_1^{-1} \circ \cdots \circ \alpha_g \circ \beta_g \circ \alpha_g^{-1} \circ \beta_g^{-1} = id.$$

$\alpha_1, \beta_1, \cdots, \alpha_g, \beta_g$ are the generators of G. We enumerate all the elements in G as

$$G = \{\gamma_0, \gamma_1, \gamma_2, \cdots\},$$

where γ_0 is the identity. The orbit of z_0 is

$$G(z_0) = \{z_0, z_1, z_2, \cdots\},$$

where $z_n = \gamma_n(z_0)$. Let

$$P_n(z_0) = \{z \in \Delta | d(z_0, z) < d(z_n, z)\},$$

d is the distance measured by Poincaré metric.

Definition 6.41 (Fundamental Polygon). *The fundamental polygon of G with respect to* z_0 *is*

$$P(z_0) = \bigcap_{n=1}^{\infty} P_n(z_0).$$

Fig. 6.7 shows the fundamental polygon of a genus two surface. If M is a compact Riemann surface, its fundamental polygon must be a compact set inside Δ, the number of edges is finite, and all of the edges do not touch $\partial\Delta$. The fundamental polygon of a non-compact Riemann surface may be very complicated, and the number of edges may be infinite.

Suppose γ is a hyperbolic Möbius transformation and $z_1, z_2 \in \partial\Delta$ are two fixed points. In Δ there is a unique geodesic through z_1 and z_2, denoted as l, then $\gamma(l) = l$, and l is called the *axis* of γ. The projection of l is a closed geodesic on M. Conversely, any closed geodesic on a compact Riemann surface is the axis of an element in its Fuchs group. We can prove that the fixed points of all elements of the Fuchs group are dense on $\partial\Delta$.

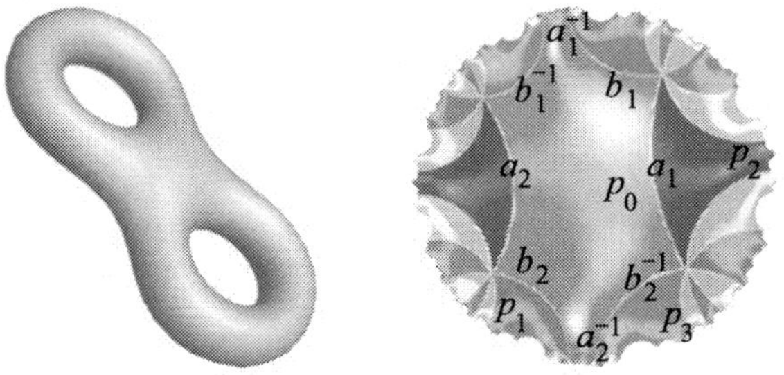

Fig. 6.7. The fundamental polygon of a genus two surface.

6.9 Teichmüller Space

This section studies the shape space of all Riemann surfaces.

6.9.1 Quasi-Conformal Map

Definition 6.42 (Quasi-Conformal Map). *Suppose* M_0 *is a Riemann surface with a conformal atlas*

$$\{(U_\alpha, \phi_\alpha)\},$$

M_1 *is another Riemann surface with a conformal atlas*

$$\{(V_\beta, \psi_\beta)\},$$

$f : M_0 \to M_1$ *is an orientation preserving homeomorphism. There exists a constant* $K \geqslant 1$, *such that if*

$$f(U_\alpha) \cap V_\beta \neq \emptyset,$$

then the local representation of f,

$$\psi_\beta \circ f \circ \phi_\alpha^{-1} : \phi_\alpha(U_\alpha \cap f^{-1}(V_\beta)) \to \psi_\beta(f(U_\alpha) \cap V_\beta),$$

is K quasi-conformal. Then we say f is a K quasi-conformal map from M_0 to M_1.

The dilatation K is independent of the choice of the local conformal coordinates.

Definition 6.43 (Beltrami Differential). *Suppose M is a Riemann surface, with conformal atlas $\{(U_\alpha, \phi_\alpha)\}$. A differential form is locally defined by*

$$\mu_\alpha(z_\alpha)\frac{d\bar{z}_\alpha}{dz_\alpha}.$$

If $U_\alpha \cap U_\beta \neq \emptyset$, then

$$\mu_\alpha(z_\alpha)\frac{d\bar{z}_\alpha}{dz_\alpha} = \mu_\beta(z_\beta)\frac{d\bar{z}_\beta}{dz_\beta},$$

where $z_\alpha = \phi_\alpha \circ \phi_\beta^{-1}(z_\beta)$. Then $\mu(z)\frac{d\bar{z}}{dz}$ is called a Beltrami differential.

If $f : M_0 \to M_1$ is K quasi-conformal, then f induces a Beltrami differential on M_0, defined as

$$\frac{\bar{\partial} f}{\partial f} = \mu(z)\frac{d\bar{z}}{dz}$$

The local representation of μ satisfies the following condition: suppose (U_α, ϕ_α) and (U_β, ϕ_β) are two local charts,

$$\mu_\beta(z_\beta) = \mu_\alpha(z_\beta(z_\alpha))\frac{\overline{\phi'_{\alpha\beta}}}{\phi'_{\alpha\beta}}.$$

We call

$$K[f] = \frac{1 + ||\mu_f||_{L^\infty(M_0)}}{1 - ||\mu_f||_{L^\infty(M_0)}}$$

the *maximal dilatation* of the quasi-conformal map f.

Assume M_0 and M_1 are two hyperbolic Riemann surfaces with universal covering spaces (Δ, π_0) and (Δ, π_1). $f : M_0 \to M_1$ is a homeomorphism, a homeomorphism $\tilde{f} : \Delta \to \Delta$ is called a *lift* of f, if

$$\pi_1 \circ \tilde{f} = f \circ \pi_0,$$

namely, the following diagram is commutative:

$$\begin{array}{ccc} \Delta & \xrightarrow{\tilde{f}} & \Delta \\ \pi_0 \downarrow & & \downarrow \pi_1 \\ M_0 & \xrightarrow{f} & M_1 \end{array}$$

Suppose G_0, G_1 are the covering transformation groups of (Δ, π_0) and (Δ, π_1) respectively, $g_0 \in G_0, g_1 \in G_1$. Then if $\tilde{f}$ is a lift, then $g_1 \circ \tilde{f} \circ g_0$ is another lift. If f is a K quasi-conformal mapping, its lift $\tilde{f}$ is also a K quasi-conformal mapping.

The lift $\tilde{f} : \Delta \to \Delta$ also induces an isomorphism

$$\chi : G_0 \to G_1, \quad g \to \tilde{f} \circ g \circ \tilde{f}^{-1}.$$

Theorem 6.44. *If f_1 and f_2 are homeomorphisms $M_0 \to M_1$, $\tilde{f}_1$ and $\tilde{f}_2$ are the corresponding lifts, then the necessary and sufficient condition for f_1 and f_2 to be homotopic is*

$$\exists g_1 \in G_1, \quad \chi_{\tilde{f}_1} = g_1 \circ \chi_{\tilde{f}_2} \circ g_1^{-1}.$$

If $\tilde{f}_1$ and $\tilde{f}_2$ are quasi-conformal, they can be extended to the boundary of Δ, $\partial\Delta$. The fixed points of the covering transformations in G_0 (or G_1) are dense in $\partial\Delta$.

Theorem 6.45. *If M_0 and M_1 are compact Riemann surfaces, with the same genus which is greater than one. $f_1, f_2 : M_0 \to M_1$ are quasi-conformal maps. f_1 and f_2 are homotopic, if and only if for any lift $\tilde{f}_1$ of f_1, there exists a lift $\tilde{f}_2$ of f_2, such that*

$$\tilde{f}_1|_{\partial\Delta} = \tilde{f}_2|_{\partial\Delta}.$$

Suppose M_0 is a compact Riemann surface with genus $g > 1$. If an automorphism $f : M_0 \to M_0$ is conformal, homotopic to identity, then f is the identity.

6.9.2 Extremal Quasi-Conformal Map

Assume M_0 and M_1 are two compact Riemann surfaces, with genus $g > 1$. Let $h : M_0 \to M_1$ be a homeomorphism preserving the orientation, let Q_h be the set of all quasi-conformal mappings $f : M_0 \to M_1$, where f is homotopic to h.

Definition 6.46 (Extremal Quasi-Conformal Map). *If $f_0 \in Q_h$, such that*

$$K[f_0] = \inf\{K[f] | f \in Q_h\},$$

then f_0 is called an extremal quasi-conformal map.

Teichmüller proved the existence and the uniqueness of the extremal quasi-conformal map, and related extremal quasi-conformal maps to the holomorphic quadratic differentials. Suppose ω is a holomorphic quadratic differential. Then $\omega = \phi(z)dz^2$. Suppose p is not a zero point of ω. we can choose a neighborhood of p and define the local coordinates

$$\zeta(z) = \int_0^z \sqrt{\phi(t)}dt,$$

where $\sqrt{\phi(t)}$ is a single valued branch, $d\zeta^2 = \phi dz^2$. Then $\zeta(z)$ is called the *natural parameter* of ω.

Definition 6.47 (Teichmüller Map). *Suppose M_0 and M_1 are two Riemann surfaces, $f: M_0 \to M_1$ is a quasi-conformal map. ω_0 and ω_1 are holomorphic quadratic differentials defined on M_0 and M_1 respectively. f maps the zero points of ω_0 to the zero points of ω_1 with same degree. In the neighborhood of regular points (non-zero points), the map can be represented under the natural parameters ζ_0, ζ_1 of ω_0, ω_1 as*

$$f: \zeta_0 \to \zeta_1 = \frac{K+1}{2}\zeta_0 + \frac{K-1}{2}\bar{\zeta}_0.$$

Then f is called a Teichmüller map, *ω_0 is called the initial holomorphic quadratic differential of f, and ω_1 is called the target holomorphic quadratic differential.*

The following theorem guarantees the uniqueness of the Teichmüller map in its homotopy class.

Theorem 6.48. *Suppose M_0 and M_1 are two compact Riemann surfaces with genus $g > 1$. If $f_0 : M_0 \to M_1$ is a Teichmüller map, then f_0 is the unique extremal quasi-conformal map in its homotopy class $[f_0]$. Namely, $\forall f \in [f_0]$,*

$$K[f] \geqslant K[f_0].$$

6.10 Teichmüller Space and Modular Space

All genus zero compact Riemann surfaces are conformally equivalent to $\bar{\mathbb{C}}$. Genus one compact Riemann surfaces can be represented as $\mathbb{C}/\Gamma$, where Γ is a lattice group

$$\Gamma = \{z \to z + m\omega_1 + n\omega_2, m,n \in Z\}, \quad \omega_1, \omega_2 \in \mathbb{C}.$$

We use $[\omega_1, \omega_2]$ to denote the conformal structure class of $\mathbb{C}/\Gamma$. It is easy to see that $[\omega_1, \omega_2] = [\tau, 1]$, where $\tau = \frac{\omega_1}{\omega_2}$. It can be shown that $[\tau, 1] = [\tau', 1]$ if and only if

$$\tau = \frac{\alpha\tau' + \beta}{\gamma\tau' + \delta}, \quad \alpha\delta - \beta\gamma = 1, \quad \alpha, \beta, \gamma, \delta \in \mathbb{Z}.$$

Therefore, all the conformal structures can be represented as

$$\{\tau \in \mathbb{C}, \mathrm{Im}\tau > 0\}/PSL(2,\mathbb{Z}),$$

where $PSL(2,\mathbb{Z})$ is called the modular group, formed by all linear rationals

$$\tau \to \frac{\alpha\tau + \beta}{\gamma\tau + \delta} \quad (\alpha, \beta, \gamma, \delta \in \mathbb{Z}, \alpha\delta - \beta\gamma = 1).$$

We can now consider high genus Riemann surfaces.

Definition 6.49 (Modular Space). *Assume M is a compact orientable topological surface with genus g. Let $\mathscr{M}_g$ be the set of all possible conformal structures on M. We say two elements of $\mathscr{M}_g$, μ_1, μ_2 are equivalent, denoted as $\mu_1 \cong \mu_2$, if there exists a conformal mapping $f: M_{\mu_1} \to M_{\mu_2}$, where M_{μ_1}, M_{μ_2} are the Riemann surfaces formed by M with conformal structures μ_1 and μ_2. The equivalence class of an element μ in $\mathscr{M}_g$ is denoted as $[\mu]$, the set of all equivalence classes is called the* Modular space of genus g Riemann surfaces, *denoted as R_g,*

$$R_g = \mathscr{M}_g / \cong.$$

Therefore R_0 has only one element, $R_1 = \mathbb{H}/PSL(2,\mathbb{Z})$. In 1857, Riemann conjectured that $R_g, g > 1$, can be described by $3g-3$ complex parameters. Modular space is very complicated to study, instead, Teichmüller space is much easier to deal with.

Definition 6.50 (Teichmüller Equivalence). *$\mu_1, \mu_2 \in \mathscr{M}_g$ are Teichmüller equivalent, denoted as $\mu_1 \sim \mu_2$, if there exists a conformal mapping $f : M_{\mu_1} \to M_{\mu_2}$ and f is homotopic to the identity map (treated as a homeomorphism).*

Definition 6.51 (Teichmüller Space). *The Teichmüller equivalence classes in $\mathscr{M}_g$ form the Teichmüller space, denoted as T_g,*

$$T_g = \mathscr{M}_g / \backsim .$$

Modular space is the quotient space of Teichmüller space:

$$R_g = T_g / Mod_g,$$

where Mod_g is the modular group. Teichmüller space is the universal covering space of the Modular space.

All the homotopy classes $[\sigma]$ of automorphisms $\sigma : M \to M$ on M form a group, which is called the *mapping class group*. The modular group Mod_g is isomorphic to the mapping class group.

6.10.1 Fricke Space Model

We use the *Fricke space model* to explain Teichmüller space. Fix a point $p_0 \in M$, and choose a set of canonical fundamental group generators $\{a_1, b_1, a_2, b_2, \cdots, a_g, b_g\}$, where a_j, b_j are loops starting and ending at p_0. Furthermore,

$$a_1 b_1 a_1^{-1} b_1^{-1} \cdots a_g b_g a_g^{-1} b_g^{-1} = 1,$$

$$a_i \times b_k = \delta_{ik}, \quad a_i \times a_k = b_i \times b_k = 0,$$

where $\alpha \times \beta$ represents the algebraic intersection number between α and β.

Suppose $\mu \in \mathscr{M}_g$ is a conformal structure on M. Let $(\mathbb{H}, \pi_0)$ be the universal covering of M_μ, G_0 be the covering transformation group. Fix one point $z_0 \in \mathbb{H}$, such that $\pi_0(z_0) = p_0$. Through z_0 lift a_j and b_j to obtain $\tilde{a}_j$ and $\tilde{b}_j$. Assume the ending points of $\tilde{a}_j$ and $\tilde{b}_j$ are z_j and w_j. Then there exist $\alpha_j, \beta_j \in G_0$, such that

$$\alpha_j(z_0) = z_j, \quad \beta_j(z_0) = w_j, \quad j = 1, 2, \cdots, g.$$

Furthermore,

$$\alpha_1 \circ \beta_1 \circ \alpha_1^{-1} \circ \beta_1^{-1} \circ \cdots \circ \alpha_g \circ \beta_g \circ \alpha_g^{-1} \circ \beta_g^{-1} = id.$$

α_j, β_j are the generators of the fundamental group $\pi_1(M, p_0)$, and they are hyperbolic Möbius transformations.

First, we use Möbius transformation to transform β_g to its conjugate form

$$\beta_g : z \to \lambda z, \quad \lambda > 1,$$

β_g has two fixed points: 0 is the repulsive fixed point and ∞ is the attractive fixed point. Then we use a scaling transformation to transform one of α_g's fixed point to 1. Under these canonical forms of the generators, we can show that α_g, β_g can be uniquely determined by $\alpha_1, \beta_1, \cdots, \alpha_{g-1}, \beta_{g-1}$. Let

$$\eta = \alpha_1 \circ \beta_1 \circ \alpha_1^{-1} \circ \beta_1^{-1} \circ \cdots \alpha_{g-1} \circ \beta_{g-1} \circ \alpha_{g-1}^{-1} \circ \beta_{g-1}^{-1}.$$

Then

$$\eta \circ \alpha_g \circ \beta_g \circ \alpha_g^{-1} \circ \beta_g^{-1} = id,$$

namely,

$$\eta \circ \alpha_g = \beta_g \circ \alpha_g \circ \beta_g^{-1}.$$

Assume

$$\eta = \frac{az+b}{cz+d}, \quad a,b,c,d \in \mathbb{R}, \quad ad-bc=1.$$

$$\alpha_g = \frac{Az+B}{Cz+D}, \quad A,B,C,D \in \mathbb{R}, \quad AD-BC=1.$$

We obtain the equation

$$\begin{pmatrix} a & b \\ c & d \end{pmatrix} \begin{pmatrix} A & B \\ C & D \end{pmatrix} = \begin{pmatrix} \sqrt{\lambda} & 0 \\ 0 & \sqrt{\lambda} \end{pmatrix} \begin{pmatrix} A & B \\ C & D \end{pmatrix} \begin{pmatrix} \frac{1}{\sqrt{\lambda}} & 0 \\ 0 & \frac{1}{\sqrt{\lambda}} \end{pmatrix},$$

also 1 is the fixed point of α_g, therefore $A+B=C+D$. Since there are 5 unknowns λ, A, B, C, D and five equations, the solution exists and is unique. Therefore the conformal structure μ is determined by $\alpha_1, \beta_1, \cdots, \alpha_{g-1}, \beta_{g-1}$.

Let γ be a hyperbolic Möbius transformation

$$\gamma : z \to \frac{az+b}{cz+d}, \quad ad-bc=1, \quad a,b,c,d \in \mathbb{R}.$$

Then the coefficient matrix

$$\begin{pmatrix} a & b \\ c & d \end{pmatrix}$$

has two eigenvalues $\lambda, \frac{1}{\lambda}, \lambda > 1$, then γ is conjugate to

$$\begin{pmatrix} \lambda & 0 \\ 0 & \lambda^{-1} \end{pmatrix},$$

namely, γ is conjugate to

$$z \to \lambda^2 z.$$

λ^2 is called the *multiplier* of γ.

Therefore, in order to represent $\alpha_1, \beta_1, \cdots, \alpha_{g-1}, \beta_{g-1}$, we can use their fixed points and multiplier, we call the following the *Fricke coordinates* of the conformal structure μ in the Teichmüller space T_g,

$$(\xi_1, \eta_1, \lambda_1, \cdots, \xi_{2g-2}, \eta_{2g-2}, \lambda_{2g-2}),$$

where ξ_k, η_k are the attractive and repulsive fixed points of α_i or β_i, λ_k is the multiplier.

Teichmüller space is path connected. This fact leads to the proof of the existence of the extremal quasi-conformal maps.

6.10.2 Geodesic Spectrum

The geodesics are the locally shortest curves on surfaces, fully determined by the Riemannian metric. The geodesic lengths reflect the global information of the surface. On general surfaces, there may be multiple geodesics in each homotopy class. For high genus surfaces with uniformization metrics, the geodesics are unique in each homotopy class.

Theorem 6.52 (Uniqueness of Geodesics). *Suppose* $(M,\mathbf{g})$ *is a closed compact surface with Riemannian metric* $\mathbf{g}$*. If Gaussian curvature is negative everywhere, then each homotopy class has a unique geodesic.*

The proof is based on the Gauss-Bonnet theorem.

Proof. Suppose $(M,\mathbf{g})$ is the surface with a metric $\mathbf{g}$, which induces a negative Gaussian curvature everywhere. We consider its universal covering space $(\tilde{M},\pi)$, where π is the projection. Then $\pi^*\mathbf{g}$ is a metric on $\tilde{M}$, which induces negative Gauss curvature everywhere.

Suppose γ_1 and γ_2 are homotopic to each other, both of them are geodesics, γ_1 does not coincide with γ_2. Then we lift them to the universal covering space $\tilde{M}$ through a common base point. Because γ_1,γ_2 are homotopic, $\tilde{\gamma}_1$, $\tilde{\gamma}_2$ share the common starting point and ending point on $\tilde{M}$. Assume the patch bounded by them is $\tilde{\sigma}$, $\partial\tilde{\sigma}=\tilde{\gamma}_1-\tilde{\gamma}_2$. According to the Gauss-Bonnet theorem,

$$\int_{\tilde{\sigma}} KdA+\int_{\tilde{\gamma}_1-\tilde{\gamma}_2} k_g ds+\alpha_1+\alpha_2=2\pi.$$

because γ_1,γ_2 are geodesics, therefore $\tilde{\gamma}_1,\tilde{\gamma}_2$ are also geodesics. k_g is zero everywhere. α_1,α_2 are the exterior angles at the common starting and ending points of $\tilde{\gamma}_1,\tilde{\gamma}_2$, both of them are less than π. Contradiction. Therefore, γ_1 coincides with γ_2. □

For the purpose of studying the conformal structures, we can always deform the surface metric to the uniformization metric, then the geodesic lengths in each homotopy class form the length spectrum.

Definition 6.53 (Length Spectrum). *Let* $(\mathrm{S},\bar{\mathbf{g}})$ *be a surface with the uniformization metric. The set of the lengths of closed geodesics on* S *is called the length spectrum of the surface* S.

The number of homotopy classes of closed curves on a compact surface is countable. Since each homotopy class contains only one geodesic curve, the length spectrum of a Riemann surface is also countable. If two surfaces are conformally equivalent, their length spectra are the same.

Theorem 6.54. M_0 *and* M_1 *are compact Riemann surfaces with the same genus* $g>1$. *The length spectra of* M_0 *and* M_1 *under the uniformization metric are* $\{l_0,l_1,l_2,\cdots\}$ *and* $\{s_0,s_1,s_2,\cdots\}$ *respectively. The corresponding homotopy classes are* $\{[\gamma_0],[\gamma_1],[\gamma_2],\cdots\}$ *and* $\{[\tau_0],[\tau_1],[\tau_2],\cdots\}$. *Then the necessary and sufficient conditions for* M_0 *and* M_1 *to be conformally equivalent are*

1. $l_k=s_k$ *for all* $k=0,1,2,\cdots$.
2. *The mapping* $h:[\gamma_k]\rightarrow[\tau_k]$ *is an isomorphism between* $\pi_1(M_0)$ *and* $\pi_1(M_1)$.

Problems

6.1. Quasi-Conformal Map Prove that there does not exist a quasi-conformal map which maps the unit disk to the whole complex plane.

6.2. Meromorphic Differential Suppose $f : S_1 \to S_2$ is a conformal map between Riemann surfaces S_1 and S_2, show that the linear space formed by the meromorphic differentials are isomorphic.

6.3. Holomorphic Differential Suppose ω is a holomorphic 1-form on a Riemann surface, $\omega = \bar{\omega}$. Prove that ω must be trivial.

6.4. Meromorphic Differential Suppose S is a Riemann surface, $p_1, p_2, \cdots, p_m$ are m points on S. Assume $C_1, C_2, \cdots, C_m$ are m complex numbers, $C_1 + C_2 + \cdots + C_m = 0$. Prove that there exists a meromorphic differential form ω, which is holomorphic on $S \backslash \{p_1, p_2, \cdots, p_m\}$, $p_1, p_2, \cdots, p_m$ are degree one poles, the residues are $C_1, C_2, \cdots, C_m$ respectively.

6.5. Holomorphic Map Let S be a compact Riemann surface with genus $g > 1$. Prove that there exists a holomorphic map $f : S \to \bar{\mathbb{C}}$, the number of covering sheets is no greater than g.

6.6. Holomorphic Map Prove that any holomorphic map $\bar{\mathbb{C}} \to \bar{\mathbb{C}}$ must be a rational function

$$R(z) = \frac{P(z)}{Q(z)}, \quad (P, Q) = 1,$$

where P and Q are polynomials.

6.7. Hyper-Elliptic Riemann Surface A compact Riemann surface S is called hyper-elliptic if there exists a divisor $D \geqslant 1$, $\deg D = 2$, $r(D^{-1}) \geqslant 2$,
(a) Prove that S is hyper-elliptic if and only if there is a non-trivial meromorphic function $f : S \to \bar{\mathbb{C}}$, the number of covering sheets is $m = 2$.
(b) Prove that all compact Riemann surfaces with genus $g \leqslant 2$ are hyper-elliptic.

6.8. Conformal Automorphism of Hyperbolic Riemann Surface Let S be a compact hyperbolic Riemann surface and $f : S \to S$ is a conformal automorphism. Prove that f must be a periodic map. Namely, there is an integer m, such that

$$\underbrace{f \circ f \circ \cdots \circ f}_{m} = id.$$

6.9. Möbius Transformation Let

$$\phi : z \to \frac{az + b}{cz + d} \quad (ad - bc = 1)$$

be a non-trivial Möbius transformation. Denote

$$A = \begin{pmatrix} a & b \\ c & d \end{pmatrix}.$$

Prove that
(a) ϕ is parabolic, if and only if $\mathrm{tr}^2(A) = 4$;
(b) ϕ is elliptic, if and only if $0 \leqslant \mathrm{tr}^2(A) < 4$;
(c) ϕ is hyperbolic, if and only if $4 < \mathrm{tr}^2(A) < \infty$;
(d) ϕ is loxodromic, if and only if $\mathrm{tr}^2(A) \notin [0, +\infty)$.

6.10. Length Spectrum Suppose S is a compact Riemann surface. Prove that if $\{l[\gamma_1], l[\gamma_2], \cdots\}$ is a length spectrum of S, then for an arbitrary positive number M, γ_n satisfying

$$l[\gamma_n] \leqslant M$$

is finite.

7

Harmonic Maps and Surface Ricci Flow

In this chapter, we introduce harmonic map theory and surface Ricci flow theory.

7.1 Harmonic Maps of Surfaces

Harmonic maps between manifolds are important tools for studying geometry and topology. They have broad applications in many fields in pure mathematics, physics, and engineering fields.

7.1.1 Harmonic Energy and Harmonic Maps

Let $(M,\mathbf{g})$, $(N,\mathbf{h})$ be surfaces with Riemannian metrics $\mathbf{g}$ and $\mathbf{h}$ respectively, and let u be a C^1 map from M to N. We denote the metrics as

$$ds_M^2 = \sum g_{\alpha\beta}(x)dx^\alpha x^\beta, \quad ds_N^2 = \sum h_{ij}(u(x))du^i du^j.$$

The pull-back of $\mathbf{h}$, $u^*(ds_N^2)$ is defined as

$$u^*(ds_N^2) = \sum_{\alpha,\beta}(\sum_{i,j} h_{ij}(u(x))\frac{\partial u^i}{\partial x^\alpha}\frac{\partial u^j}{\partial x^\beta})dx^\alpha dx^\beta.$$

We find two orthonormal vector fields $\mathbf{e}_1, \mathbf{e}_2$ and their dual 1-forms ω_1, ω_2, then diagonalize $u^*(ds_N^2)$ such that

$$u^*(ds_N^2) = \lambda_1\omega_1^2 + \lambda_2\omega_2^2.$$

Then the symmetric functions of λ_1, λ_2 are important invariants. One of them is the *trace*, which gives the energy density.

Definition 7.1 (Energy Density). *The trace*

$$|du|^2 = Tr_{ds_M^2}(u^* ds_N^2) = \lambda_1 + \lambda_2$$

is called the energy density *of u and its coordinate representation is given by*

$$|du|^2 = \sum_{i,j,\alpha,\beta} g^{\alpha\beta}(x)h_{ij}(u(x))\frac{\partial u^i}{\partial x^\alpha}\frac{\partial u^j}{\partial x^\beta}.$$

Then the *energy density* of u is independent of the choice of coordinate system of M.

Definition 7.2 (Harmonic Maps). *The* energy functional $E(u)$ *is given by*

$$E(u) = \int_M |du|^2 dv_M,$$

where $dv_M = \sqrt{\det \mathbf{g}}dx$ *is the area element of M. The critical points of E in the space of maps are called* harmonic maps.

If the target surface $N = \mathbb{R}^3$ and $h_{ij} = \delta_{ij}$, then the harmonic energy has a simpler form

$$E(u) = \sum_{i=1}^{3}\int |\nabla u^i|^2 dv_M.$$

Let $M = [0,1]$, and $u : [0,1] \to N$ be a curve in N. Then the harmonic energy is

$$E(u) = \int_0^1 |\frac{du}{dt}|^2 dt.$$

The critical points are geodesics with constant speed parameterization. From this point of view, harmonic maps can be treated as the generalization of geodesics.

7.1.2 Harmonic Map Equation

Suppose N is isometrically embedded in $\mathbb{R}^3$. Then the harmonic energy has the form of

$$E(u) = \sum_{i=1}^{3} \int_M |\nabla u^i|^2 dv_M.$$

In this way we can get *harmonic map equation*

$$\Delta u^i - g^{\alpha\beta} A^i_{u(x)}\left(\frac{\partial u}{\partial x^\alpha}, \frac{\partial u}{\partial x^\beta}\right) = 0, \tag{7.1}$$

where A is the second fundamental form of N defined by $A_u(X,Y) = (D_X Y)^\perp$. Harmonic map equation (7.1) can be written as

$$(\Delta u)^{T_u N} \equiv 0,$$

the tangential component of Δu is zero. In local coordinates on N, equation (7.1) becomes

$$\Delta u^i + g^{\alpha\beta}\Gamma^i_{jl}(u(x))\frac{\partial u^j}{\partial x^\alpha}\frac{\partial u^l}{\partial x^\beta} = 0, \quad i = 1,2,$$

where Γ^i_{jk} denotes the Christoffel symbols of N.

Harmonic energy is *conformally invariant*. Suppose M is equipped with two Riemannian metrics $\mathbf{g} = g_{\alpha\beta}dx^\alpha dx^\beta$, and $\tilde{\mathbf{g}} = e^{2\lambda}\mathbf{g}$, where $\lambda \in C^\infty(M)$. Then

$$|\tilde{du}|^2 = e^{-2\lambda}|du|^2$$

and

$$\sqrt{\det\tilde{\mathbf{g}}} = e^{2\lambda}\sqrt{\det\mathbf{g}},$$

hence $\tilde{E}(u) = E(u)$. This shows that the harmonic energy solely depends on the conformal structure of M, not a specific metric.

7.1.3 Radó's Theorem

From the theories of elliptic partial differential equations, we can verify the following theorem.

Theorem 7.3 (Radó). *Assume $\Omega \subset \mathbb{R}^2$ is a convex domain with a smooth boundary $\partial\Omega$ and D is a topological disk with a Riemannian metric $\mathbf{g}$. Given any homeomophism ϕ : $\mathbb{S}^1 \to \partial\Omega$, there exists a unique harmonic map $u : D \to \Omega$, such that $u = \phi$ on $\partial D = \mathbb{S}^1$ and u is a diffeomorphism.*

The harmonic map can be computed by minimizing the harmonic energy $E(u) = \sum_i \int_D |\nabla u_i|^2 dxdy$. A harmonic map is not conformal in general. In fact, a conformal map minimizes energies over all orientation preserving homeomorphisms: $D \to \Omega$. which are one to one and continuous on ∂D. Therefore, the boundary condition plays a crucial role.

If Ω is not convex, then the above result is not true, since folding may occur near the concave part of $\partial\Omega$ if a small neighborhood of a point in ∂D is mapped by ϕ onto a large concave part of $\partial\Omega$.

7.1.4 Hopf Differential

Let a map $u : M \to N$ be harmonic. We choose normal coordinates centered at $u(x) \in N$, such that $\Gamma^k_{jk}(u(x)) = 0$. Then the harmonic map equation on the local representation reduces to

$$\Delta u^i(x) = 0.$$

On the other hand,

$$\Delta u^i = 4\frac{\partial^2 u^i}{\partial z \partial \bar{z}}.$$

If we define

$$\phi(z) = \left\langle \frac{\partial u}{\partial z}, \frac{\partial u}{\partial z} \right\rangle,$$

or equivalently,

$$\phi(z) = \sum_{i,j} h_{ij}(u(z)) \frac{\partial u^i}{\partial z}\frac{\partial u^j}{\partial z},$$

hence

$$\frac{\partial \phi}{\partial \bar{z}} = 0.$$

$\phi(z)$ is independent of the choice of the coordinates and ∇h_{ij} are zeros at $u(z)$ in the normal coordinates. Therefore $\phi(z)$ is holomorphic.

Definition 7.4 (Hopf Differential). $\Phi = \phi(z)dz^2$ *is called the* Hopf differential.

If u is a harmonic map, then its Hopf differential is a holomorphic quadratic differential.

Lemma 7.5. *If u is conformal, then its Hopf differential is zero.*

Proof. Since

$$\phi(z) = \frac{1}{4}\left(\left|\frac{\partial u}{\partial x}\right|^2 - \left|\frac{\partial u}{\partial y}\right|^2 - 2i\left\langle \frac{\partial u}{\partial x}, \frac{\partial u}{\partial y} \right\rangle \right),$$

$\phi(z) \equiv 0$ if and only if u is conformal. □

Proposition 7.6. *A harmonic map $u : \mathbb{S}^2 \to N$ is automatically conformal.*

Proof. There exists no nonzero holomorphic quadratic differential on $\mathbb{S}^2$. A harmonic map u induces a holomorphic quadratic differential, which must be zero. Therefore, u must be a conformal map. □

Corollary 7.7. *There exists no nontrivial harmonic map from $\mathbb{S}^2$ to a surface N with positive genus, regardless of the metric on N.*

Proof. Suppose $u : \mathbb{S}^2 \to N$ is harmonic. Then it is conformal. Then we lift the map $\tilde{u} : \mathbb{S}^2 \to \tilde{N}$, where $\tilde{N}$ is the universal covering space of N. Then we get a conformal map from $\mathbb{S}^2$ to the unit disk Δ, then $\tilde{u}$ is a constant. Contradiction. □

In fact, a nontrivial harmonic map between a torus and an arbitrary surface exists.

7.1.5 Complex Form

It is more convenient to discuss harmonic maps using complex forms. Let M,N be compact oriented surfaces without boundary and u a map from M to N. Suppose $z = x+iy$, $u = u_1 + iu_2$ be complex coordinates of M, N respectively. Then the metrics of M,N are of the form $\lambda(z)|dz|^2, \rho(u)|du|^2$ respectively. We denote

$$\frac{\partial}{\partial z} = \frac{1}{2}\left(\frac{\partial}{\partial x} - i\frac{\partial}{\partial y}\right),$$

$$\frac{\partial}{\partial \bar{z}} = \frac{1}{2}\left(\frac{\partial}{\partial x} + i\frac{\partial}{\partial y}\right).$$

Then by direct computation, we get

$$|du|^2 = 2\frac{\rho}{\lambda}\left(\left|\frac{\partial u}{\partial z}\right|^2 + \left|\frac{\partial u}{\partial \bar{z}}\right|^2\right),$$

the Jacobian of u is

$$J(u) = \frac{\rho}{\lambda}\left(\left|\frac{\partial u}{\partial z}\right|^2 - \left|\frac{\partial u}{\partial \bar{z}}\right|^2\right).$$

The harmonic map equation is

$$u_{z\bar{z}} + (\log\rho)_u u_z u_{\bar{z}} = 0.$$

7.1.6 Bochner Formula

Let K_M, K_N be the Gaussian curvature of M,N respectively. Then

$$K_M = -\frac{1}{2}\Delta_M \log\lambda, \quad K_N = -\frac{1}{2}\Delta_N \log\rho,$$

$\Delta_M = \dfrac{4}{\lambda}\dfrac{\partial^2}{\partial z \partial \bar{z}}$. Let

$$|\partial u|^2 = \frac{\rho}{\lambda}|u_z|^2, \quad |\bar{\partial} u|^2 = \frac{\rho}{\lambda}|u_{\bar{z}}|^2.$$

The Bochner formula are

$$\Delta_M \log|\partial u| = K_M - K_N J(u), \tag{7.2}$$

$$\Delta_M \log|\bar{\partial} u| = K_M + K_N J(u). \tag{7.3}$$

The functions $|\partial u|, |\bar{\partial} u|$ are either identically zero or they have isolated zeros with well-defined order.

Theorem 7.8. *Suppose M,N are closed. If $|\partial u|$ is not identically zero, then*

$$\sum_{\substack{p\in M \\ |\partial u|(p)=0}} n_p = -\deg(u)(2g_N - 2) + (2g_M - 2). \tag{7.4}$$

If $|\bar{\partial}u|$ is not identically zero, then

$$\sum_{\substack{p\in M\\|\bar{\partial}u|(p)=0}} m_p = \deg(u)(2g_N-2)+(2g_M-2). \tag{7.5}$$

where n_p, m_p are the orders of $|\partial u|$, $|\bar{\partial}u|$ at p respectively.

Proof. Since the zeros of $|\partial u|$ are isolated and M is compact, there exists a finite number of zeros $p_1,\cdots,p_k$ in M. For $\varepsilon>0$ small enough, let $M_\varepsilon = M/\cup_{j=1}^k D_\varepsilon(p_j)$, where $D_\varepsilon(p_j)$ is the disk of radius ε centered at p_j. From the Bochner formula we have

$$\int_{M_\varepsilon} \Delta\log|\partial u| dv_M = -\int_{M_\varepsilon} K_N J(u) dV_M + \int_{M_\varepsilon} K_M dv_M.$$

By Stokes theorem, we obtain

$$\int_{M_\varepsilon} \Delta\log|\partial u| dv_M = -\sum_{j=1}^k \int_{C_\varepsilon} \frac{\partial}{\partial r}\log|\partial u| r d\theta,$$

where $\frac{\partial}{\partial r}$ is the radial derivative and $C_\varepsilon(p_j)=\partial D_\varepsilon(p_j)$. Assume

$$|\partial u| = |z|^{n_{p_j}} g_j(z)$$

in $D_s(p_j)$, where $g_j(z)$ is a C^∞ positive function. Hence for each j,

$$\int_{C_\varepsilon(p_j)} \frac{\partial}{\partial r}\log|\partial u| r d\theta = 2\pi\cdot n_{p_j} + O(\varepsilon).$$

Let $\varepsilon\to 0$, we get

$$\int_M \Delta\log|\partial u| dv_M = -2\pi\sum_{j=1}^k n_{p_j} = -2\pi \sum_{\substack{p\in M\\|\partial u|(p)=0}} n_p.$$

On the other hand,

$$-\int_M K_N J(u) dv_M + \int_M K_M dv_M = -2\pi\deg(u)\cdot(2-2g_N)+2\pi(2-2g_M).$$

Therefore,

$$\sum_{\substack{p\in M\\|\partial u|(p)=0}} n_p = -\deg(u)(2g_N-2)+(2g_M-2).$$

The second statement can be derived in a similar way. □

Corollary 7.9. *Suppose $N=\mathbb{S}^2$, M and N are equipped with arbitrary metrics. If $u: M\to \mathbb{S}^2$ is harmonic and $\deg(u) > g_M - 1$, then u is holomorphic.*

Proof. Otherwise, $\bar{\partial}u \not\equiv 0$, then

$$0 \leqslant \sum_{|\bar{\partial}u|(p)=0} m_p = \deg(u)(0-2) + (2g_M - 2) < -2(g_M - 1) + (2g_M - 2) = 0,$$

contradiction. So $\bar{\partial}u \equiv 0$ and so u must be holomorphic. □

Corollary 7.10. *There is no harmonic map from T^2 to $\mathbb{S}^2$ with* $\deg(u) = 1$.

Proof. Suppose u is such a harmonic map. Since $\deg(u) = 1 > \text{genus}(T^2) - 1$, u is holomorphic. Hence $J(u) \geqslant 0$. u is a branched covering with degree 1, therefore, u is a diffeomorphism. Contradiction. □

Orientation preserving harmonic maps are necessarily branched coverings, any zero of $J(u)$ is a nontrivial branch point of u.

Proposition 7.11. *Let $u : \Omega \subset M \to N$ be harmonic on an open connected set Ω of M. If $J(u) \geq 0$ and $J(u)$ is not identically zero in Ω, then the zeros of $J(u)$ are isolated. Moreover, if there is a number l such that $\#(u^{-1}(q)) \leqslant l$ for any regular value $q \in N$, then any zero of $J(u)$ is a nontrivial branch point of u (where $J(u)$ is the Jacobian of u).*

7.1.7 Existence and Regularity

In general, for any diffeomorphism between two surfaces, there exists a harmonic map homotopic to it. The harmonic map may not necessarily be unique.

Theorem 7.12. *Suppose M and N are compact surfaces without boundary, with arbitrary Riemannian metrics. $h : M \to N$ is a diffeomorphism. Then there exists a harmonic diffeomorphism $u : M \to N$, such that u is isotopic to h. Furthermore, u is of least energy among all diffeomorphisms isotopic to h.*

The same assertion holds with 'isotopic' replaced by 'homotopic'. Homotopic diffeomorphisms are isotopic by Baer's theorem. This theorem was proved by Jost and Schoen, a detailed proof can be found in [34].

A harmonic map in general is regular. The following theorem gives a general criterion under which a harmonic map between surfaces is a diffeomorphism.

Theorem 7.13. *Suppose* $\mathbf{g}_M = \mathbf{g}_N = \mathbf{g}$, $\deg(u) = 1$, *u harmonic, and $K_N < 0$ everywhere. Then u is a diffeomorphism.*

Proof. Let

$$w = \log \frac{|\partial u|}{|\bar{\partial} u|}.$$

Then

$$\Delta w = -K_N |\phi| \sinh w.$$

From equation (7.5), we see that $|\partial u|$ has no zeros. It follows that w is bounded below and achieves its minimum value (unless $\bar{\partial}u \equiv 0$, in which case the theorem is clearly true). Let p_0 be the minimum point of w. Then $\Delta w(p_0) \geqslant 0$. Hence $-K_N(u(p_0))|\phi(p_0)| \sinh w(p_0) \geqslant 0$. Since w is finite at p_0, we have $|\bar{\partial}(p_0)| > 0$. It follows that $\phi(p_0) \neq 0$, and hence $w(p_0) \geqslant 0$. By the strong maximum principle, we have $w(p_0) > 0$. Hence $J(u) > 0$. It follows that u is a covering map and therefore a diffeomorphism since it is of degree 1. □

7.1.8 Uniqueness

For harmonic diffeomorphisms of hyperbolic surfaces, the Hopf differential determines the image surface uniquely up to an isometry.

Theorem 7.14. *Let u_1, u_2 be harmonic maps of degree one, from $(M, \mathbf{g})$ to (M, ρ_1) and (M, ρ_2) respectively. Suppose $K_{\rho_1} = K_{\rho_2} = -1$. If the Hopf differential $\Phi_{u_1} \equiv \Phi_{u_2}$, then $(u_1 \circ u_2^{-1})^* \rho_1 = \rho_2$. Namely, (M, ρ_1) is isometric to (M, ρ_2).*

Proof. Let
$$w_1 = \log \frac{|\partial u_1|}{|\bar{\partial} u_1|}, \quad w_2 = \log \frac{|\partial u_2|}{|\bar{\partial} u_2|}.$$
By assumption,
$$\phi(z) = 4\rho_1(u_1(z)) u_{1_z} \bar{u}_{1_z} = 4\rho_2(u_2(z)) u_{2_z} \bar{u}_{2_z}. \tag{7.6}$$
Hence
$$\Delta w_i = |\phi| \sinh w_i, \quad i = 1, 2.$$
Note that the poles of w_1, w_2 occur at the zeros of ϕ. Hence
$$\frac{\partial u_i}{\bar{\partial} u_i} = |a(z)| \cdot |z - p|^{-m_p},$$
where $a(z)$ is smooth and $a(p) \neq 0$. Therefore,
$$w_i(z) = m_p \log \frac{1}{|z - p|} + \text{regular function}.$$
Hence $v = w_1 - w_2$ is smooth. We show $v \equiv 0$. Suppose v is positive somewhere, and let $\Omega_+ = \{x \in M, v(x) > 0\}$. In Ω_+,
$$\Delta v = |\phi|(\sinh w_1 - \sinh w_2) > 0. \tag{7.7}$$
Moreover, Δv is smooth, hence v is subharmonic and $v \equiv 0$ on $\partial\Omega_+$. By the strong maximum principle, $v \equiv 0$ in Ω_+, which is a contradiction. Therefore $v \leq 0$ everywhere. Similarly we conclude $v \geqslant 0$ everywhere. Thus $v \equiv 0$ and $w_1 \equiv w_2$. From equation (7.6) we get $|\partial u_1| \cdot |\bar{\partial} u_1| = |\partial u_2| \cdot |\bar{\partial} u_2|$. Hence we get
$$|\partial u_1| \equiv |\partial u_2|, \quad |\bar{\partial} u_1| \equiv |\bar{\partial} u_2|.$$
Therefore the energy densities of u_1 and u_2 are equal as functions.

Let $u : (M, z) \to (M, \rho)$, $\rho = \rho(u)|du|^2$. Then
$$u^* \rho = \rho(|u_z|^2 + |u_{\bar{z}}|^2)|dz|^2 + \frac{1}{2}\text{Re}(\Phi).$$
We get
$$\begin{aligned} u_1^* \rho_1 &= \lambda(z)(|\partial u_1|^2 + |\bar{\partial u_1}|^2)|dz|^2 + \tfrac{1}{2}\text{Re}(\Phi_{u_1}) \\ u_2^* \rho_2 &= \lambda(z)(|\partial u_2|^2 + |\bar{\partial u_2}|^2)|dz|^2 + \tfrac{1}{2}\text{Re}(\Phi_{u_2}) \end{aligned}$$
From equation (7.7) and the hypothesis, we get
$$u_1^* \rho_1 \equiv u_2^* \rho_2$$
on M. □

The following global uniqueness result of Al'ber and Hartman is for general Riemannian manifolds.

Theorem 7.15. *Let $u : M \to N$ be a harmonic map between compact Riemannian manifolds without boundaries. Suppose N has negative sectional curvature. Then u is unique harmonic map in its homotopy class unless $u(M)$ is a point or a closed geodesic.*

If the sectional curvature of N is non-positive, then for any two homotopic harmonic $u_0, u_1 : M \to N$, there exists a family $u_t : M \to N$ of harmonic maps, with the property that the curves $u_t(x)$, for fixed $x \in M$, $t \in [0,1]$ varying, constitute a family of parallel geodesics, parameterized proportionally to arc length. In particular, all maps u_t have the same energy.

A detailed proof of this theorem can be found in [49].

7.2 Surface Ricci Flow

In this section, we introduce several important concepts in the theory of the *surface Ricci flow*.

7.2.1 Conformal Deformation

The Gauss-Bonnet theorem bridges Gaussian curvature with the topology of a surface. We can consider a special Riemannian metric such that the Gaussian curvature is constant everywhere, the constant is one of $\{+1, 0, -1\}$. The uniformization theorem provides the existence and the uniqueness of such a metric, as follows.

Theorem 7.16 (Poincaré-Klein-Koebe Uniformization Theorem). *Let $(S, \mathbf{g})$ be a closed 2-dimensional surface with a Riemannian metric $\mathbf{g}$. Then there is a unique metric $\bar{\mathbf{g}}$ conformal to $\mathbf{g}$ with constant Gaussian curvature, the constant is one of $\{+1, 0, -1\}$ depending on the topology of the surface.*

Such a metric is called the *uniformization metric*. According to the Gauss-Bonnet theorem, the sign of the constant Gaussian curvatures is determined by the Euler number of the surface.

Fig. 7.1 illustrates the uniformization theorem. The universal covering space of any closed surface with the uniformization metric can be isometrically embedded onto one of the three canonical spaces:

1. the *unit sphere* $\mathbb{S}^2$ for genus zero surfaces with $\chi > 0$;
2. the *plane* $\mathbb{R}^2$ for genus one surfaces with $\chi = 0$;
3. the *hyperbolic space* $\mathbb{H}^2$ for high genus surfaces with $\chi < 0$.

Definition 7.17 (Conformal Deformation). *Let $(M, \mathbf{g})$ be a two dimensional smooth Riemannian manifold. If $\tilde{\mathbf{g}}$ is another Riemannian metric on M, we say that $\tilde{g}$ is* conformal to $\mathbf{g}$ *if and only if there exists a positive function $\rho \in C^\infty(M)$, such that*

$$\tilde{\mathbf{g}} = \rho \mathbf{g}.$$

We also say $\tilde{\mathbf{g}}$ is a conformal deformation *of* $\mathbf{g}$.

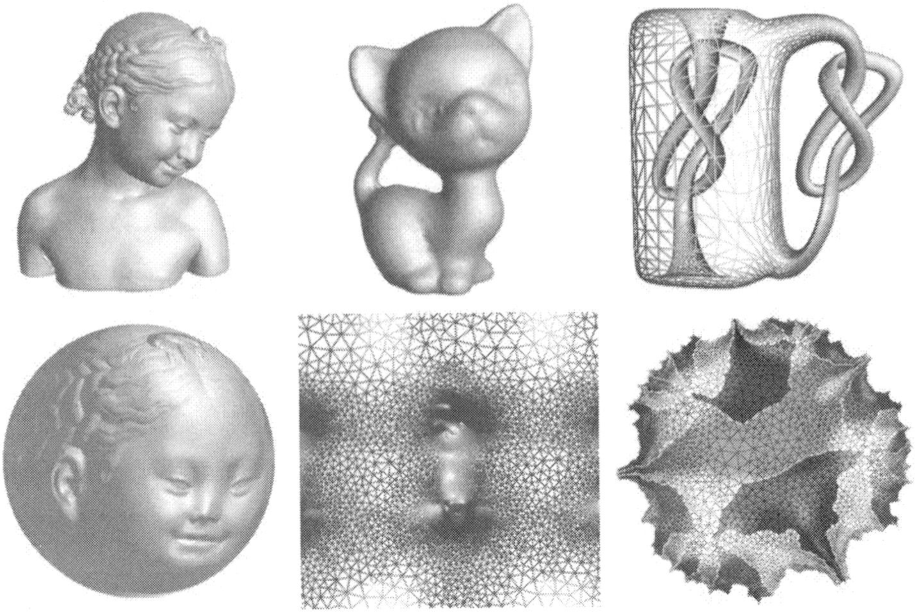

Fig. 7.1. Uniformization Theorem: all surfaces with Riemannian metric can be conformally embedded onto three canonical spaces: the sphere, the plane, and the hyperbolic space.

We choose isothermal coordinates of the surface,

$$\mathbf{g} = e^{2\lambda(u,v)}(du^2 + dv^2).$$

The Gaussian curvature is

$$K(u,v) = -\frac{1}{e^{2\lambda(u,v)}}(\frac{\partial^2}{\partial u^2} + \frac{\partial^2}{\partial v^2})\lambda.$$

We use $\Delta_g = \frac{1}{e^{2\lambda}}\left(\frac{\partial^2}{\partial u^2} + \frac{\partial^2}{\partial v^2}\right)$. Then

$$K = -\Delta_g \lambda$$

After conformal deformation,

$$\tilde{\mathbf{g}} = e^{2\tilde{\lambda}}(du^2 + dv^2),$$

then

$$\tilde{K} = -\Delta_{\tilde{g}}\tilde{\lambda}.$$

Let $\tau = \tilde{\lambda} - \lambda$. Then we get

$$\begin{aligned}
\tilde{K} &= -\Delta_{\tilde{g}}\tilde{\lambda} \\
&= -\frac{1}{e^{2(\lambda+\tau)}}(\Delta\lambda + \Delta\tau) \\
&= -\frac{1}{e^{2\tau}}\frac{1}{e^{2\lambda}}(\Delta\lambda + \Delta\tau) \\
&= \frac{1}{e^{2\tau}}(K - \Delta_g \tau).
\end{aligned}$$

The equation is

$$\tilde{K} = \frac{1}{e^{2\tau}}(K - \Delta_g \tau) \tag{7.8}$$

Similarly, if the surface has boundaries, then the geodesic curvature along the boundary will become

$$\tilde{k} = \frac{1}{e^{\tau}}(k - \partial_{\mathbf{n}} u), \tag{7.9}$$

where $\mathbf{n}$ is the outwards normal on the tangent planes along the boundary.

Equations 7.8 and 7.9 can be solved using the surface Ricci flow.

7.2.2 Surface Ricci Flow

Suppose S is a smooth surface with a Riemannian metric $\mathbf{g}$. The Ricci flow is the process to deform the metric $\mathbf{g}(t)$ according to its induced Gaussian curvature $K(t)$, where t is the time parameter:

$$\frac{dg_{ij}(t)}{dt} = -2K(t)g_{ij}(t). \tag{7.10}$$

Ricci flow was introduced by Hamilton in [50] inspired by the heat flows on manifolds. Suppose $T(t)$ is a temperature field on the surface. The heat diffusion equation is

$$\frac{dT(t)}{dt} = -\Delta_g T(t),$$

where Δ is the Laplace-Beltrami operator induced by the Riemannian metric. The temperature field becomes more and more uniform with the increase of t, and it will become a constant eventually.

In a physical sense, the curvature evolution induced by the Ricci flow is exactly the same as heat diffusion on the surface, as follows:

$$\frac{K(t)}{dt} = -\Delta_{\mathbf{g}(t)} K(t), \tag{7.11}$$

where $\Delta_{\mathbf{g}(t)}$ is the Laplace-Beltrami operator induced by the metric $\mathbf{g}(t)$. We can simplify the Ricci flow equation (7.10). Let $g(t) = e^{2u(t)}g(0)$. Then Ricci flow is

$$\frac{du(t)}{dt} = -2K(t). \tag{7.12}$$

It is very easy to see that equation 7.12 preserves the conformal class, that is, $\mathbf{g}(t)$ is a solution, then $\mathbf{g}(t)$ and $\mathbf{g}(t')$ are conformal. To study the convergence of Ricci flow, Hamilton introduced the *normalized Ricci flow* 7.13 so that the surface area is preserved during the flow.

$$\frac{du(t)}{dt} = -2(K(t) - 2\pi\frac{\chi(S)}{A(t)}). \tag{7.13}$$

The following theorems postulate that the Ricci flow defined in equation (7.10) is convergent and leads to the conformal uniformization metric.

Theorem 7.18 (Hamilton 1988 [51]). *For a closed surface of non-positive Euler characteristic, the normalized Ricci flow will converge to a metric such that the Gaussian curvature is constant everywhere.*

Theorem 7.19 (Chow 1991[52]). *For a closed surface of positive Euler characteristic, the normalized Ricci flow will converge to a metric such that the Gaussian curvature is constant everywhere.*

The corresponding metric $\mathbf{g}(\infty)$ is the *uniformization metric*. Moreover, at any time t, the metric $\mathbf{g}(t)$ is conformal to the original metric $\mathbf{g}(0)$.

The Ricci flow can be easily modified to compute a metric with a *prescribed curvature* $\bar{K}$, and then the flow becomes

$$\frac{dg_{ij}(t)}{dt} = 2(\bar{K} - K)g_{ij}(t). \tag{7.14}$$

With this modification, any target curvatures $\bar{K}$, which are admissible with the Gauss-Bonnet theorem, can be induced from the solution metric $\mathbf{g}(\infty)$.

Problems

7.1. Harmonic Function Suppose S is a compact closed surface with a Riemannian metric. Prove that all harmonic functions on S must be constants.

7.2. Laplace Operator on Surface Let S be a surface,

$$\mathbf{r}(u^1,u^2) = (x(u^1,u^2), y(u^1,u^2), z(u^1,u^2)).$$

Use the Gauss equation

$$\frac{\partial^2 \mathbf{r}}{\partial u^i \partial u^j} = \sum_{k=1}^{2} \Gamma_{ij}^k \frac{\partial \mathbf{r}}{\partial u^k} + \Omega_{ij}\mathbf{n},$$

where $\mathbf{n}$ is the unit normal of the surface;

$$\begin{pmatrix} \Omega_{11} & \Omega_{12} \\ \Omega_{21} & \Omega_{22} \end{pmatrix} = \begin{pmatrix} L & M \\ M & N \end{pmatrix}$$

is the second fundamental form. Introduce the Laplace operator

$$\Delta_s = \sum_{i,j=1}^{2} g^{ij} \left(\frac{\partial^2}{\partial u^i \partial u^j} - \sum_{k=1}^{2} \Gamma_{ij}^k \frac{\partial}{\partial u^k} \right),$$

where

$$\begin{pmatrix} g^{11} & g^{12} \\ g^{21} & g^{22} \end{pmatrix} = \frac{1}{EG - F^2} \begin{pmatrix} G & -F \\ -F & E \end{pmatrix}$$

is the inverse matrix of $\begin{pmatrix} E & F \\ F & G \end{pmatrix}$. (a) Prove that under isothermal coordinates,

$$\Delta_s = \frac{1}{E(u,v)} \left(\frac{\partial^2}{\partial u^2} + \frac{\partial^2}{\partial v^2} \right).$$

(b) Prove another formula of Laplace operator

$$\Delta_s = \sum_{i,j=1}^{2} \frac{1}{\sqrt{g}} \frac{\partial}{\partial u^i} \left(\sqrt{g} g^{ij} \frac{\partial}{\partial u^j} \right)$$

where $g = EG - F^2$.

7.3. Show that the Laplace operator is independent of the choice of the coordinates.

7.4. Show that a harmonic map from a regular surface patch to a planar convex domain must be a diffeomorphism.

8

Geometric Structure

According to Felix Klein's Erlanger program (1872), a classical geometry is the study of properties of a space X invariant under a group G of transformations of X.

For example, *Euclidean geometry* is the geometry of Euclidean space $\mathbb{R}^3$, which is invariant under its group $Euc(\mathbb{R}^3)$ of isometries (i.e., rigid motions, congruence transformations). The following concepts are in Euclidean geometry: parallelism of lines, angles between lines, distance between points, area, and volume. All the concepts are derived from the metric structure of the Euclidean space. *Affine geometry* arises when we talk about points, lines and the relation of parallelism. When we only concern point and lines and the relation of incidence between them, then *projective geometry* should be applied.

If M is a manifold, we would like to impart to it different geometries, such as affine geometry, projective geometry, and so on. The feasible means are the so-called *affine structure* and *projective structure*.

8.1 (X,G) Geometric Structure

Suppose X is a space with a geometry on it and G is the group of transformations of X, which preserves the geometry. In the following we define the basic concept of (X,G) structure.

Definition 8.1 ((X,G) atlas). *Suppose M is a manifold. An (X,G) atlas on M is an atlas $\{(U_\alpha,\phi_\alpha)\}$, such that the local coordinates are defined in X, $\phi_\alpha : U_\alpha \to X$. If $U_\alpha \cap U_\beta \neq \emptyset$, then the chart transition function*

$$\phi_\beta \circ \phi_\alpha^{-1} : \phi_\alpha(U_\alpha) \to \phi_\beta(U_\beta)$$

belongs to G.

Definition 8.2 ((X,G) Structure). *An (X,G) structure on M is a maximal (X,G)-atlas and (X,G) manifold is a manifold together with an (X,G) structure.*

Definition 8.3 ((X,G) Map). *Suppose that M and N are two (X,G) manifolds and $f : M \to N$ is a map. Then f is an (X,G) map if for each pair of charts*

$$\phi_\alpha : U_\alpha \to X, \quad \psi_\beta : V_\beta \to X,$$

the composition $\psi_\beta \circ f \circ \phi_\alpha^{-1}$ restricted to $\phi_\alpha(U_\alpha \cap f^{-1}(V_\beta))$ is locally (X,G).

It is obvious that the set of (X,G) automorphisms $M \to M$ forms a group, which we denote by $\mathrm{Aut}_{(X,G)}(M)$.

Every surface admits a metric of constant curvature. A sphere $\mathbb{S}^2$ is a covering space of the real projective space $\mathbb{RP}^2$, therefore it has a real projective structure. A plane of $\mathbb{R}^2$ has a natural Euclidean structure, which is also a real projective structure. Using the Klein model of hyperbolic geometry, every hyperbolic structure (i.e., Riemannian metric of constant curvature -1) determines a real projective structure. Therefore, all surfaces have a real projective structure. The following theorem is fundamental.

Theorem 8.4. *Every surface admits a real projective $\mathbb{RP}^2$ structure.*

8.2 Development and Holonomy

There is a globalization of the coordinate charts of a geometric structure using universal covering space and the fundamental group. Let M be an (X,G) manifold, choose a universal covering space $\mathbf{p} : \tilde{M} \to M$ and let $\pi(M)$ be the fundamental group of M. It is easy to see that the covering projection $\mathbf{p}$ induces an (X,G) structure on $\tilde{M}$ upon which π acts by (X,G) automorphisms. Because $\tilde{M}$ is simply connected, then there exists an (X,G) map $f:\tilde{M}\to X$. The (X,G) map f is called a *develop map* for $\tilde{M}$. Furthermore, suppose both $f,f':\tilde{M}\to X$. Then there exists an (X,G) automorphism $\phi:\tilde{M}\to\tilde{M}$, and an element $g\in G$, such that the following diagram commutes,

$$\begin{array}{ccc} \tilde{M} & \xrightarrow{f'} & X \\ {\scriptstyle\phi}\downarrow & & \downarrow{\scriptstyle g} \\ \tilde{M} & \xrightarrow{f} & X \end{array}$$

Theorem 8.5 (Development Theorem). *Let M be an (X,G) manifold with universal covering space $\mathbf{p}:\tilde{M}\to M$ and group of deck transformations $\pi=\pi_1(M)$. Then there exists a pair $(\mathbf{dev},h)$ such that $\mathbf{dev}:\tilde{M}\to X$ is an (X,G) map and $h:\pi\to G$ is a homomorphism such that, for each $\gamma\in\pi$,*

$$\begin{array}{ccc} \tilde{M} & \xrightarrow{\mathbf{dev}} & X \\ {\scriptstyle\gamma}\downarrow & & \downarrow{\scriptstyle h(\gamma)} \\ \tilde{M} & \xrightarrow[\mathbf{dev}]{} & X \end{array}$$

commutes. Furthermore, if $(\mathbf{dev}',h')$ is another such pair, there exists $g\in G$ such that $\mathbf{dev}'=g\circ\mathbf{dev}$ and $h'(\gamma)=gh(\gamma)g^{-1}$ for each $\gamma\in\pi$.

We call such a pair $(\mathbf{dev},h)$ a *development pair*, and the homomorphism h the *holonomy representation*. The developing map is a globalization of the coordinate charts of the manifold and the holonomy representation is a globalization of the coordinate changes.

8.3 Affine Structures on Surfaces

We introduce a basic result of Benzécri that a closed surface admits an affine structure if and only if its Euler characteristic vanishes.

Every non-compact orientable surface admits an immersion into $\mathbb{R}^2$ and such an immersion determines an affine structure with trivial holonomy. Suppose M is of genus g and with a single boundary. We glue a disk along ∂M to form a closed surface $\bar{M}$. We can construct a canonical homology basis of $\bar{M}$ $\{a_1,b_1,a_2,b_2,\cdots,a_g,b_g\}$, the union of the base curves form a graph, which does not intersect ∂M. Then the graph is a deformation retract of M; we can easily immerse the graph into $\mathbb{R}^2$, therefore, M can be immersed into $\mathbb{R}^2$ as well. For general non-compact surfaces, we can construct immersion in the same way.

Theorem 8.6 (Benzécri 1955). *Let M be a closed 2-dimensional affine manifold. Then $\chi(M)=0$.*

Benzécri gave an elementary proof, which can be found in [35]. Milnor observed that this result follows from a more general theorem on flat vector bundles [53]. Let E be a

2-dimensional oriented vector bundle over M, whose total space is the quotient of $\tilde{M} \times \mathbb{R}^2$ by the diagonal action of π by deck transformations on $\tilde{M}$ and via $h(\gamma)$ on $\mathbb{R}^2$, (i.e., the flat vector bundle over M associated to the linear holonomy representation). Suppose $(p,v) \in \tilde{M} \times \mathbb{R}^2$. Then

$$(p,v) \sim (\gamma(p), h(\gamma)(v)), \quad \forall \gamma \in \pi,$$

where γ is an element in the fundamental group π, we also use γ to represent the corresponding deck transformation, $h(\gamma)$ is the corresponding affine transformation. Then the vector bundle is defined as

$$\tilde{M} \times \mathbb{R}^2 / \sim .$$

This bundle has a natural flat structure, since the coordinate changes for this bundle are locally constant linear maps. Now an oriented $\mathbb{R}^2$ bundle over a space M is classified by its Euler class which lies in $H^2(M,\mathbb{Z}) \cong \mathbb{Z}$. If ξ is an oriented $\mathbb{R}^2$ bundle over M which admits a flat structure, Milnor showed that its Euler class

$$|e(\xi)| < g.$$

If M is an affine manifold, then the bundle E is isomorphic to the tangent bundle of M and hence has Euler number $e(TM) = 2 - 2g$. Therefore, the only closed orientable surface whose tangent bundle has a flat structure is a torus. Furthermore, Milnor showed that any $\mathbb{R}^2$ bundle over M whose Euler number satisfies the above inequality has a flat connection.

8.4 Spherical Structure

Suppose M is a closed genus zero Riemann surface. Then M is conformally equivalent to the unit sphere $\mathbb{S}^2$. If $\phi : M \to \mathbb{S}^2$ is a degree one harmonic map, then ϕ must be conformal. ϕ induces a spherical structure (Fig. 8.1) on M.

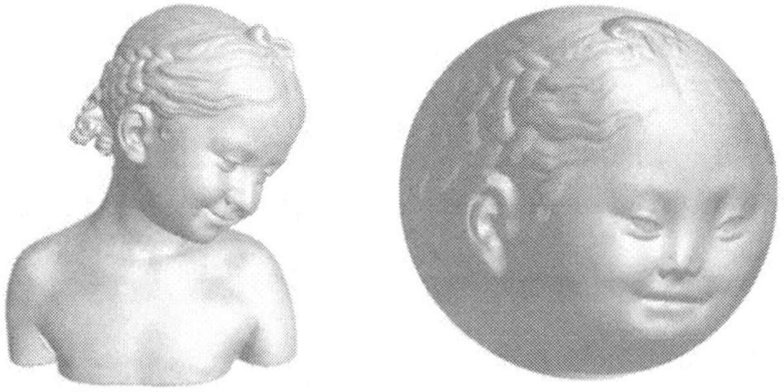

Fig. 8.1. Spherical structure.

Assuming the sphere $\mathbb{S}^2$ is embedded in $\mathbb{R}^3$, we can compute the harmonic map $\mathbf{f}$: $M \to \mathbb{S}^2$ in the following way. First we compute the Gauss map of M, $\mathbf{g} : M \to \mathbb{S}^2$, $\mathbf{r} \to \mathbf{n}$, which is a degree one map. Initialize $\mathbf{f}$ as $\mathbf{g}$, $\mathbf{f} \leftarrow \mathbf{g}$, we treat $\mathbf{f} = (f_1, f_2, f_3)$ as the map to $\mathbb{R}^3$. Then we minimize the harmonic energy of the map,

$$E(\mathbf{f}) = \sum_{i=1}^{3} \int_M |\nabla f_i|^2 dA,$$

the minimization can be carried out by the following heat flow method:

$$\frac{d\mathbf{f}}{dt} = -(\Delta \mathbf{f} - (\Delta \mathbf{f})^{\perp}),$$

where Δ is the Laplace-Beltrami operator on M, $(\Delta \mathbf{f})^{\perp}$ is the component of $\Delta \mathbf{f}$ along the normal at $\mathbf{f}(p) \in \mathbb{S}^2, p \in M$. Because the conformal map is not unique, two conformal maps from M to $\mathbb{S}^2$ differ by a Möbius transformation of the sphere, we need to add some normalization condition to ensure the uniqueness of the solution. A common method is to fix the images of three points. This variational method leads to the conformal spherical maps from a genus zero closed surface to the unit sphere.

For non-compact orientable surfaces, they can be immersed onto the Euclidean plane. By stereo-graphic projection, the Euclidean plane can be mapped onto the sphere, so non-compact orientable surfaces also admit spherical structures.

8.5 Euclidean Structure

If a closed surface admits a Euclidean structure, then we can define a flat Riemannian metric on it, which induces zero Gauss curvature everywhere. According to the Gauss-Bonnet theorem, the surface must be a torus.

Given a torus, it is straightforward to compute a Euclidean structure (Fig. 8.2) There are two direct methods: one is to use a holomorphic 1-form; the other is to use surface Ricci flow.

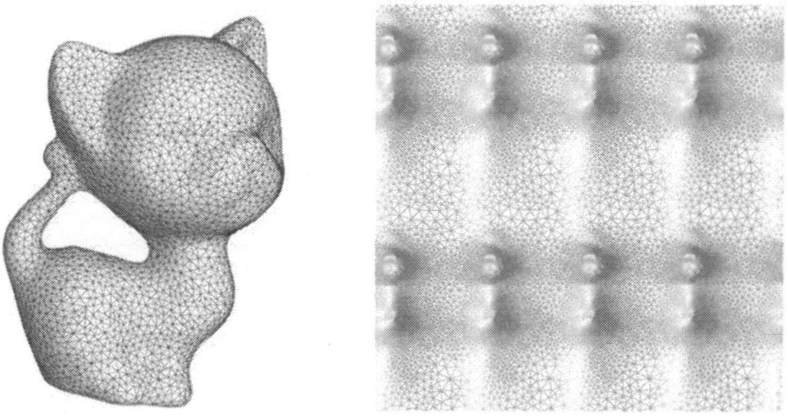

Fig. 8.2. Euclidean structure.

Holomorphic 1-form Method

The first method is to compute a holomorphic one-form ω of the torus M with an initial metric $\mathbf{g}_0$. Because ω has no zero points, we can define a new Riemannian metric $\mathbf{g}$,

$$ds^2 = \omega\bar{\omega},$$

where $\bar{\omega}$ is the complex conjugate to ω, then $\mathbf{g}$ is a flat Riemannian metric conformal to $\mathbf{g}_0$.

Let $\{U_\alpha\}$ be a family of open sets, $M \subset \cup U_\alpha$. We construct $\phi_\alpha : U_\alpha \to \mathbb{C}$ in the following way: first we choose a base point $p_\alpha \in U_\alpha$. For any point $p \in U_\alpha$, we choose an arbitrary path γ from p_α to p. Then we define

$$\phi_\alpha(p) = \int_\gamma \omega.$$

Lemma 8.7. *The atlas $\{(U_\alpha, \phi_\alpha)\}$ constructed by the holomorphic 1-form ω is a Euclidean atlas.*

Proof. Suppose $U_\alpha \cap U_\beta \neq \emptyset$, $p,q \in U_\alpha \cap U_\beta$ are two arbitrary points. Then they have two different local coordinates $\phi_\alpha(p)$, $\phi_\beta(p)$ and $\phi_\alpha(q), \phi_\beta(q)$. We want to show that all the coordinate changes are only translations, namely,

$$\phi_\alpha(p) - \phi_\beta(p) = \phi_\alpha(q) - \phi_\beta(q). \tag{8.1}$$

First, we choose two paths in U_α,

$$\gamma_\alpha^p : p_\alpha \to p, \quad \gamma_\alpha^q : p_\alpha \to q.$$

Then we choose two paths in U_β,

$$\gamma_\beta^p : p_\beta \to p, \quad \gamma_\beta^q : p_\beta \to q.$$

Then

$$(\phi_\alpha(p) - \phi_\beta(p)) - (\phi_\alpha(q) - \phi_\beta(q)) = \left(\int_{\gamma_\alpha^p} \omega - \int_{\gamma_\alpha^q} \omega\right) - \left(\int_{\gamma_\beta^p} \omega - \int_{\gamma_\beta^q} \omega\right).$$

The four curves $\gamma_\alpha^p - \gamma_\alpha^q - \gamma_\beta^p + \gamma_\beta^q$ form a closed loop, and bound a domain D,

$$\partial D = \gamma_\alpha^p - \gamma_\alpha^q - \gamma_\beta^p + \gamma_\beta^q.$$

By the Stokes formula, the above equation equals

$$\int_D d\omega = 0$$

because ω is holomorphic. Therefore, the chart transitions are translations. Namely, $\{(U_\alpha, \phi_\alpha)\}$ is a Euclidean atlas. □

We can generalize this method to arbitrary compact surfaces. Suppose M is a compact surface with a Riemannian metric. Then it is a Riemann surface with a conformal structure induced by the metric. Suppose ω is a holomorphic 1-form on M. Then ω induces a flat metric with cone singularities,

$$\tilde{\mathbf{g}} = \omega\bar{\omega},$$

the zeros of ω become cone singularities and the Gauss curvature are pulse functions. Furthermore,

$$\forall p, \quad \omega(p) = 0, \quad K_{\tilde{g}}(p) = 2n\pi, \quad n \in \mathbb{Z}^-.$$

From the above discussion, we can see that the elements in the holonomy group are only translations.

Theorem 8.8. *Suppose M is a compact Riemann surface. $\mathbf{g}$ is a flat metric with cone singularities, the curvatures at the singularities are $2n\pi, n \in \mathbb{Z}^-$. The holonomy group is rotation free, if and only if $\mathbf{g}$ is induced by a holomorphic 1-form ω.*

Proof. Using the method similar to the proof of lemma 8.7, we can show that a holomorphic 1-form induces a flat metric with cone singularities, which induces a Euclidean structure, the holonomy group consists of only translations.

Now, assume we have a Euclidean structure which induces a rotation free holonomy group. dz is a holomorphic 1-form on $\mathbb{R}^2$, and also on $\tilde{M}$. Because all deck transformations are translation, therefore dz is well defined on $\tilde{M}/\pi(M)$, which is a holomorphic 1-form on M. □

Ricci Flow Method

The second method is to use surface Ricci flow,

$$\frac{d\mathbf{g}(t)}{dt} = -K_{\mathbf{g}(t)}\mathbf{g}(t),$$

under the constraint

$$\int_M dA_{\mathbf{g}(t)} = \int_M dA_{\mathbf{g}_0}.$$

Then the solution is a flat metric on the surface M. Assume $(\tilde{M}, \mathbf{p})$ is the universal covering space of M. Then the projection map $\mathbf{p}$ induces the flat metric to $\tilde{M}$. We can directly isometrically embed $\tilde{M}$ onto the Euclidean plane $\mathbb{R}^2$. This gives a natural Euclidean structure on M.

8.6 Hyperbolic Structure

For closed surfaces with negative Euler numbers, they admit hyperbolic structures. Suppose the surface is equipped with an initial metric $\mathbf{g}_0$. We can compute the uniformization hyperbolic metric using the surface Ricci flow method. In particular, the uniformization metric is conformal to the initial metric $\mathbf{g}_0$ and induces -1 Gauss curvature everywhere. The Ricci flow is given by

$$\frac{d\mathbf{g}(t)}{dt} = -K_{\mathbf{g}(t)}\mathbf{g}(t),$$

under the constraint

$$\int_M dA_{\mathbf{g}} = \int_M dA_{\mathbf{g}_0}.$$

Then the solution $\mathbf{g}(\infty)$ is the hyperbolic metric on the surface M. Assume $(\tilde{M}, \mathbf{p})$ is the universal covering space of M. Then the projection map $\mathbf{p}$ induces the hyperbolic metric on $\tilde{M}$. We can directly isometrically embed $\tilde{M}$ onto the Poincaré disk Δ.

The deck transformations of $\tilde{M}$ are Möbius transformations. Therefore, the hyperbolic structure induces a *complex projective structure* $\mathbb{CP}^1$ of M.

Let Σ be a closed Riemann surface with genus g, and $\phi(z)dz^2$ be a holomorphic quadratic differential on M. We can show that each holomorphic quadratic differential on M corresponds to a holomorphic developing map and induces a complex structure and vice versa. Therefore all possible complex projective structures have one to one correspondence to $(\Sigma, \phi(z)dz^2)$. We know that the Teichmüller space for a genus g closed surface is $3g-3$ complex dimensional. Holomorphic quadratic differentials are $3g-3$ complex dimensional. Therefore, the complex projective structures of genus g closed surfaces form a $6g-6$ complex dimensional space.

If we use the Klein model to represent the hyperbolic space, then all the Möbius transformations are real projective transformations. Then the hyperbolic structure induces a *real projective structure* $\mathbb{RP}^2$ of M as shown in Fig. 8.3.

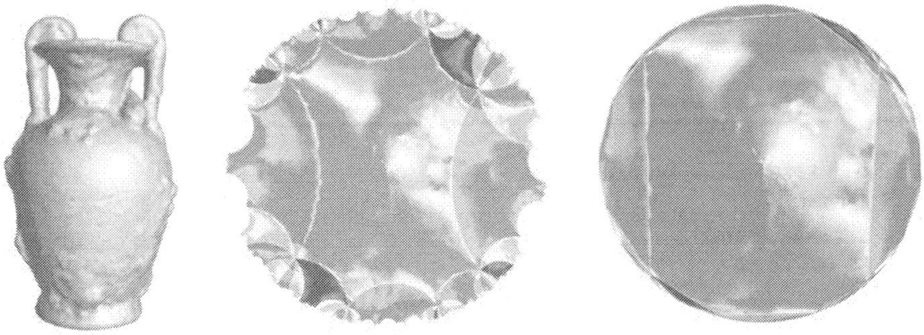

Fig. 8.3. Hyperbolic structure, induced complex projective structure,and real projective structure of a genus two surface.

8.7 Real Projective Structure

Let $\mathbb{RP}^2$ be the real projective plane and $PGL(3,\mathbb{R})$ the group of projective transformations $\mathbb{RP}^2 \to \mathbb{RP}^2$. M is a compact surface. A real projective structure on M is a maximal atlas of coordinate charts locally modeled on $\mathbb{RP}^2$ with coordinate changes lying in $PGL(3,\mathbb{R})$. If M is an $\mathbb{RP}^2$ surface, then a geodesic on M is a curve which in local coordinates maps to a projective line in $\mathbb{RP}^2$.

An $\mathbb{R}P^2$ manifold M can be developed into $\mathbb{R}P^2$ as follows. The coordinate atlas globalizes to define a local diffeomorphism of the universal covering $\tilde{M} \to \mathbb{R}P^2$, called the *developing map*. The deck transformations of the universal covering space $\tilde{M}$ define automorphisms of $\tilde{M}$. The resulting *holonomy homomorphism* $\pi(M) \to PGL(3,\mathbb{R})$ corresponds to the coordinate changes in the atlas for the $\mathbb{R}P^2$ structure. The pair of the developing map and the holonomy homomorphism is unique up to the $PGL(3,\mathbb{R})$ action by composition and conjugation respectively.

$\mathbb{R}P^2(M)$ is closely related to the space $Hom(\pi(M), PGL(3,\mathbb{R}))$ of homomorphisms $\pi(M) \to PGL(3,\mathbb{R})$. The group $PGL(3,\mathbb{R})$ acts on this space by conjugation. We define its orbit space

$$X(M) = Hom(\pi_1(M), PGL(3,\mathbb{R}))/PGL(3,\mathbb{R}).$$

Taking the holonomy homomorphism of a projective structure defines a map

$$hol : \mathbb{R}P^2(M) \to X(M),$$

which is essentially a local homeomorphism for closed surfaces with $\chi(M) < 0$.

Theorem 8.9. *Let M be a closed surface with $\chi(M) < 0$. Then the deformation space $\mathbb{R}P^2(M)$ is a Hausdorff real analytic manifold of dimension $-8\chi(M)$.*

Another approach is through conformal structure and complex vector bundle. For a large and important class of manifolds M with $\mathbb{R}P^2$ structure, the *convex ones*, $M = \Omega/\Gamma$, with Ω a convex domain in some $\mathbb{R}^2 \subset \mathbb{R}P^2$ and $\Gamma \subset PGL(3,\mathbb{R})$ an appropriate subgroup, Loftin [54] proved that

Theorem 8.10. *The data of a compact Riemann surface of genus $g > 1$, together with a holomorphic section of K^3, the tricanonical bundle, is equivalent to the data of an oriented, compact surface with convex $\mathbb{R}P^2$ structure.*

The Teichmüller space is $3g-3$ complex dimensional, the tricanonical bundle is $5g-5$ complex dimensional. Therefore, the space of convex $\mathbb{R}P^2$ structure on a oriented compact surface of genus $g > 1$ is $16g-16$ dimensional.

The more thorough result is as follows. The space of projective equivalence classes of real projective structures on a closed orientable surface of genus $g > 1$ is a countable disjoint union of open cells of dimension $16g-16$. The method is to decompose a real projective structure into convex subsurfaces along closed geodesics. More references can be found in [54] and [35].

Problems

8.1. Euclidean Structure
Prove that if a closed surface admits a Euclidean structure, then it must be a torus.

8.2. Euclidean Structure
Prove that any closed surface with one point removed admits a Euclidean structure.

8.3. Holomorphic Quadratic Differential
Prove that holomorphic quadratic differentials induce Euclidean structure with cone singularities. What is the holonomy induced by the Euclidean structure?

8.4. Affine Structure
(a) Given a planar triangulation and a point. The point location algorithm finds the triangle covering the point in a linear time. Design an algorithm for point location using barycentric coordinates.
(b) Suppose a triangulation is defined on a surface with an affine structure. Generalize your algorithm to the surface point location problem.

8.5. Projective Structure
(a) Prove that all surfaces admit real projective structure.
(b) Define a real projective line on a surface.
(c) Define the cross ration for arbitrary four points on a projective line on a surface.

Part II Algorithms

The concepts in algebraic topology, such as homotopy groups, homology groups and cohomology groups can be directly computed using discrete meshes without any approximation.

Most concepts in conformal geometry require solving geometric elliptic partial differential equations. It is well known that the solutions to elliptic partial differential equations are smooth, unique and stable. The convergence and stability can be rigorously established in the finite element field. In [55], Luo proved that the discrete holomorphic 1-forms converges to the smooth holomorphic 1-form in L^2 sense and the period matrix entries converges to the real one when the discrete surfaces converges to the smooth surface.

All the computational conformal geometric algorithms are based on the *geometric variational principle*: first, we define specific functionals for the desired mappings, differential forms or Riemannian metrics; then, we optimizes the functionals, the global optima of the functionals are the desired mappings, forms or metrics.

There are three categories of algorithms according to the solution types: *mapping, form or metrics*.

1. Algorithms in the mapping category aim at finding some special mappings, such as conformal mappings or harmonic mappings;
2. Algorithms in form category computes harmonic or holomorphic 1-forms. For high genus surfaces, the global mappings do not exist directly. Instead, holomorphic forms are computed by optimizing special harmonic energies defined on the forms;
3. Algorithms in metric category finds special Riemmanian metrics. In order to compute the uniformization metrics, we find some functionals whose global critical points are the desired metrics. In practice, the variational methods are robust and stable.

The metric method is more general than the form method; the form method (treated as derivatives) is more general than the mapping method. The form method and the mapping method are linear methods in general; whereas the metric method is highly non-linear. Different methods have advantages and disadvantages depending on the practical applications.

9

Topological Algorithms

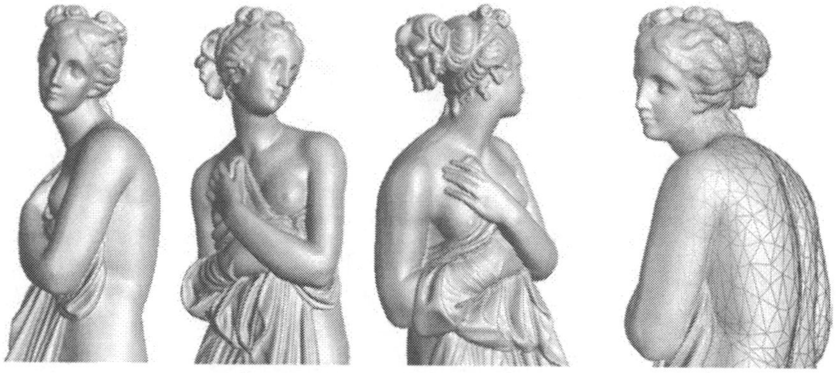

Fig. 9.1. Half-Edge data structure to represent a mesh.

9.1 Triangular Meshes

In the digital geometry processing field, surfaces are generally represented as triangular meshes, which are directly supported by common graphics hardware. Any other geometric representations, such as *splines*, *implicit surfaces*, and *level sets*, need to be converted to triangular meshes for display purposes.

The popularity of triangular meshes can be explained in terms of both theory and practice. In theory, any surface with C^1 continuity can be triangulated. Simplicial homology and simplicial cohomology theories in topology are directly based on the triangular meshes. In practice, all surfaces in real life can be digitized using 3D scanners. The acquired point clouds can be easily converted to triangulated meshes. Most digital geometry processing softwares support triangular meshes as the default internal data structure. More importantly, modern graphics hardware supports triangular meshes directly.

A triangular mesh is exactly a two-dimensional simplicial complex in algebraic topology. Intuitively, a mesh is a set of triangular faces coherently glued together. All of the topological information is implied by the connectivity. In general, a mesh is embedded in $\mathbb{R}^3$; therefore, all vertices have Euclidean coordinates and all faces are planar triangles. The lengths of all edges are induced by the Euclidean metric of $\mathbb{R}^3$. A mesh is flat everywhere except at the vertices, where the curvatures can be defined and measured.

Topological problems can be solved accurately on meshes without any approximation. Differential forms, curvatures, geodesics, and conformal mappings can only be solved on meshes by approximation.

9.1.1 Half-Edge Data Structure

A *half-edge* data structure is commonly used (Fig. 9.1) in geometric softwares to represent triangular meshes.

A mesh $M = (V, E, F)$ consists of a list of *vertices* V, *edges* E, and *faces* F. Suppose $V = \{v_0, v_1, \cdots, v_n\}$ is the list of vertices of a mesh M. In the following discussion, we use $\{v_1, v_2, \cdots, v_k\}$ to represent the set of vertices and $[v_1, v_2, \cdots, v_k]$ the set of ordered vertices. For example, $\{v_i, v_j\}$ represents a non-oriented edge connecting v_i and v_j;$[v_i, v_j]$ represents an oriented edge, starting from v_i and ending at v_j. We call an oriented edge a half-edge, denoted as $h_{ij} = [v_i, v_j]$ where v_i is called the source vertex and v_j is the target vertex.

Suppose three vertices $\{v_i, v_j, v_k\}$ form a face, then all the permutations of v_i, v_j, v_k can be divided into two classes. If two permutations differ by an even number of swaps, then the two permutations are equivalent; if two permutations differ by an odd number of swaps, then they are not equivalent.

Definition 9.1 (Face Orientation). *Each equivalence class of the permutation of the vertices define one orientation of the face.*

For example, the orientations of $[v_i, v_j, v_k]$ and $[v_j, v_k, v_i]$ are the same; the orientations of $[v_i, v_j, v_k]$ and $[v_j, v_i, v_k]$ are opposite.

We then define the boundary relations among oriented faces, oriented edges, and vertices.

Definition 9.2 (Boundary). *The boundary of an oriented face* $[v_i, v_j, v_k]$ *is*

$$\partial[v_i, v_j, v_k] = [v_i, v_j] + [v_j, v_k] + [v_k, v_i].$$

The boundary of an oriented edge $[v_i, v_j]$ *is*

$$\partial[v_i, v_j] = v_j - v_i.$$

Each edge $e_{ij} = \{v_i, v_j\}$ has two orientations, $h_{ij} = [v_i, v_j]$ and $h_{ji} = [v_j, v_i]$. We say the half-edges are dual to each other.

Definition 9.3 (Dual Half-Edge). *Suppose an oriented edge is $h_{ij} = [v_i, v_j]$. We call half-edge $h_{ji} = [v_j, v_i]$ the* dual half-edge *of h_{ij}.*

Given an edge $\{v_i, v_j\}$, if it attaches to two faces $\{v_i, v_j, v_k\}$ and $\{v_i, v_j, v_l\}$, then it is called an interior edge; if it only attaches to one face $\{v_i, v_j, v_k\}$, then it is called a boundary edge.

Definition 9.4 (Interior Edge). *Suppose an edge in the mesh attaches to two faces, then the edge is called an* interior edge. *The half-edges attached to it are called* interior half-edges.

Definition 9.5 (Boundary Edge). *Suppose an edge in the mesh attaches to only one face, then the edge is called a* Boundary Edge. *The half-edge attach to it is called a* Boundary Half-Edge.

Each face $\{v_i, v_j, v_k\}$ has two orientations $[v_i, v_j, v_k]$ and $[v_j, v_i, v_k]$. Each orientation gives a set of boundary half-edges: $[v_i, v_j, v_k]$ gives $\{[v_i, v_j], [v_j, v_k], [v_k, v_k]\}$; $[v_i, v_j, v_k]$ gives $\{[v_j, v_i], [v_k, v_j], [v_i, v_k]\}$. Suppose we choose an orientation for each face, then we define the half-edge set of the mesh as the union of the boundary half-edges of all the oriented faces. We use F to represent the set of all oriented faces,

$$H(M) = \bigcup_{[v_i, v_j, v_k] \in F} \partial[v_i, v_j, v_k].$$

If we can choose the orientations of all faces in a consistent way, such that any interior edge attaches to a pair of dual half-edges in $H(M)$, then the mesh is orientable.

Definition 9.6 (Mesh Orientation). *Given a mesh M, if one can choose the orientation of each face, the oriented face set is F, then for any interior edge $e_{ij} = \{v_i, v_j\}$, $[v_i, v_j]$ and $[v_j, v_i]$ are in $H(M)$, then M is orientable.*

There are many non-orientable surfaces, such as the Möbius band, Klein bottle, and so on. If they are triangulated and converted to meshes, then the corresponding meshes are non-orientable.

In practice, we mainly consider the mesh embedded in $\mathbb{R}^3$. If the mesh is closed, then the orientation can be easily defined in the following intuitive way. The mesh divides the whole space of $\mathbb{R}^3$ into two volumes: one is finite, called the *inside volume*, the other is infinite, called the *outside volume*. The *normal* of an oriented face $f = [v_i, v_j, v_k]$ is the unit vector perpendicular to the face plane and towards the outside. The normal is expressed as

$$\mathbf{n} = \frac{(v_j - v_i) \times (v_k - v_i)}{|(v_j - v_i) \times (v_k - v_i)|}. \tag{9.1}$$

Namely, the vectors $v_j - v_i$, $v_k - v_i$ and the normal satisfy the right-hand rule. The orientation of the mesh gives the two sides of the mesh: the front side and the back side. The side the normals point to is the front side and the opposite side is the back side.

If the mesh is not closed, then all of the boundary edges have only one half-edge attaching to them. Each boundary of the mesh consists of a loop of consecutive half-edges.

If we stand on the front side of the mesh and walk on the boundary along the half-edge directions, then the surface is always on our left-hand side, and the hole is on our right-hand side (Fig. 9.2).

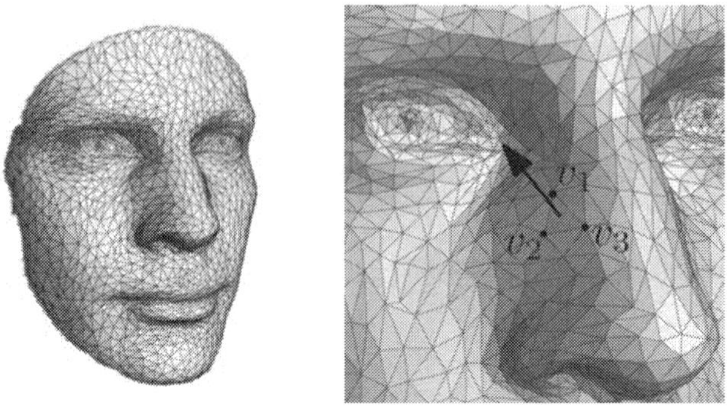

Fig. 9.2. Mesh orientation and surface normal. Three vertices are ordered counter-clock-wisely, the normal points to the front side. Each interior edge attaches to two half-edges with the opposite orientations. Each boundary edge attaches to one half-edge. If we walk along the boundary, the surface is on our left-hand side, the hole is on our right hand-side.

For the convenience of geometric processing, all the adjacent simplexes are connected together coherently using pointers in the data structure. In other words, the topological adjacency information is encoded by the combinatorial structure of the pointers. In practice, it is important to preserve the consistency among the pointers and it is very efficient to execute the following operations using the half-edge data structure, such as

- Traverse all the vertices connecting to a given vertex in counter-clock-wise order.
- Traverse all the edges connecting to a given vertex in counter-clock-wise order.
- Traverse all the faces connecting to a given vertex in counter-clock-wise order.
- Traverse all the half-edges pointing to a given vertex in counter-clock-wise order.
- Traverse all the half-edges pointing away from a given vertex in counter-clock-wise order.
- Traverse all the half-edges of a boundary of the mesh in clock-wise order.

All classes in the half-edge data structure are connected by the half-edge class. Therefore, the half-edge class plays a vital role in the half-edge data structure. The following discussion goes through all the pointers in half-edge data structure (Fig. 9.3).

1. Each half-edge points to the face it belongs to, $h \to f$.
2. Each face points to its first half-edge, $f \to h$.
3. Each half-edge points to the next half-edge in the same face, ordered counter-clock-wisely, $h \to h$.
4. Each half-edge points to the previous half-edge in the same face, ordered counter-clock-wisely, $h \to h$.

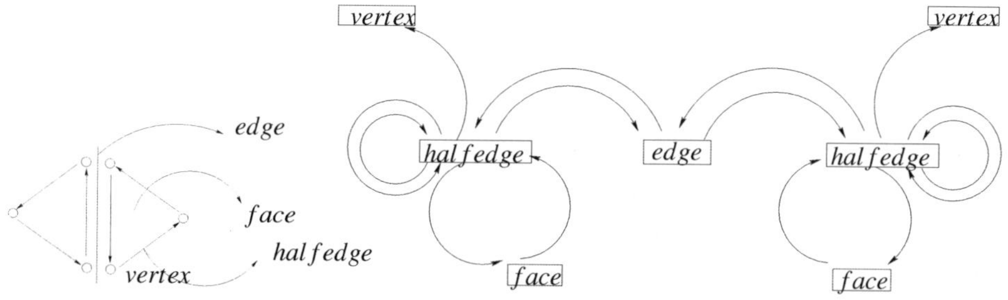

Fig. 9.3.

5. Each half-edge points to the edge it attaches to, $h \rightarrow e$.
6. Each edge points to all half-edges attaching to it, $e \rightarrow h$.
7. Each half-edge points to its target vertex, $h \rightarrow v$.
8. Each vertex points to the first half-edge, which targets at the vertex, $v \rightarrow h$.

There are no other direct connections. For example, no pointer connects two vertices, or a vertex and a face. All of the indirect adjacency information can be obtained through the pointers. Here, we discuss two simple examples. More complicated cases can be handled in a similar way.

9.1.2 Code Samples

The Standford bunny surface in Fig. 9.4 is represented as a triangular mesh using the half-edge data structure.

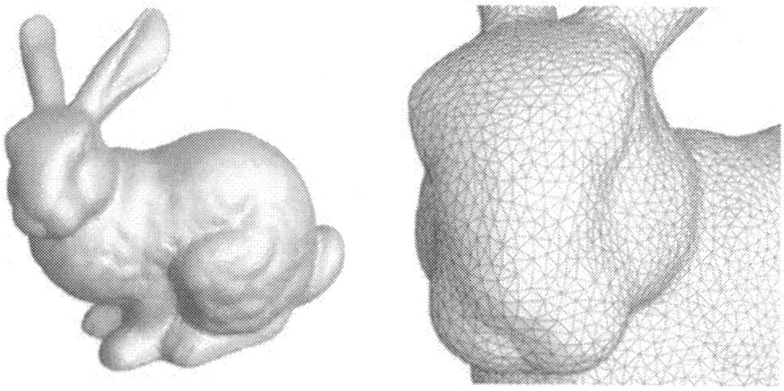

Fig. 9.4. The Standford bunny surface is represented as a triangular mesh using the half-edge data structure.

Basic Class Definitions

The following show the pseudo-code for the half-edge data structure. First, we illustrate the primary definitions of basic classes: vertex, edge, half-edge, face, and mesh.

```
class Vertex{
    int   vertex_id;
    Point position;
    HalfEdge* pHalfEdge;
};


class Face {
    int        face_id;
    HalfEdge * pHalfedge;
};

class Edge {
    double     length;
    HalfEdge * pLeftHalfedge;
    HalfEdge * pRightHalfedge;
};

class HalfEdge {
    HalfEdge * pPrevHalfedge;
    HalfEdge * pNextHalfedge;
    Edge     * pEdge;
    Face     * pFace;
    Vertex   * pVertex;
};


class Mesh{
    List<Vertex> vertices;
    List<Face>   faces;
};
```

The following methods get the source vertex and the target vertex of a half-edge.

```
Vertex * HalfEdge::source( ) const
{
    return ph->pPrevHalfedge->pVertex;
};

Vertex * HalfEdge::target( ) const
{
    return ph->pVertex;
};
```

Simple Iterators

Suppose we want to go through all the half-edges connecting to a given interior vertex v in counter-clock-wise order. The following algorithm performs such an operation:

```
HalfEdge * HalfEdge::dual( const HalfEdge * he ) const
{
    Edge     * pe     = pEdge;
    assert( pe!= NULL );
    HalfEdge * ph;
    if ( pe->pLeftHalfedge != this )
        ph = pe->pLeftHalfedge;
    else
        ph = pe->pRightHalfedge;
    return ph;
};
```

The input is a mesh *M* represented as the half-edge data structure and an interior vertex *v*; the output is a list of half-edges targeting at *v* and sorted counter-clock-wisely.

```
void VertexInHalfEdge( Vertex * v, List<HalfEdge*> lst_hes )
{
    HalfEdge * ph = v->pHalfEdge;
    lst_hes.add( ph );
    HalfEdge * ph =ph->dual();
    assert( ph != NULL );
    ph = ph->pPrevHalfedge;

    while( ph != phead )
    {
        lst_hes.add( ph );
        ph = ph->dual();
        assert( ph!=NULL );
        ph = ph->pPrevHalfedge;
    };
}
```

Mesh Construction

The following explains the basic steps for constructing a mesh. For example, the program can read an input 3D mesh file and build the half-edge data structure internally. A general mesh file includes :

1. a list of vertices, their ids, positions, normals, texture coordinates, and so on;
2. a list of faces, each face is given by three vertex ids, sorted counter-clock-wisely.

The following algorithm explains the details to build a face from three vertices. The basic task is to connect all the objects coherently and check the consistency. Assume the input is the three vertex ids, the output is the pointer to the newly created face.

1. Generate a face object,

```
    Face * face = new Face;
    assert( face );
```

2. Locate vertices in the vertex list,

```
Vertex * v[3];
for( int i = 0;i < 3; i ++ )
{
    v[i] = vertices.get(id[i]);
    assert( v[i] );
}
```

3. Generate half-edges and connect them,

```
HalfEdge * he[3];
for( int i = 0; i < 3; i ++ )
{
    he[i] = new HalfEdge;
    assert( he[i] );
    he[i]->pVertex = v[i];
    he[i]->pFace   = face;
}

for( int i = 0; i < 3; i ++ )
{
    he[i]->next = he[(i+1)%3];
    he[i]->prev = he[(i+2)%3];
}
```

4. Attach half-edge to edge,

```
for( int i = 0; i < 3; i ++ )
{
    int id_1 = he[i]->pVertex->id;
    int id_2 = he[i]->prev->pVertex->id;

    Edge * pe = edges.find(id_1,id_2);

    if( e != NULL )
    {
        assert( pe->pLeftHalfEdge != NULL  );
        pHalfEdge * pdh = pe->pLeftHalfEdge;
        assert( pdh->source()  == he[i]->target() );
        assert( pdh->target()  == he[i]->source() );
        e->pRightHalfEdge = he[i];
    }
    else
    {
            pe = new Edge;
            assert( pe );
            e->LeftHalfedge  = he[i];
            edges.add( pe );
```

```
        }
    }
```

5. Return the face pointer $face$.

The following algorithm generates a mesh, represented as the half-edge data structure. The input consists of a list of vertex symbols and a list of face symbols.

Algorithm 1: Create a mesh from a vertex list and a face list

input : A vertex list, a face list
output: A mesh represented as half-edge data structure

Generate a mesh object;
$Mesh* \ M \leftarrow new \ Mesh$;

forall *elements in vertex list* **do**
    generate a Vertex object v;
    store the vertex object in the mesh vertex list;
    $M \rightarrow vertices.add(v)$;
end

forall *elements in face list* **do**
    generate a Face object f using the face construction procedure;
    store the face object in the mesh face list;
    $M \rightarrow faces.add(f)$;
end

Slice a mesh along an edges graph

The following two concepts are very useful in practice. One is the *corner* of a face and the other is the *wedge*.

Suppose $f = [v_i, v_j, v_k]$ is a face, it has three corners with v_i, v_j and v_k as the apexes. We represent each corner using the corresponding half-edge. Half-edge $[v_i, v_j]$ represents the corner at the vertex v_j. A corner in the face f with vertex v_j is also denoted as (f, v_j).

A wedge is a collection of corners adjacent to each other (Fig. 9.5).

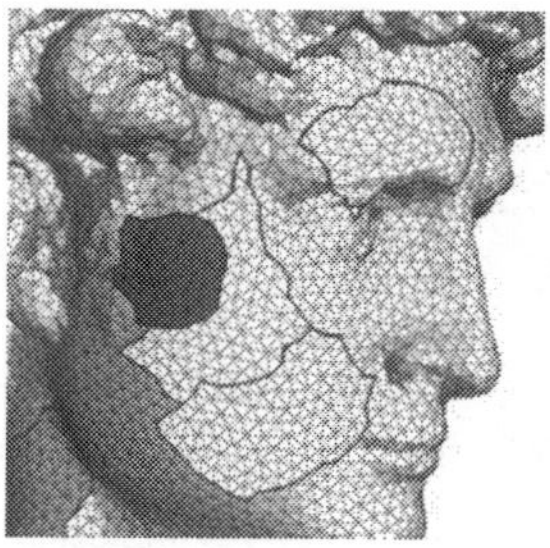

wedges

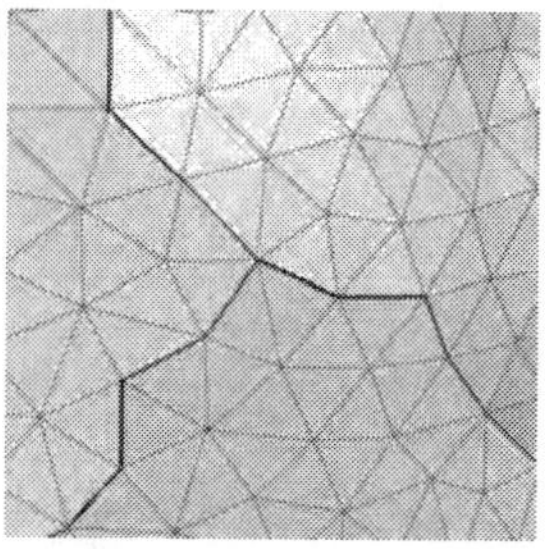

3 wedges around a vertex

Fig. 9.5.

Definition 9.7 (Wedge). *Suppose γ is a collection of edges. Given a vertex v, all of the corners are ordered counter-clock-wisely. A maximal sequence of adjacent corners form a* wedge, *such that no two adjacent corners in a wedge share an edge in γ.*

The wedge concept is useful for the high level operation, such as slicing a mesh open along a closed curve and so on. The following algorithm explains the algorithm for slicing a mesh along a curve consisting of a list of consecutive half-edges.

Algorithm 2: Slice a mesh along a curve

input : A mesh M, a list of half-edges γ
output: A mesh $\tilde{M}$ obtained by slicing M along γ

1. For each vertex $v \in M$, if v is not in γ, generate a wedge which includes all the corners attaching to v.
2. For each vertex on $v \in \gamma$, suppose all the half-edges adjacent to v are sorted counter-clock-wisely

$$h_0, h_1, h_2, \cdots, h_{n-1}.$$

Suppose $i_0 < i_1 \cdots < i_k$, h_{i_k} is in γ or $h_{i_k}^{-1}$ is in γ. Then we construct wedges W_k :

$$W_k = \{h_j | i_k \leqslant j < i_{k+1}\}$$

From the above construction, each corner c belongs to a unique wedge $w(c)$.
3. Construct the mesh $\tilde{M}$, such that the vertex set of M' is the set of all wedges and for any face $f_{ijk} \in M$, its three corners are c_i, c_j, c_k. Then there is a face $\tilde{f}$ in $\tilde{M}$, $f' = [w(c_i), w(c_j), w(c_k)]$.

9.2 Cut Graph

A cut graph on a mesh is a set of edges, such that its complement is a topological disk, which includes all the faces. All the topological information of the mesh is encoded in the cut graph.

Definition 9.8 (Cut Graph). *A cut graph* (Fig. 9.6, 9.7) *G in a mesh M consists of a set of edges of M, such that M/G is a simply connected mesh.*

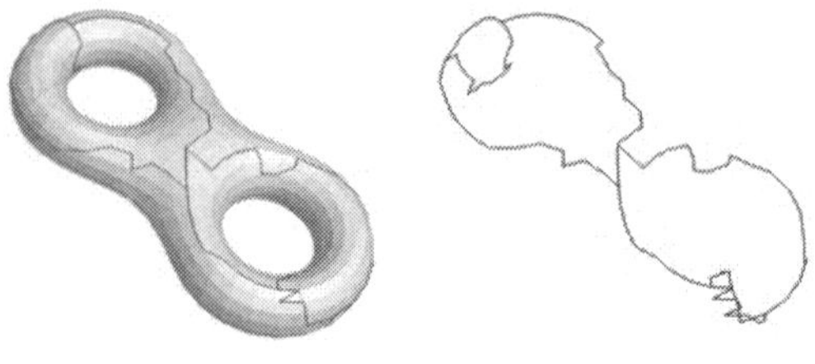

Fig. 9.6. Cut graph of a genus two surface.

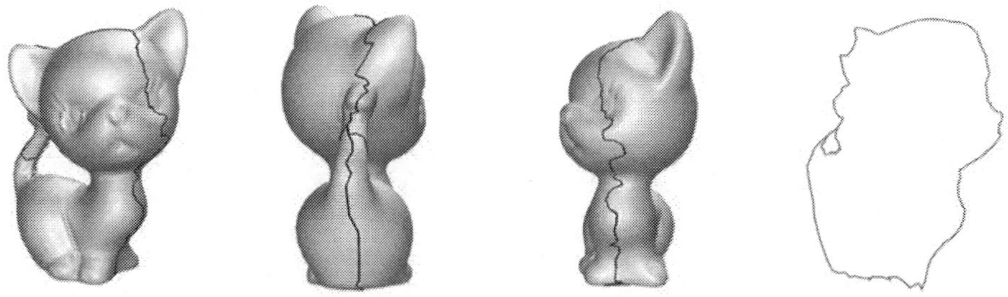

Fig. 9.7. Cut graph of a genus one mesh

A simply connected mesh has no handle and only one boundary, namely, it is a topological disk. In order to compute the topology of the mesh, we need to compute the cut graph.

First, we construct a dual mesh of M. A dual mesh $\bar{M}$ is the mesh constructed in the following way:

1. Each face f in M corresponds to a unique vertex of $\bar{v}(f) \in \bar{M}$.
2. Each vertex v in M corresponds to a unique face of $\bar{f}(v) \in \bar{M}$. Suppose the faces adjacent to v are $f_1, f_2, \cdots, f_n \in M$ sorted counter-clock-wisely. Then
$$\bar{f}(v) = [\bar{v}(f_1), \bar{v}(f_2), \cdots, \bar{v}(f_n)].$$
3. Each edge $e \in M$ adjacent to faces $f_i, f_j \in M$ corresponds to an edge $\bar{e}$,
$$\bar{e} = [\bar{v}(f_i), \bar{v}(f_j)].$$

The following simple algorithm computes a cut graph of a mesh with arbitrary topology.

Algorithm 3: Compute a cut graph

input : A mesh M
output: A cut graph G of M

1. Compute the dual mesh $\bar{M}$ of M;
2. Generate a minimal spanning tree $\bar{T}$ of the vertices of $\bar{M}$;
3. $G = \{e | \bar{e} \notin \bar{T}\}$.

The idea of the algorithm is to remove a topological disk D from M. D is as big as possible, such that by removing D, only a one-dimensional graph consisting of vertices and edges is left. Therefore, the cut graph G equals to M/D. A spanning tree of vertices of $\bar{M}$ corresponds to a topological disk consisting of all faces of M.

The cut graph is not unique, it depends on the construction of the minimal spanning tree $\bar{T}$. The spanning tree could be generated using depth first search or breadth first search, it also depends on the chosen root vertex.

The cut graph can be further improved by pruning the branches using the following algorithm:

Algorithm 4: Prune the cut graph

input : A cut graph G of M
output: A pruned cut graph G'
$G' \leftarrow G$;
while *there is valence one node v in* G' **do**
  remove v and the segment attached to it
end
return G'.

9.3 Fundamental Domain

The fundamental domain of a mesh is a topological disk which consists of all faces of M. Fundamental domain is important for many applications, such as surface parameterization and surface matching. The following algorithm describes the algorithm to compute a fundamental domain.

Algorithm 5: Compute a fundamental domain

input : A mesh M
output: A fundamental domain D of M
Compute a cut graph G of M;
Slice the mesh along G using the wedge structure to get a mesh D.

The algorithm is straight forward: the boundary of the fundamental domain ∂D is determined by the cut graph G.

We locate all of the *branching vertices* in G, whose valence is not equal to two. Then a *segment* in G is a set of consecutive edges in G, which connect two branching vertices. Namely, the branching vertices separate the cut graph into a collection of segments, $S = \{s_1, s_2, \cdots, s_n\}$. We give each segment an orientation, then the boundary of the fundamental domain ∂D can be written as

$$\partial D = s_{i_1}^{d_1} s_{i_2}^{d_2} s_{i_3}^{d_3} \cdots s_{i_m}^{d_m}, \tag{9.2}$$

where $i_k \in \{1, 2, \cdots, n\}, d_k \in \{+1, -1\}, 1 \leqslant k \leqslant m$. The boundary ∂D goes through each segment s_k twice with opposite directions.

Suppose the loop generators of the cut graph are $\gamma_1, \gamma_2, \cdots, \gamma_p$. Then ∂D is also a loop on G, therefore,

$$\partial D = \gamma_{i_1}^{d_1}, \gamma_{i_2}^{d_2}, \cdots, \gamma_{i_q}^{d_q}, \tag{9.3}$$

where $i_k \in \{1, 2, \cdots, p\}, d_k \in \{+1, -1\}$. Equation 9.3 shows the relation among the loop generators $\gamma_1, \gamma_2, \cdots, \gamma_k$. Equation 9.3 can be obtained from equation 9.2.

9.4 Basis of Homotopy Group

A set of Homotopy Group Basis can be shown in Fig. 9.8. The set of homotopy group generators of the surface is also the set of homotopy group generators of its cut graph. Therefore, we can just compute the homotopy of the cut graph, which is a one-dimensional graph. This dimension reduction method can be generalized to computing the topologies of higher dimensional manifolds.

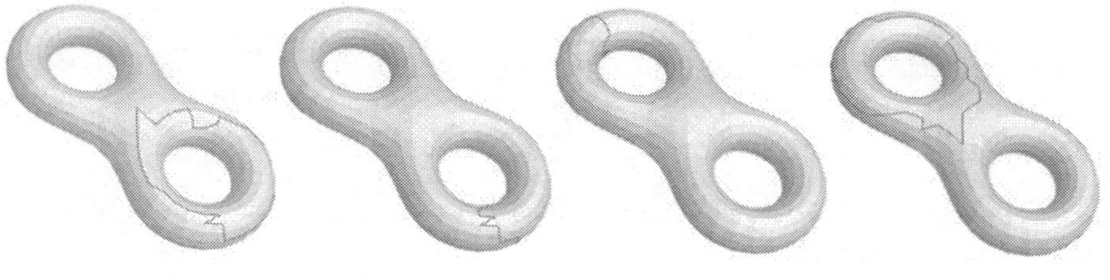

Fig. 9.8. A set of homotopy group basis of a genus two surface.

Proposition 9.9. *The set of homotopy group generators of G is also a set of homotopy group generators of M.*

Proof. Let $D = M/G$. From the construction, D is a topological disk. Pick one point $p \in D$. Let $M_p = M/p$. Then M is the union of D and M_p,

$$M = M_p \cup D.$$

The intersection of D and M_p is a topological annulus $D_p = D - p$,

$$D_p = D \cap M_p.$$

Suppose the generators of fundamental group $\pi_1(M_p)$ are $\{\gamma_1, \gamma_2, \cdots, \gamma_n\}$. The annulus has one fundamental group generator, and the fundamental group of the disk D is trivial. According to Seifert-van Kampen theorem 2.16, $\pi_1(D \cup M_p) = \{\gamma_1, \gamma_2, \cdots, \gamma_n | R_1\}$, the equivalence relation in R_1 is caused by the generator of $\pi_1(D_p)$. Suppose ∂D is a loop in M_p, and can be represented as

$$\gamma_1 = s_1 s_2 \cdots s_m,$$

each s_k is one of γ_j or γ_j^{-1}. Therefore, the generators of M_p are equal to the generators of M.

Furthermore, G is the deformation retract of M_p, therefore the generators of $\pi_1(G)$ are also a set of generators of $\pi_1(M)$. □

Therefore, the problem boiled down to computing the independent loops in the cut graph G, which form the homotopy group generators of G.

Algorithm 6: Compute homology group basis

input : A mesh M
output: Independent loops of G
1. Compute a cut graph G of the mesh M;
2. Choose a root node v of G;
3. Use breadth first search to construct the minimal spanning tree T of G. For each leaf v_i, compute a unique path γ_i from the root to v_i;
4. Suppose
$$G - T = \{e_1, e_2, \cdots, e_k\}.$$
where e_k connects leaves v_i and v_j. Then we get a loop
$$l_k = \gamma_i [v_i, v_j] \gamma_j^{-1}.$$
Then $\{l_1, l_2, \cdots, l_k\}$ are all of the independent loops in G, also the basis for the homology group basis $H_1(M, \mathbb{Z})$.

9.5 Gluing Two Meshes

This section explains the algorithm to merge two meshes along some parts of their boundaries.

Suppose $M_1 = \{V_1, E_1, F_1\}$ and $M_2 = \{V_2, E_2, F_2\}$. We are also given the equivalence relations among the boundary vertices, represented as pairs,

$$\sim := \{(v_i^1, v_j^2) | v_i^1 \in \partial M_1, v_j^2 \in \partial M_2\},$$

the pair (v_i^1, v_j^2) means that v_i^1 should be glued to v_j^2.

Then we construct the vertex list and face list of the glued mesh $M = (V, E, F)$ in the following way:
$$V = V_1 \cup V_2 / \sim .$$
We denote the equivalence class of $v_i \in M_1$ or $v_i \in M_2$ as $[v_i]$. The face list is

$$F = \{[[v_i], [v_j], [v_k]] | [v_i, v_j, v_k] \in M_1 \ or \ [v_i, v_j, v_k] \in M_2\}.$$

There are some natural requirements to ensure the glued mesh is still a manifold. In practice, it is very useful to glue an open mesh with itself with opposite orientation. The resulting mesh is a closed, symmetric mesh, which is called *the double covering mesh* or *doubled mesh* (Fig. 9.9).

The following algorithm describes the details for double covering an open mesh:

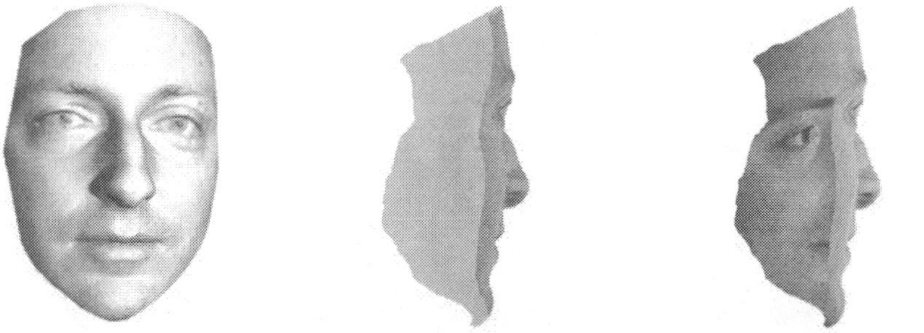

(a) Front view of the single surface

(b) Back view of the single surface

(c) Doubled surface

Fig. 9.9. Double covering an open mesh.

Algorithm 7: Double covering of an open mesh

input : A mesh M with boundaries
output: The double covering $\bar{M}$ of M

1. Generate the mesh M^- with opposite orientation: If $v \in M$, then $v^- \in M^-$; If $[v_i, v_j, v_k] \in M$, then $[v_j^-, v_i^-, v_k^-] \in M^-$;
2. Define equivalence relation

$$v_k \sim v_k^- \; if \; v_k \in \partial M, \quad v_k^- \in \partial M^-.$$

3. Glue M and M^- along their corresponding boundaries

$$\bar{M} = M \cup M^- / \sim .$$

Any computational algorithms for closed meshes can be generalized to open meshes by the double covering technique. In the later discussion, we will follow this approach to compute Riemann mapping, harmonic forms, and holomorphic forms for open meshes.

9.6 Universal Covering Space

The construction of a finite portion of the universal covering space (Fig. 9.10) is to coherently glue copies of the fundamental domains. The most difficult part is to guarantee the consistency.

Given a closed mesh M, first we compute a cut graph G. Then we locate all of the branching vertices in G, which separate the cut graph into segments. Each segment is assigned an orientation, which is arbitrarily chosen. All of the oriented segments are denoted as $\{s_1, s_2, \cdots, s_n\}$. We slice the mesh M along the cut graph G to get the fundamental domain of M, denoted as D. The boundary of D consists of oriented segments,

$$\partial D = s_{i_1}^{d_1} s_{i_2}^{d_2} \cdots s_{i_m}^{d_m},$$

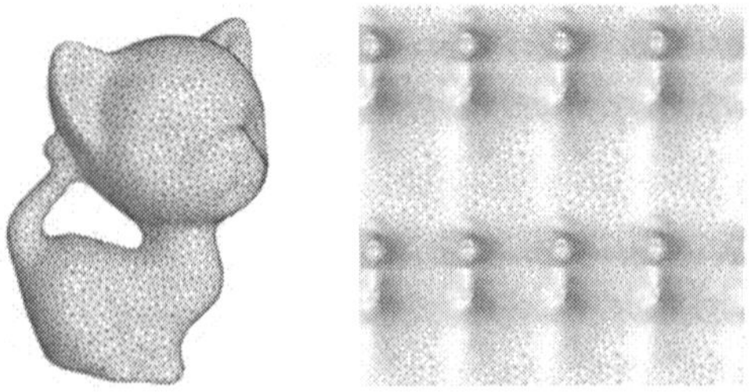

Fig. 9.10. Universal covering space of a genus one surface.

where $i_k \in \{1,2,\cdots,n\}, d_k \in \{+1,-1\}, 1 \leqslant k \leqslant m$.

We denote the finite portion of the universal covering space of M as $\bar{M}$. We initialize the $\bar{M}$ as one copy of the fundamental domain D. Then we merge a new copy of D with the current $\bar{M}$ in the following way. Suppose an oriented segment s_k^{+1} is on the boundary of $\bar{M}$, $s_k^{+1} \subset \partial\bar{M}$, then s_k^{-1} is on the boundary of D. We merge D with $\bar{M}$ by gluing s_k^{-1} to s_k^{+1}, denoted as

$$\bar{M} = \bar{M} \cup_{s_k} D.$$

Suppose the boundary of the new $\bar{M}$ is

$$\partial\bar{M} = s_{i_1}^{d_1} s_{i_2}^{d_2} \cdots s_{i_n}^{d_n}.$$

Find a consecutive pair of segments $s_{i_k}^{d_k}, s_{i_{k+1}}^{d_{k+1}}$, such that $i_k = i_{k+1}, d_k = -d_{k+1}$, namely

$$(s_{i_k}^{d_k})^{-1} = s_{i_{k+1}}^{d_{k+1}}.$$

Then stitch the boundary along $s_{i_k}^{d_k}$ and $s_{i_{k+1}}^{d_{k+1}}$. We repeat this stitching process, until there is no such consecutive opposite pairs of segments in the boundary of $\bar{M}$.

We can repeat this process to glue more copies of the fundamental domain to enlarge the universal covering space until it is big enough for our purpose.

Algorithm 8: Computing a finite portion of the universal covering space

input : A mesh M
output: A finite portion of the universal covering space of M, $\bar{M}$

1. Compute a cut graph G of M.
2. Separate G into segments $\{d_1, d_2, \cdots, d_m\}$ by the branching vertices.
3. Slice M along G to form a fundamental domain D.
4. Label the boundary of D as a sequence of oriented segments,

$$\partial D = s_1, s_2, \cdots s_n,$$

each s_i is d_k^+ or d_k^-, where $1 \leqslant k \leqslant m$.

5. Put the sequence of segments into a queue Q, set $\bar{M} = D$,

$$\partial \bar{M} = \partial D.$$

6. **while** *the queue is not empty and $\bar{M}$ is not big enough* **do**
   a. remove the first element in the queue, without loss of generality, assume the removed element is $s_1 \in \partial \bar{M}$.

   b. generate a copy of the fundamental domain D, arrange the boundary of D as

$$\partial D = s_1^- d_2, d_3, \cdots, d_m.$$

   c. glue $\bar{M}$ and D along $s_1 \in \partial \bar{M}$ and $s_1^- \in \partial D$, where s_1 and s_1^- are dual to each other. Update the boundary of $\bar{M}$ as

$$\partial \bar{M} = s_2, s_3, \cdots, s_n d_2, d_3, \cdots, d_m.$$

   d. go through ∂m, locate a consecutive pairs s_i, s_{i+1}, such that $s_{i+1} = s_i^{-1}$. Stitch $\bar{M}$ along the pair, if s_i or s_{i+1} is in the queue, remove it from the queue.

   e. repeat d until no such pair can be found in $\partial \bar{M}$.

   f. insert the new half-segments in $\partial \bar{M}$ to the queue.

end

This algorithm is straightforward and easy to understand. In practice, this algorithm consumes too much memory. The data structure has a lot of redundant information. A more sophisticated data structure can be carefully designed to remove the redundancy and reduce the memory requirement dramatically. Furthermore, if the problem to solve is purely topological, we can use simplified mesh instead of the original high resolution mesh to further reduce the memory requirement.

9.7 Curve Lifting

Any curve on the original surface can be lifted to the universal covering space (Fig. 9.11). A closed loop with nontrivial homotopy type can be lifted to an open path in the universal covering space. A lot of topological problems can be solved on the universal covering space in a much easier way than on the original surface.

Suppose γ is a closed loop (a closed 1-chain), represented as a list of consecutive half-edges, denoted as

$$\{[v_1, v_2], [v_2, v_3], [v_3, v_4], \cdots, [v_{n-1}, v_n], [v_n, v_1]\}.$$

$\bar{M}$ is the universal covering space of M, $\pi : \bar{M} \to M$ is the projection. Locally the projection is a one-to-one map. Let $\bar{v}_1 \in \pi^{-1}(v_1) \subset \bar{M}$ be a pre-image of v_1. We denote the one-ring neighborhoods of $v_1, \bar{v}_1$ as $N(v_1), N(\bar{v}_1)$ respectively. Then the restriction of $\pi|_{N(\bar{v}_1)} : N(\bar{v}_1) \to N(v_1)$ is a one-to-one map. Therefore, we can uniquely find the pre-image of v_2 in $N(\bar{v}_1)$, $\bar{v}_2$. Then we uniquely locate the pre-image of v_3, $\bar{v}_3$ in the one ring neighbor of $\bar{v}_2$. We continue this process step by step. At the *k-th* step, we can uniquely determine the pre-image of v_k, until we reach the starting vertex v_1 again. Then we get another pre-image of v_1, denoted as $\bar{v}'_1$. In general, $\bar{v}'_1$ may not necessarily be identical to $\bar{v}_1$. Therefore, a loop on the original surface is lifted to a path on the universal covering space.

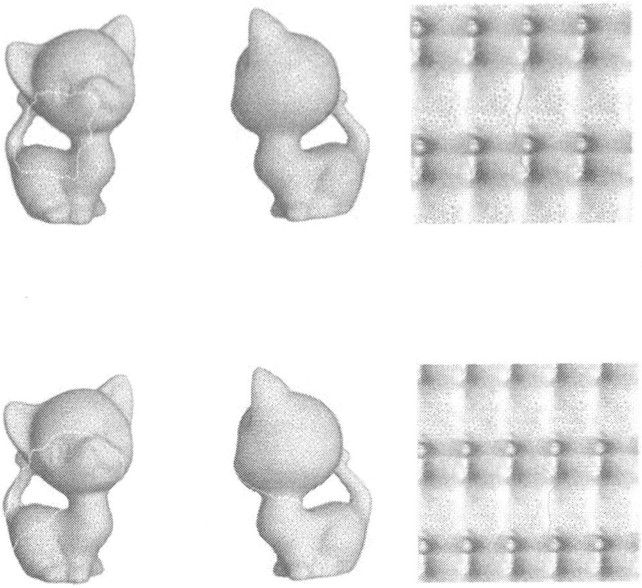

Fig. 9.11. curve lifting

The following algorithm describes the details of lifting a loop on the surface to an open path on the universal covering space.

Algorithm 9: Compute a lift of a curve in the universal covering space

input : A mesh M, a 1-chain $\gamma \subset M$ represented as a list of half-edges

$$\gamma = \{[v_0, v_1], [v_1, v_2], \cdots, [v_{n-3}, v_{n-2}], [v_{n-2}, v_{n-1}], [v_{n-1}, v_n]\}.$$

v_n may equal to v_0.

output: A lift of γ to a 1-chain $\bar{\gamma} \subset \bar{M}$ on the universal covering space.

1. Construct a finite portion of the universal covering space of M, denoted as $\bar{M}$;
2. Represent the 1-chain as a list of vertices $\{v_0, v_1, \cdots, v_{n-1}\}$;
3. Locate a pre-image $\bar{v}_0 \in \bar{M}$, such that $\pi(\bar{v}_0) = v_0$;
4. Go through the list of vertices. In the one ring neighborhood of $\bar{v}_k$, locate the unique pre-image of v_{k+1}, denoted as $\bar{v}_{k+1}$;
5. Construct the lift of γ, $\bar{\gamma}$ in the universal covering space, represented as a 1-chain

$$\bar{\gamma} = \{[\bar{v}_0, \bar{v}_1], [\bar{v}_1, \bar{v}_2], \cdots, [\bar{v}_{n-2}, \bar{v}_{n-1}], [\bar{v}_{n-1}, \bar{v}_n]\}.$$

9.8 Homotopy Detection

Suppose γ_1 and γ_2 are two closed loops on a surface. A commonly asked question is whether they are homotopic. General homotopy detection is NP-hard. In the following, we offer one solution to the challenging problem utilizing the powerful tool of the universal covering space.

First, we select one point p_1 on γ_1 and p_2 on γ_2. We then connect p_1 and p_2 by a path γ. If γ_1 is homotopic to γ_2, then

$$\Gamma = \gamma_1 \circ \gamma \circ \gamma_2^{-1} \circ \gamma^{-1}$$

is homotopic to zero. If we lift Γ to the universal covering space, then its lift $\bar{\Gamma}$ should be a closed loop. The major difficulty is that the interior of $\bar{\Gamma}$ may be very large and includes many copies of the fundamental domain. Suppose $\bar{\Gamma}$ goes through n copies of the fundamental domain, the interior region inside $\bar{\Gamma}$ may include Ce^n copies of the fundamental domain. Therefore, the problem is NP hard in nature.

The homotopy detection is explained in detail in algorithm 10.

Algorithm 10: Homotopy dection

input : A mesh M, two 1-chains γ_1, γ_2
output: Whether γ_1 is homotopic to γ_2

1. Represent γ_1 by a circular list of vertices

$$\gamma_1 = \{v_1, v_2, \cdots, v_n\}.$$

2. Represent γ_2 by a circular list of vertices

$$\gamma_2 = \{w_1, w_2, \cdots, w_m\}.$$

3. Use Dijkstra's algorithm to compute a shortest path connecting v_1 and w_1

$$\gamma = \{v_1, d_1, d_2, \cdots, d_k, w_1\}.$$

4. Construct the loop of $\Gamma = \gamma_1 \circ \gamma \circ \gamma_2^{-1} \circ \gamma^{-1}$,

$$\Gamma = \{v_1, v_2, \cdots, v_n, v_1, d_1, \cdots, d_k, w_1, w_m, w_{m-1}, \cdots, w_1, d_k, d_{k-1}, \cdots, d_1\}.$$

5. Construct a finite portion of the universal covering space $\bar{M}$.

6. Lift Γ to the universal covering space $\bar{\Gamma}$. If Γ is a loop, then γ_1 is homotopic to γ_2, otherwise, they are not homotopic equivalent.

The algorithm is simple and brute force. In practice, we can simplify the mesh first. During the simplification process, we preserve the homotopy type of γ_1 and γ_2 and perform the homotopy detection using the simplified mesh.

9.9 The Shortest Loop

In graph theory, computing the shortest path is a common task. It has broad applications in many fields, such as network routing, path planning etc. On a surface with non-trivial topologies, computing the shortest loop in each homotopy class is a challenging problem.

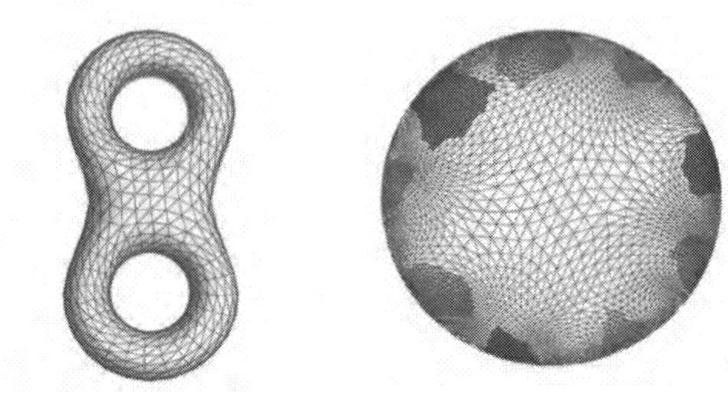

Fig. 9.12. Shortest cycle in the homotopy class

On a smooth surface with a Riemannian metric, the shortest loop of each homotopy class must be geodesics (Fig. 9.12). If the surface has handles, then the shortest loop corresponding to the homotopy generators of the handle must be greater than zero. The uniqueness of the shortest loop is determined by the Gaussian curvature. The existence of closed geodesics is an interesting but difficult problem. It can be tackled by studying the fixed point of automorphisms of the circle bundle of the surface.

In the discrete case, we can use 1-chains to approximate curves on the mesh, and study the shortest loop on the mesh in each homotopy class.

First, we fix a vertex v, and study the shortest loop through v in a given homotopy class. Assume $\bar{M}$ is the universal covering space of M and $\pi : \bar{M} \to M$ is the projection. The pre-images of v are

$$\pi^{-1}(v) = \{\bar{v}_0, \bar{v_1}, \bar{v}_2, \cdots\}.$$

Suppose γ is a 1-chain through v. Then we lift γ to the universal covering space of M, $\bar{\gamma} \subset \bar{M}$,

$$\partial \bar{\gamma} = \bar{v}_k - \bar{v}_0.$$

Suppose γ' is another loop through v homotopic to γ. Then we lift γ' to $\bar{\gamma}'$,

$$\partial \bar{\gamma}' = \bar{v}_k - \bar{v}_0.$$

Namely, all loops through v homotopic to γ are lifted to paths in $\bar{M}$, connecting $\bar{v}_0$ and $\bar{v}_k$. The shortest loop through v corresponds to the shortest path connecting $\bar{v}_0$ and $\bar{v}_k$, which can be easily computed using conventional Dijkstra's algorithm. Second, we move the base vertex to traverse all the vertices on M, and choose the shortest one. Algorithm 11 describes the details.

Algorithm 11: Shortest loop through a given point

input : A base vertex v, a loop γ through v on a mesh M
output: the shortest loop through v and homotopic to γ

1. Compute a finite portion of the universal covering space $\bar{M}$. Compute the pre-images of v

$$\pi^{-1}(v) = \{\bar{v}_0, \bar{v}_1, \bar{v}_2, \cdots\}.$$

2. Lift γ to $\bar{\gamma} \subset \bar{M}$, such that

$$\partial \bar{\gamma} = \bar{v}_k - \bar{v}_0.$$

3. Compute the shortest path $\bar{\Gamma}$ in the universal covering space $\bar{M}$,

$$\partial \Gamma = \bar{v}_k - \bar{v}_0,$$

connecting $\bar{v}_k$ and $\bar{v}_0$.
4. The projection of the shortest path

$$\Gamma = \pi(\bar{\Gamma})$$

is the shortest loop through v homotopic to γ.

The following algorithm 12 computes the shortest loop in a given homotopy class.

Algorithm 12: Shortest loop in a given homotopy class

input : A closed loop γ on a mesh M
output: The shortest loop on M homotopic to γ

foreach *vertex w on M* **do**
  Find the nearest vertex v on γ;
  Find the shortest path connecting v and w, denoted as γ_0;
  Construct a loop

$$\Gamma_w = \gamma_0 \circ \gamma \circ \gamma_0^{-1}.$$

  Compute the shortest loop through w, homotopic to Γ_w;
end
Select the loop with the minimal length:

$$\Gamma = \min_{w \in M} \Gamma_w.$$

9.10 Canonical Homotopy Group Generator

In practice, it is highly preferred to compute canonical homotopy group generators. Many computations can be dramatically simplified by using a set of canonical homotopy group generators. This section introduces the algorithm to find a set of canonical homotopy group generators.

Suppose a closed surface M of genus g, a set of loops

$$\{a_1, b_1, a_2, b_2, \cdots, a_g, b_g\}$$

form a set of canonical group generators, if

$$a_i \times b_i = 1, \quad a_i \times a_j = 0, \quad b_i \times b_j = 0, \quad a_i \times b_j = 0,$$

where $a_i \times b_j$ represents the (algebraic) intersection number between a_i and b_j.

In practice, it is ideal that (a_i, b_i) are completely separated from other (a_j, b_j)'s. If the tessellation of the mesh is too coarse, then it is impossible. By mesh subdivision, we can make a mesh refined enough to guarantee the existence of such kind of separated generator pairs. The following is a simple algorithm for surface subdivision.

Basically, we insert a new vertex on each edge of the mesh, and split each face to four smaller faces. Suppose $[v_1, v_2, v_3]$ is a face in the original mesh. We split three edges and insert three new vertices, v_{12} in the middle of edge e_{12}, v_{23} in the edge e_{23}, and v_{31} in the edge e_{31}. Then we form four faces

$$[v_1, v_{12}, v_{31}], \ [v_2, v_{23}, v_{12}], \ [v_3, v_{31}, v_{23}], \ [v_{12}, v_{23}, v_{31}].$$

The positions of the new vertices can be calculated using interpolations. All of the vertex positions can be updated using simple rules. In the following algorithm, we emphasize the connectivity change for the purpose of topological computations, and omit the details of determining the vertex positions.

Algorithm 13: Mesh subdivision

input : A mesh M
output: A subdivided mesh

1.Suppose the vertex list of M is

$$V = \{v_1, v_2, \cdots, v_n\},$$

the edge list of M is

$$E = \{e_1, e_2, \cdots, e_m\},$$

the face list of M is F. Generate the vertex list of $\bar{M}$,

$$\bar{V} = V \cup E = \{v_1, v_2, \cdots, v_n, e_1, e_2, \cdots, e_m\}.$$

2. Suppose $f \in M$, $f = [v_1, v_2, v_3]$, edges

$$e_1 = [v_1, v_2], \quad e_2 = [v_2, v_3], \quad e_3 = [v_3, v_1],$$

split f to four faces

$$\bar{f} = \{[v_1, e_1, e_3], [v_2, e_2, e_1], [v_3, e_3, e_2], [e_1, e_2, e_3]\}.$$

The face list of $\bar{M}$ is

$$\bar{F} = \cup_{f \in F} \bar{f}.$$

3. The subdivided mesh $\bar{M}$ is represented as the vertex list and the face list,

$$\bar{M} = \{\bar{V}, \bar{F}\}.$$

We can subdivide a mesh several times to make it refined enough, such that a canonical homotopy group generators can be found and the pairs (a_i, b_i) are separated from (a_j, b_j), $i \neq j$.

The following is the intuitive idea for the computation. First we compute a closed loop a_1. Then we slice the mesh along a_1. We get an open mesh M_1, such that the boundary of M_1 is

$$\partial M_1 = a_1 - a_1^{-1}.$$

Then we choose a vertex $v_1 \in a_1$ and its dual vertex $v_1^{-1} \in a_1^{-1}$. Then we compute a shortest path from v_1 to v_1^{-1}. In the computation, we make sure the shortest path does not go through the boundary edges. The shortest path corresponds to a closed loop in M, denoted as b_1.

Then we slice the mesh along $a_1 \cup b_1$ to get a new mesh M_2,

$$\partial M_2 = a_1 b_1 a_1^{-1} b_1^{-1}.$$

We compute the cut graph G_2 of M_2. Then we remove all boundary edges of M_2 from G_2, and find a loop a_2 in $G_2 \backslash \partial M_2$. Then we slice M_2 along a_2, to get M_2',

$$\partial M_2' = a_2 + a_2^{-1} + a_1 b_1 a_1^{-1} b_1^{-1}.$$

Then we choose $v_2 \in a_2$ on M_2', $v_2^{-1} \in a_2^{-1}$, compute the shortest path connecting v_2, v_2^{-1} without intersecting the boundary except at the start and end points. The shortest path corresponds to a loop in M, which is b_2. We then repeat this procedure to compute all the other pairs (a_j, b_j). In fact, each pair (a_j, b_j) corresponds to a handle. That is, locating a handle means to find the pair (a_j, b_j) on it.

Algorithm 14: Locate a handle

input : A mesh M
output: A pair of loops a, b which trasversely intersect at only one point

1. Compute a cut graph G of M.
2. Remove all boundary edges from G,

$$G \leftarrow G - \partial M.$$

If G is a tree, return the empty set.
3. Find the shortest loop in G, denoted as a.
4. Slice M along a, to get a new mesh

$$\bar{M} \leftarrow M - \{a\}, \quad \partial \bar{M} = \partial M \cup a^+ \cup a^-.$$

5. Choose a vertex $v^+ \in a^+$ and a vertex $v^- \in a^-$, compute a shortest path γ connecting v^+ and v^{-1}, such that the shortest path has no other intersections with the boundary of the mesh

$$\gamma \cap \partial \bar{M} = v^+ \cup v^-.$$

6. All the half edges on $\bar{M}$ have one-to-one correspondence with the half-edges on M, therefore γ corresponds to a closed loop on M, denoted as b. Then (a, b) is the output.

Locating a set of canonical homotopy group generators is simply finding all the handles.

Algorithm 15: Compute a Set of Canonical Homotopy Group Generator

input : A closed mesh M with resolution high enough, the genus of M is g
output: A set of canonical homotopy group generator $\{a_1, b_1, a_2, b_2, \cdots, a_g, b_g\}$

$M_1 \leftarrow M$;
repeat
  Locate a handle of M_k to obtain (a_k, b_k);
  $M_{k+1} \leftarrow M_k \backslash \{a_k, b_k\}$
until *no more handles can be found*;
Output $\{a_1, b_1, a_2, b_2, \cdots, a_g, b_g\}$.

Further Readings

Half-edge data structures have been implemented in the following open source [56] and [57], which support sophisticated geometric processing operations. Hoppe proposed progressive mesh in [58], which is an efficient and robust data structure designed for mesh compression purposes. In practice, progressive mesh is essential for many geometric processing applications, such as mesh compression. The following works [59] and [60] are good surveys in mesh compression.

Vegter and Yap [61], Dey and Schipper [62] gave linear time algorithms to compute polygonal schemas whose removal cuts the surface into a disk. Erickson and Har-Peled [63] showed that computing graphs of shortest length whose removal cut the surface into a disk is NP-hard. Verdiére and Lazarus [64] gave an algorithm for computing a system of loops on a surface which is shortest among the homotopy class of a given system. Erickson and Whittlesey [65] gave a greedy algorithm to compute the shortest system of loops, among all systems of loops, relaxing the homotopy condition. A survey in the area of computational topology can be found in [66], where you can find more references and more research topics.

Problems

9.1. Half-Edge Data Structure
Implement the half-edge data structure to represent triangular meshes.
(a) Support file input and output.
(b) Verify the consistency of the mesh file. In the input file, if there exists three faces sharing one edge, report error information; if there exists two faces sharing the same half-edge, report error information.
(c) Find all of the boundary loops, store the consecutive boundary half-edges in lists.
(d) Compute the Euler number of the mesh, and the number of handles, the number of boundaries.

9.2. Non-Orientable Meshes
(a) Triangulate a Möbius band, generate an input mesh file.
(b) How to detect the orientability of the input mesh file in your mesh program?

9.3. Homotopy Group Generators
(a) Implement the algorithm to compute the fundamental domain in your mesh program.
(b) Implement the algorithm to compute the cut graph in your mesh program.
(c) Implement the algorithm to compute the homotopy group generators.

9.4. Universal Covering Space
(a) Implement the double covering algorithm.
(b) Implement the algorithm to construct a finite portion of the universal covering space.
(c) Implement the algorithm for curve lifting.
(d) Implement the algorithm for shortest loop in a given homotopy group.
(e) Design a data structure to represent the finite portion of the universal covering space, which is a graph. Each node represents a fundamental domain, each arc represents a pair of boundary segments to be glued together.

9.5. Volumetric Half-Face Data Structure
Generalize half-edge data structure for surfaces to half-face data structure for volumetric solids. The basic data structures are vertex, half-edge, edge, half-face, face, and tetrahedron. All the data structures are connected coherently by pointers.

9.6. Handle Detection
Geometric meshes reconstructed from scanned point clouds are always noisy. There are in general many small handles on the surface. Design an algorithm to locate the small handles and fill the small holes.

9.7. Mesh Subdivision
(a) Implement the mesh subdivision algorithm.
(b) Compute the homotopy group generators for both the original mesh and the subdivided mesh.

9.8. Covering Space
(a) Calculate the homotopy group generators of a closed torus mesh, denoted as $\{a, b\}$.
(b) Construct a covering mesh, whose homotopy group generator is only a.

9.9. Double Covering
(a) Implement the mesh double covering algorithm.
(b) Compute the Euler number of the original mesh and the doubled mesh.
(c) Compute the homotopy group generators of the original mesh and the doubled mesh.

10

Algorithms for Harmonic Maps

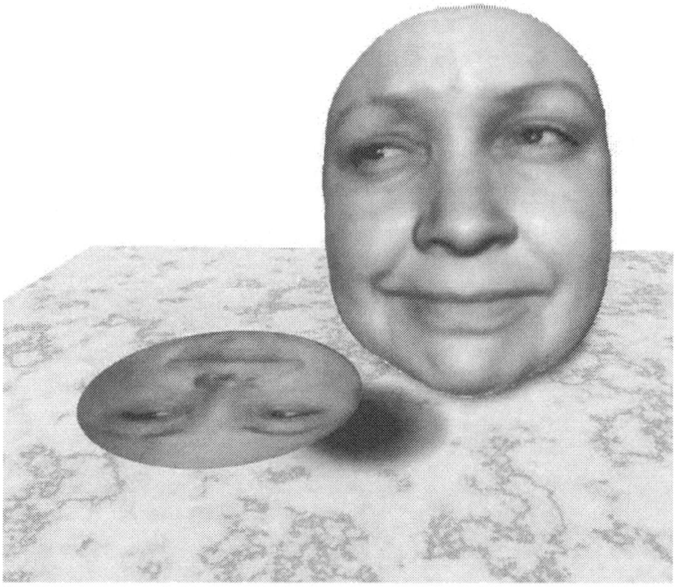

Fig. 10.1. Harmonic mapping for a face surface.

This chapter focuses on the algorithms for computing harmonic maps between genus zero surfaces. The input surface is either a topological disk or a topological sphere (Fig. 10.1—10.3), represented as triangular meshes. Our algorithm depends only on the Riemannian metric of the input surface, therefore it is invariant to the changes of the embeddings. It is also insensitive to the resolutions and the tessellations of the input mesh. The algorithms are mainly based on the heat flow method.

10.1 Piecewise Linear Functional Space, Inner Product and Laplacian

In digital geometry processing, smooth surfaces are approximated by discrete meshes. Smooth functions defined on surfaces are approximated by piecewise-linear functions. Suppose M is a triangular mesh,

Definition 10.1 (Function Space). *All piecewise linear functions defined on M form a linear space, denoted by $C^{PL}(M)$.*

Namely, the function space $C^{PL}(M)$ is the 0-forms of the mesh.

Definition 10.2 (Inner Product). *Suppose a set of string constants k_{ij} are assigned for each edge $\{v_i, v_j\}$. The inner product on $C^{PL}(M)$ is defined as the quadratic form*

$$\langle f, g\rangle = \frac{1}{2} \sum_{[v_i,v_j]\in M} k_{ij}(f(v_j) - f(v_i))(g(v_j) - g(v_i)). \tag{10.1}$$

The string energy is defined as the norm on $C^{PL}(M)$.

Definition 10.3 (String Energy). *Suppose $f \in C^{PL}(M)$, the string energy is defined as*

$$E(f) = \langle f, f\rangle = \sum_{[v_i,v_j]\in M} k_{ij}(f(v_j) - f(v_i))^2. \tag{10.2}$$

By changing the string constants k_{ij} in the energy formula, we can define different string energies.

Fig. 10.4 shows the definition of harmonic edge weight k_{ij}, which is the summation of cotangents of the two angles against the edge.

Definition 10.4 (Harmonic Energy). *Suppose edge $[v_i, v_j]$ has two adjacent faces T_α, T_β, $T_\alpha = \{v_i, v_j, v_k\}$ with three corner angles $\{\theta_i, \theta_j, \theta_k\}$. Define parameters*

$$a^\alpha_{v_i,v_j} = \frac{1}{2}\frac{(v_i - v_k)\cdot(v_j - v_k)}{|(v_i - v_k)\times(v_j - v_k)|} = \cot\theta_k, \tag{10.3}$$

$$a^\alpha_{v_j,v_k} = \frac{1}{2}\frac{(v_j - v_i)\cdot(v_k - v_i)}{|(v_j - v_i)\times(v_k - v_i)|} = \cot\theta_i, \tag{10.4}$$

$$a^\alpha_{v_k,v_i} = \frac{1}{2}\frac{(v_k - v_j)\cdot(v_i - v_j)}{|(v_k - v_j)\times(v_i - v_j)|} = \cot\theta_j. \tag{10.5}$$

T_β is defined similarly. If $[v_i, v_j]$ is an interior edge adjacent to T_α and T_β,

$$k_{ij} = a^\alpha_{v_i,v_j} + a^\beta_{v_i,v_j}. \tag{10.6}$$

If $[v_i, v_j]$ is a boundary edge and only adjacent to T_α, then

$$k_{v_i,v_j} = a^\alpha_{v_i,v_j}.$$

Then the string energy obtained is called the harmonic energy.

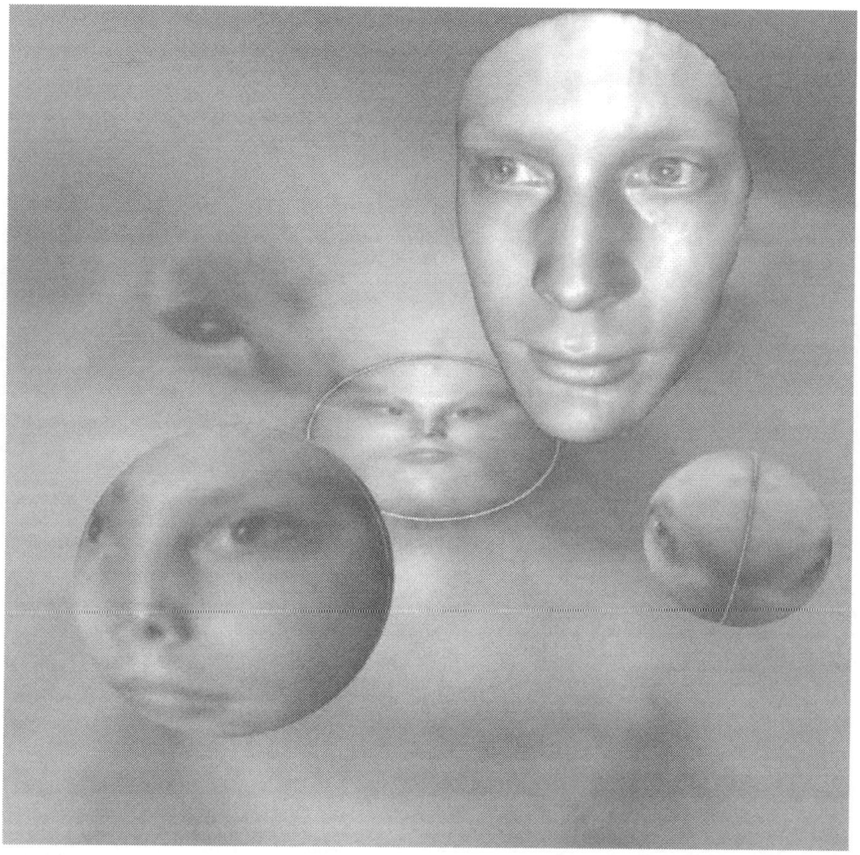

Fig. 10.2. The human face is double covered to form a topological sphere. Then the double covering surface is conformally mapped to the unit sphere, the sphere is conformally mapped onto the plane by the stereo-graphic projection.

The string energy is always a quadratic form. By carefully choosing the string coefficients, we make sure the quadratic form is *positive definite*. This will guarantee the convergence of the optimization process.

Definition 10.5 (Laplace Operator). *The piecewise Laplacian is a linear operator*

$$\Delta_{PL} : C^{PL}(M) \to C^{PL}(M)$$

on the space of piecewise linear functions on M, defined by the formula

$$\Delta_{PL} f(v_i) = \sum_{[v_i,v_j]\in M} k_{ij}(f(v_j) - f(v_i)). \tag{10.7}$$

If f minimizes the string energy, then f satisfies the condition

$$\Delta_{PL} f \equiv 0.$$

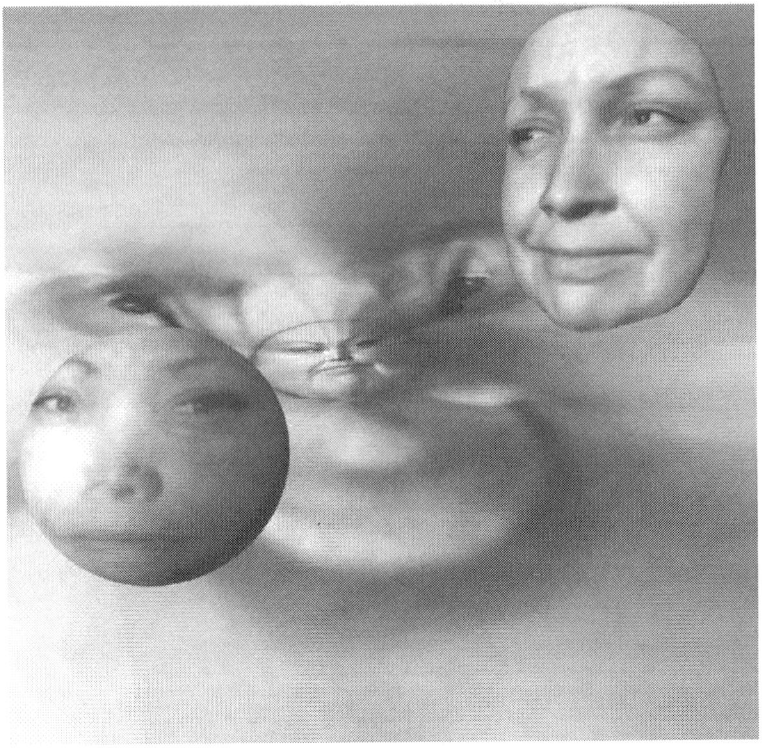

Fig. 10.3. Conformal mapping of a double covered human face surface to the unit sphere. The sphere is conformally mapped to the plane by a stereographic projection.

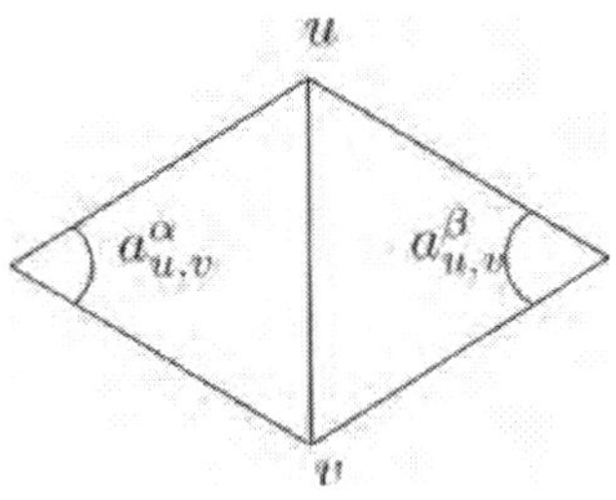

Fig. 10.4. Cotangent formulae for harmonic edge weight.

Suppose M_1, M_2 are two meshes and the map $\mathbf{f}: M_1 \to M_2$ is a map between them, $\mathbf{f}$ can be treated as a map from M_1 to $\mathbb{R}^3$ also.

Definition 10.6 (Harmonic Energy of Maps). *For a map $\mathbf{f}: M_1 \to \mathbb{R}^3$, $\mathbf{f} = (f_1, f_2, f_3)$, $f_i \in C^{PL}(M_1), i = 1,2,3$, we define the energy as the norm of $\mathbf{f}$:*

$$E(\mathbf{f}) = \sum_{i=1}^{3} E(f_i). \tag{10.8}$$

The Laplacian is defined in a similar way.

Definition 10.7 (Laplace Operator). *For a map* $\mathbf{f}: M_1 \to \mathbb{R}^3$ *, the piecewise Laplacian of* $\mathbf{f}$ *is*

$$\Delta_{PL}\mathbf{f} = (\Delta_{PL}f_1, \Delta_{PL}f_2, \Delta_{PL}f_3). \tag{10.9}$$

Definition 10.8 (Harmonic Map). *A map* $\mathbf{f}: M_1 \to M_2$ *is harmonic, if and only if* $\Delta_{PL}\mathbf{f}$ *only has a normal component, and its tangential component is zero*:

$$\Delta_{PL}(\mathbf{f}) = (\Delta_{PL}\mathbf{f})^{\perp}. \tag{10.10}$$

From harmonic map theory, suppose $f: M \to \mathbb{R}$ is a function defined on the mesh. Then the conventional harmonic energy is defined as

$$E(f) = \int_M |\nabla f|^2 d\sigma,$$

where $d\sigma$ is the area element of the surface and ∇f is the gradient of the function. A harmonic function is a harmonic map from M to $\mathbb{R}$, therefore a harmonic function has the property

$$\Delta f = 0.$$

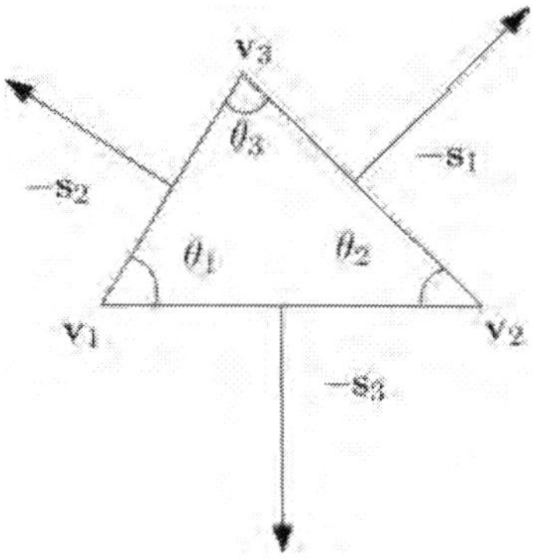

Fig. 10.5. A triangular face.

Lemma 10.9. *The conventional harmonic energy and the discrete harmonic energy are consistent.*

Proof. Let $[v_1, v_2, v_3]$ be a triangular face as shown in Fig. 10.5, the normal of the face is $\mathbf{n}$,

$$\begin{aligned} \mathbf{s}_1 &= \mathbf{n} \times (\mathbf{v}_3 - \mathbf{v}_2), \\ \mathbf{s}_2 &= \mathbf{n} \times (\mathbf{v}_1 - \mathbf{v}_3), \\ \mathbf{s}_3 &= \mathbf{n} \times (\mathbf{v}_2 - \mathbf{v}_1). \end{aligned}$$

Then

$$\mathbf{s}_1+\mathbf{s}_2+\mathbf{s}_3=\mathbf{n}\times[(\mathbf{v}_3-\mathbf{v}_2)+(\mathbf{v}_1-\mathbf{v}_3)+(\mathbf{v}_2-\mathbf{v}_1)]=0.$$

Therefore,

$$\langle \mathbf{s}_i,\mathbf{s}_i\rangle=-\sum_{i\neq j}\langle \mathbf{s}_i,\mathbf{s}_j\rangle.$$

Let $\mathbf{r}$ be an arbitrary interior point on the face,

$$\mathbf{r}=\lambda_1\mathbf{v}_1+\lambda_2\mathbf{v}_2+\lambda_3\mathbf{v}_3,$$

where $(\lambda_1,\lambda_2,\lambda_3)$ are the barycentric coordinates,

$$\lambda_i=\frac{1}{2}\frac{\langle \mathbf{r},\mathbf{s}_i\rangle}{A},\quad i=1,2,3,$$

where A is the area of the face.

Suppose f is a linear function defined on the face. Then

$$f(\mathbf{r})=\sum_{i=1}^{3}\lambda_i f(\mathbf{v}_i)=\frac{1}{2A}\sum_{i=1}^{3}\langle \mathbf{r},f(\mathbf{v}_i)\mathbf{s}_i\rangle.$$

Therefore, the gradient of f is

$$\nabla f=\frac{1}{2A}\sum_{i=1}^{3}f(\mathbf{v}_i)\mathbf{s}_i.$$

The conventional harmonic energy is defined as

$$E(f)=\langle\nabla f,\nabla f\rangle A=-\frac{1}{4}\sum_{i\neq j}\frac{\langle \mathbf{s}_i,\mathbf{s}_j\rangle}{A}(f_i-f_j)^2.$$

Furthermore,

$$\frac{\langle \mathbf{s}_i,\mathbf{s}_j\rangle}{2A}=-\cot\theta_k.$$

The conventional harmonic energy and the discrete harmonic energy differ by a constant. □

In the classical finite element literature, it has been proven that the discrete solutions to the discrete Laplace equation converge to the smooth solution when the meshes are subdivided to the limit. One subtle requirement is that the corner angles on the subdivided meshes should have a lower bound.

10.2 Newton's Method for Open Surface

This section explains the discrete harmonic maps for topological disks. A harmonic Map for a human face surface is shown in Fig. 10.6.1

Suppose M is a topological disk, according to Radó's theorem 7.3, the harmonic map from M to a convex planar domain is a diffeomorphism. A harmonic map can be directly computed by optimizing the harmonic energy with the *Dirichlet boundary condition*. Suppose M is a topological disk, $\mathbf{f}$ maps the mesh onto the plane,

$$\mathbf{f}: M \to \mathbb{R}^2.$$

Then the harmonic energy is

$$E(\mathbf{f}) = \sum_{[v_i,v_j]\in M} k_{ij} |\mathbf{f}(v_i) - \mathbf{f}(v_j)|^2,$$

where the boundary condition

$$\mathbf{f}|_{\partial M} = \mathbf{g},$$

where $\mathbf{g} : \partial M \to \mathbb{R}^2$ is given. The harmonic energy optimization is a linear problem and can be solved directly using Newton's method.

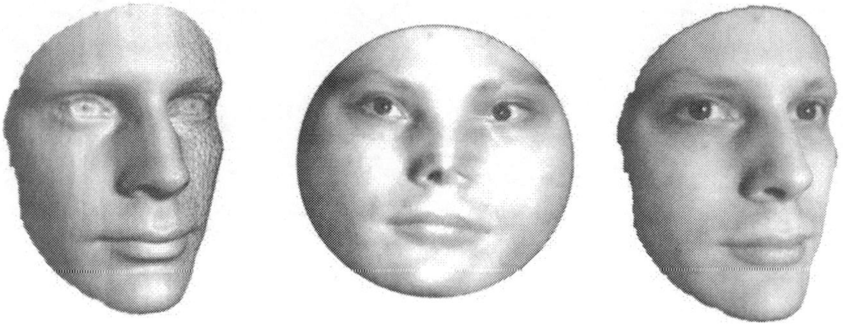

Fig. 10.6. Harmonic mapping for a face surface.

Algorithm 16: Harmonic map for a topological disk

input : A simply connected mesh with a single boundary
output: A harmonic map, mapping the mesh to the unit disk

Traverse the boundary of M, store boundary vertices to a list

$$\partial M = \{v_0, v_1, \cdots, v_{n-1}\}.$$

where v_0 and v_n are identical.

$$s \leftarrow \sum_{i=0}^{n-1} l_{v_i,v_{i+1}},$$

where $l_{v_i,v_{i+1}}$ is the edge length of edge $[v_i, v_{i+1}]$;
forall $v_i \in \partial M$ **do**

$$s_i \leftarrow \sum_{j=1}^{i} l_{v_{j-1},v_j}$$

$$\theta_i \leftarrow 2\pi \frac{s_i}{s}$$

$$\mathbf{f}(v_i) \leftarrow (\cos\theta_i, \sin\theta_i)$$

end

Optimize the harmonic energy

$$E(f) = \sum_{[v_i,v_j]\in M} k_{ij}(f(v_j) - f(v_i))^2,$$

using Newton's method with fixed boundary condition.

The energy optimization is equivalent to solving the following linear system,

$$\Delta_{PL}\mathbf{f}(v_i) = \mathbf{0}, \quad \forall v_i \notin \partial M,$$

with the boundary vertices fixed.

10.3 Non-Linear Heat Diffusion for Closed Surfaces

It is well known that any closed genus zero surface can be mapped conformally onto the sphere (Fig.10.7). The mapping is one-to-one and angle preserving. The pull-back metric on the surface is conformal to the original induced Euclidean metric. The area distortion is described by the conformal factor function. All such conformal mappings differ by a Möbius transformation on the sphere, which form a six dimensional group. Furthermore, any harmonic maps from the surface to the unit sphere is conformal.

Fig. 10.7. Conformal spherical mapping.

In this section, we will explain the discrete algorithm for computing harmonic maps from a genus zero closed mesh to the unit sphere. The algorithm is based on the nonlinear heat diffusion method. Our algorithm for computing conformal mappings is based on the fact that for genus zero closed surfaces, a map is conformal if and only if it is harmonic. Our method is as follows: we first find a degree one map $\mathbf{h}$ between M and the unit sphere $\mathbb{S}^2$, the map may not be a homeomorphism. The map will be smoothed out automatically during the process. Then we evolve $\mathbf{h}$ to minimize its harmonic energy until it becomes a harmonic map. The evolution of the map is according to a nonlinear heat diffusion process,

$$\frac{d\mathbf{f}(t)}{dt} = -\Delta\mathbf{f}(t), \tag{10.11}$$

where Δ is the Laplace-Beltrami operator on the surface determined by the induced Euclidean metric. But $\mathbf{f}(M)$ is constrained to be on the unit sphere $\mathbb{S}^2$, so we project $-\Delta\mathbf{f}$ onto the tangent space of the sphere. This projection causes the non-linearity, otherwise we can formulate the optimization as a linear system.

In detail, let $\mathbf{f}: M_1 \to \mathbb{S}^2$ be the map, $\mathbf{f}(v)$ be the image of the vertex $v \in M$. The normal on the target surface at $\mathbf{f}(v)$ is $\mathbf{n}(\mathbf{f}(v))$. (If the target surface is the unit sphere $\mathbb{S}^2$, then $\mathbf{n}(\mathbf{f}(v))$ equals to $\mathbf{f}(v)$.) Define the normal component of the Laplacian as follows.

Definition 10.10. *The normal component of the Laplacian is*

$$(\Delta\mathbf{f}(v))^{\perp} = \langle \Delta\mathbf{f}(v), \mathbf{n}(\mathbf{f}(v)) \rangle \mathbf{n}(\mathbf{f}(v)), \tag{10.12}$$

where $\langle , \rangle$ *is the inner product in* $\mathbb{R}^3$.

The tangential component of the Laplacian is

$$(\Delta\mathbf{f}(v))^{\parallel} = \Delta\mathbf{f}(v) - \Delta\mathbf{f}(v)^{\perp}.$$

The nonlinear heat diffusion equation is

$$\frac{d\mathbf{f}(v,t)}{dt} = -(\Delta\mathbf{f}(v))^{\parallel}. \tag{10.13}$$

Then $\mathbf{f}(v,\infty)$ is the harmonic map.

The solutions to the harmonic map (conformal map) from M to the unit sphere are not unique. Suppose $\mathbf{f}_1, \mathbf{f}_2 : M \to \mathbb{S}^2$ are two harmonic maps. Then

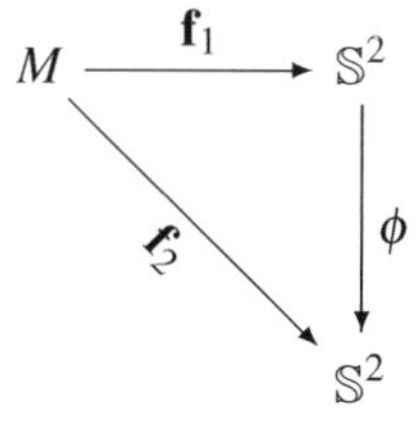

$\phi = \mathbf{f}_2 \circ \mathbf{f}_1^{-1} : \mathbb{S}^2 \to \mathbb{S}^2$ is a Möbius transformation of the unit sphere, or equivalently, the extended complex plane $\bar{\mathbb{C}}$.

Definition 10.11 (Möbius Transformation). *Mapping* $\mathbf{f}: \bar{\mathbb{C}} \to \bar{\mathbb{C}}$ *is a Möbius transformation if and only if*

$$\phi : z \to \frac{az+b}{cz+d}, \quad a,b,c,d \in \mathbb{C},\ ad - bc = 1. \tag{10.14}$$

The Möbius transformations of the unit sphere form a 6-dimensional Möbius transformation group. To ensure the convergence of the algorithm and the uniqueness of the solution, constraints need to be added. In practice we use the following *zero mass-center constraint*.

Definition 10.12. *Mapping* $\mathbf{f}: M \to \mathbb{S}^2$ *satisfies the zero mass-center condition if and only if*

$$\int_{\mathbb{S}^2} \mathbf{f} d\sigma = 0, \tag{10.15}$$

where $d\sigma$ *is the area element on* M.

All conformal maps from M to $\mathbb{S}^2$ satisfying the zero mass-center constraint are unique up to a Euclidean rotation (the Euclidean rotation group in $\mathbb{R}^3$ is 3-dimensional). We use the Gauss map as the initial map.

Definition 10.13 (Gauss Map). *A Gauss map* $g: M \to \mathbb{S}^2$ *is defined as*

$$g(v) = \mathbf{n}(v), \quad v \in M, \tag{10.16}$$

$\mathbf{n}(v)$ *is the unit normal at* v.

The following algorithm describes the details for computing a discrete conformal map from a topological sphere to the unit sphere.

Algorithm 17: Spherical conformal mapping

input : Mesh M, step length δt, energy difference threshold δE
output: A harmonic map $\mathbf{f}: M \to S^2$, which satisfies the zero mass-center constraint.

Compute a degree one map, such as Gauss map $\mathbf{g}: M \to \mathbb{S}^2$;
Initialize $\mathbf{f} \leftarrow \mathbf{g}$, compute harmonic energy E_0;
repeat
  forall *vertex* $v \in M$ **do**
    Compute the Laplacian $\Delta \mathbf{f}$;
    Compute the normal component;

$$\Delta \mathbf{f}^{\perp} = < \Delta \mathbf{f}, \mathbf{f} > \mathbf{f}$$

    Compute the tangential component;

$$\Delta \mathbf{f}^{\parallel} = \Delta \mathbf{f} - \Delta \mathbf{f}^{\perp}$$

    Update $\mathbf{f}(v)$ by

$$\mathbf{f}(v) = \mathbf{f}(v) - \delta t \times \Delta \mathbf{f}^{\parallel}.$$

  end
  Compute Möbius transformation $\varphi: \mathbb{S}^2 \to \mathbb{S}^2$, such that the mass center of $\varphi \circ \mathbf{f}$ is sphere center;
  $\mathbf{f} \leftarrow \varphi \circ \mathbf{f}$;
  $E_0 \leftarrow E$;
  Compute the harmonic energy $E(\mathbf{f})$.
until *Harmonic energy difference* $|E - E_0|$ *is less than* δE;
Return $\mathbf{f}$

The step of normalization using Möbius transformation is non-linear and expensive to compute. In practice we use the following simple procedure instead:

Algorithm 18: Normalization

input : Mesh M, a mapping to the sphere $\mathbf{f}: M \to \mathbb{S}^2$
output: Normalized mapping $\tilde{\mathbf{f}}$, whose mass center is at the sphere center

Compute the mass center of $\mathbf{f}$:

$$\mathbf{c} \leftarrow \int_{\mathbb{S}^2} \mathbf{f} d\sigma.$$

where $d\sigma$ is the area element on the original mesh M. **forall** $v \in M$ **do**

$$\tilde{\mathbf{f}}(v) \leftarrow \mathbf{f}(v) - \mathbf{c}$$

end
forall $v \in M$ **do**

$$\tilde{\mathbf{f}}(v) \leftarrow \frac{\tilde{\mathbf{f}}(v)}{|\tilde{\mathbf{f}}(v)|}.$$

end

This approximation method is good enough for our purpose in practice. By choosing the step length carefully, the energy can be decreased monotonically during the process. The initial map could be the Gauss map or any other map as long as its degree equals to one. In practice, the quality of the initial map affects the efficiency and stability of the whole process prominently. Fig. 10.8 shows a real example of conformal brain mapping using this algorithm. The brain surface is reconstructed from MRI images and represented as triangular meshes. Fig. 10.9 shows the conformal spherical mapping results for a hand model and a foot model. It is easy to see that the extruding part (like the fingers) will induce higher area distortion.

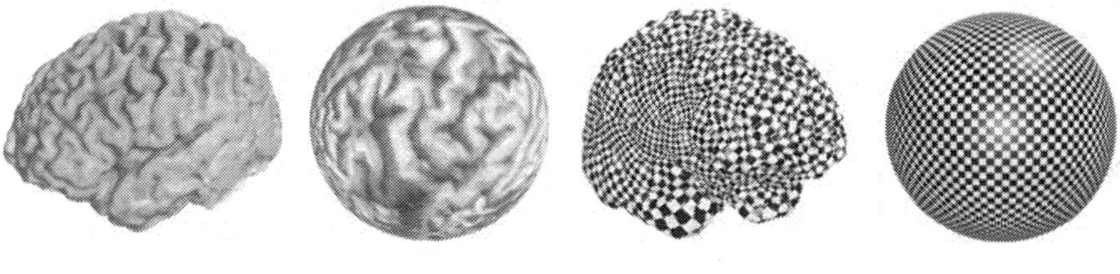

Fig. 10.8. Conformal spherical brain mapping.

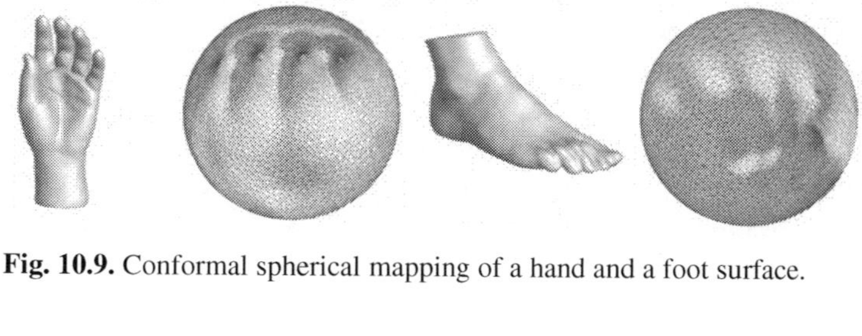

Fig. 10.9. Conformal spherical mapping of a hand and a foot surface.

10.4 Riemann Mapping

Conformal maps are harmonic, but harmonic maps are not necessarily conformal (Fig. 10.10). Riemann mapping theorem claims the existence of conformal mappings between a topological disk and the unit planar disk. This section introduces the algorithm for computing the Riemann mapping for simply connected surfaces with single boundaries.

The major idea is as follows. Suppose M is a simply connected surface and ∂M is a single loop. First, we double cover the surface to form a closed symmetric surface $\bar{M}$, which is a topological sphere. Then we apply the non-linear heat diffusion algorithm to find a conformal map from $\bar{M}$ to the unit sphere. Second, we use a Möbius transformation to adjust the conformal map such that ∂M is mapped to the equator of the unit sphere. Finally, we use the stereo-graphic projection to map the unit sphere onto the plane, which conformally maps the lower hemisphere to the unit disk.

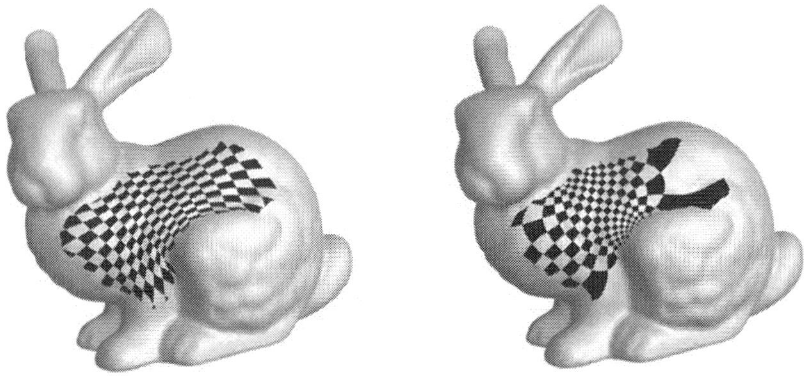

Fig. 10.10. Comparison between a harmonic map and a conformal map. The left frame shows a harmonic map, the right frame shows a conformal map.

A *stereo-graphic projection* is a conformal map, which maps the unit sphere $\mathbb{S}^2$ onto the whole complex plane $\mathbb{C}$:

$$\phi : \mathbb{S}^2 \to \mathbb{C}.$$

Geometrically, we can imagine that we put a light source at the north pole of the sphere, and lay down the screen as the tangent plane at the south pole. Then the whole surface of the sphere is projected onto the screen:

$$\phi : (x,y,z) \to \left(\frac{2x}{2-z}, \frac{2y}{2-z} \right),$$

the south pole is mapped to the origin, the north pole is mapped to infinity.

A Möbius transformation is a conformal map from the complex plane to itself $\tau : \mathbb{C} \to \mathbb{C}$. By using a Möbius transformation, we can map an arbitrary three points $z_0, z_1, z_2 \in \mathbb{C}$ to $0, 1, \infty$,

$$\tau_1 = \frac{(z-z_0)(z_1-z_2)}{(z-z_2)(z_1-z_0)}.$$

Suppose w_0, w_1, w_2 is another set of three points, τ_2 maps them to $0, 1, \infty$. The composition map $\tau_2^{-1} \circ \tau_1$ transforms $\{z_0, z_1, z_2\}$ to $\{w_0, w_1, w_2\}$ individually. If we require both $\{z_0, z_1, z_2\}$ are $\{w_0, w_1, w_2\}$ on the equators, then all the Möbius transformation form a 3 dimensional group and map each hemisphere to itself. Therefore, all the conformal mappings from the unit disk to itself form a 3 dimensional group. All the Riemann mappings from a topological disk to the unit disk differ by a Möbius transform on the unit disk. Therefore, we need to add constraints to make the mapping unique. For example, we can fix the images of three boundary points; we can fix the images of one interior vertex and one boundary vertex.

The following algorithm computes the Riemann mapping from a topological disk to the unit disk.

Algorithm 19: Riemann mapping

input : A genus zero mesh M with a single boundary.
output: $\mathbf{h} : M \rightarrow \mathbb{H}^2$, $\mathbf{h}$ is conformal

Double covering M to $\bar{M}$;
Compute a harmonic map $\mathbf{f} : \bar{M} \rightarrow \mathbb{S}^2$;
Compute the stereographic projection $\phi : \mathbb{S}^2 \rightarrow \mathbb{C}$;
Select three points v_0, v_1, v_2 on the boundary of M;
Use a Möbius transformation τ to map v_0, v_1, v_2 to $0, 1, i$;
$\mathbf{h} \leftarrow \tau \circ \phi \circ f$.

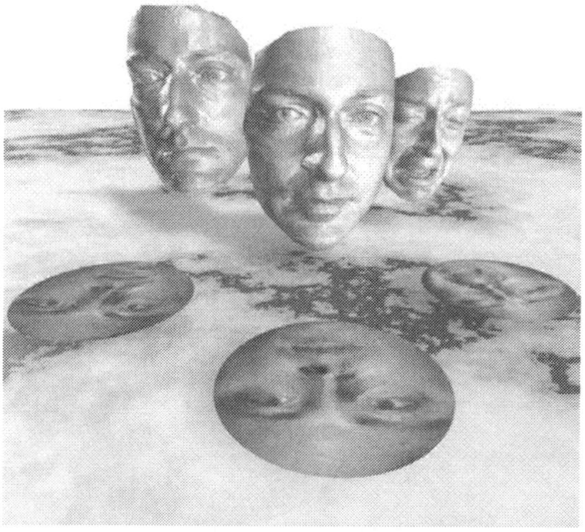

Fig. 10.11. Riemann mapping for three face surfaces of the same person with different expressions.

Fig. 10.11 illustrates three Riemann mappings of three face surfaces of the same person with different expressions. The face surfaces are topological disks and mapped onto the unit disk on the plane. The correspondences between the face and its planar image can be visualized by the texture mapping. The mappings are conformal, hence, there is no local shape distortion.

10.5 Least Square Method for Solving Beltrami Equation

In the theory part, in order to study quasi-conformal maps, we introduced the Beltrami equation. Suppose $f : D \to \mathbb{C}$ is a map, where $D \subset \mathbb{C}$ is a domain in the complex plane. The *Beltrami equation* of f is defined as

$$\frac{\partial f}{\partial \bar{z}} = \mu(z)\frac{\partial f}{\partial z}.$$

If $\mu(z)$ is zero everywhere, then we get the conventional Cauchy-Riemann equation.

In the following discussion, we use real coordinates (x,y) to represent the complex number $x+iy$. Suppose D is a triangular mesh and f is a piecewise linear function. We first choose one face $[v_1,v_2,v_3]$ embedded on the plane $\mathbb{R}^2$, the planar coordinates of v_k are (x_k,y_k) as shown in Fig. 10.5. $f : (x,y) \to (u(x,y),v(x,y))$, from the proof of Lemma 10.9. We can obtain the formula of the gradients of u and v as follows,

$$\begin{aligned}\nabla u &= u_1\mathbf{s}_1 + u_2\mathbf{s}_2 + u_3\mathbf{s}_3,\\ \nabla v &= v_1\mathbf{s}_1 + v_2\mathbf{s}_2 + v_3\mathbf{s}_3.\end{aligned}$$

where $\mathbf{s}_1,\mathbf{s}_2,\mathbf{s}_3$ are illustrated in Fig. 10.5.

Therefore, $\frac{\partial u}{\partial x},\frac{\partial u}{\partial y}$ are linear functions of u_1,u_2,u_3; $\frac{\partial v}{\partial x},\frac{\partial v}{\partial y}$ are linear functions of v_1,v_2,v_3. Let $\mu(x,y)$ be a piecewise constant function,

$$\mu(x,y) = (a,b).$$

Then the Beltrami equation is equivalent to

$$\begin{aligned}\mathbf{p}\cdot\nabla u + \mathbf{q}\cdot\nabla v &= 0,\\ -\mathbf{q}\cdot\nabla u + \mathbf{p}\cdot\nabla v &= 0,\end{aligned}$$

where

$$\mathbf{p} = \begin{pmatrix} a-1 \\ b \end{pmatrix}, \quad \mathbf{q} = \begin{pmatrix} -b \\ a+1 \end{pmatrix}.$$

The Beltrami equation is transformed to the following linear equation system:

$$\begin{pmatrix} \lambda_1 & \lambda_2 & \lambda_3 & \gamma_1 & \gamma_2 & \gamma_3 \\ -\gamma_1 & -\gamma_2 & -\gamma_3 & \lambda_1 & \lambda_2 & \lambda_3 \end{pmatrix} \begin{pmatrix} u_1 \\ u_2 \\ u_3 \\ v_1 \\ v_2 \\ v_3 \end{pmatrix} = 0,$$

where

$$\lambda_k = \mathbf{p}\cdot\mathbf{s}_k, \gamma_k = \mathbf{q}\cdot\mathbf{s}_k, k = 1,2,3.$$

For each face, we can construct a linear system, and we pack the linear equations together to form a linear system:

$$\begin{pmatrix} \Lambda & \Gamma \\ -\Gamma & \Lambda \end{pmatrix} \begin{pmatrix} \mathbf{u} \\ \mathbf{v} \end{pmatrix} = 0 \tag{10.17}$$

Suppose the vertex set of the mesh is $\{v_1, v_2, \cdots, v_n\}$, and the face set of the mesh is $\{f_1, f_2, \cdots, f_m\}$. Then

$$\Lambda_{m\times n} = (\lambda_{ij}), \quad \Gamma_{m\times n} = (\gamma_{ij}),$$

λ_{ij} represents the λ value of vertex v_j in the face f_i. Similarly, γ_{ij} represents the γ value of v_j in the face f_i.

$$\mathbf{u} = \begin{pmatrix} u_1 \\ u_2 \\ \vdots \\ u_n \end{pmatrix}, \quad \mathbf{v} = \begin{pmatrix} v_1 \\ v_2 \\ \vdots \\ v_n \end{pmatrix}.$$

In order to ensure the existence and the uniqueness of the solution, we need to prescribe the (u, v) coordinates of at least two vertices. The solution to the linear system can be solved using the standard least square method, namely by minimizing the following quadratic form:

$$\begin{pmatrix} \mathbf{u}^T & \mathbf{v}^T \end{pmatrix} \begin{pmatrix} \Lambda^T & -\Gamma^T \\ \Gamma^T & \Lambda^T \end{pmatrix} \begin{pmatrix} \Lambda & \Gamma \\ -\Gamma & \Lambda \end{pmatrix} \begin{pmatrix} \mathbf{u} \\ \mathbf{v} \end{pmatrix}.$$

Fig. 10.12 illustrates the conformal parameterization results using least square method solving Beltrami equation. We set four anchor points for each face, and set the boundary free. In the general case, if the topology of the surface is complicated, this method may produce immersions instead of embedding.

Fig. 10.12. Parameterizations for three face surfaces of the same person with different expressions using the least square method of solving Beltrami equation.

10.6 General Surface Mapping

Surface matching is a fundamental problem in computer graphics, computer vision and visualization. Harmonic maps are valuable for matching surfaces. The basic approach is as follows. First, we map surfaces to a canonical planar domain; then, we find a mapping from the planar domain to itself, which induces the matching between the two surfaces. By using this method, we convert the 3D matching problem to the planar matching problem, which is much easier to deal with (Fig. 10.13).

Suppose S_1, S_2 are two surfaces embedded in $\mathbb{R}^3$, $\phi_1 : S_1 \rightarrow D$, $\phi_2 : S_2 \rightarrow D$ map S_1, S_2 to a common planar domain $D \subset \mathbb{R}^2$ respectively. ϕ_1, ϕ_2 are one to one, onto, differentiable and the inverses are also differentiable. We want to find an optimal map from D to itself, $\bar{\mathbf{f}} : D \rightarrow D$, such that the induced map

$$\mathbf{f} = \phi_2{}^{-1} \circ \bar{\mathbf{f}} \circ \phi_1 : S_1 \rightarrow S_2$$

is the desired map between S_1 and S_2.

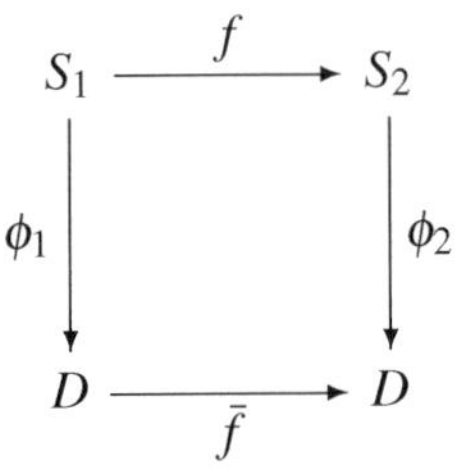

The mappings $\phi_1, \phi_2, \bar{\mathbf{f}}$ should be chosen to reflect the geometric characteristics of the surfaces S_1, S_2. Namely, ϕ_1 and ϕ_2 should be determined by the geometries of the surfaces. The conformal mapping technique is very suitable for this purpose. As shown in Fig. 10.1 and 10.11, human face surfaces are mapped onto the unit planar disk, the mapping preserves angles, therefore, all the geometric shapes are preserved on the planar images. Such kind of mappings $\phi_\mathbf{k} : S_k \rightarrow D$ form a three-dimensional group, therefore, the map can be controlled by specifying three vertex images on the boundary.

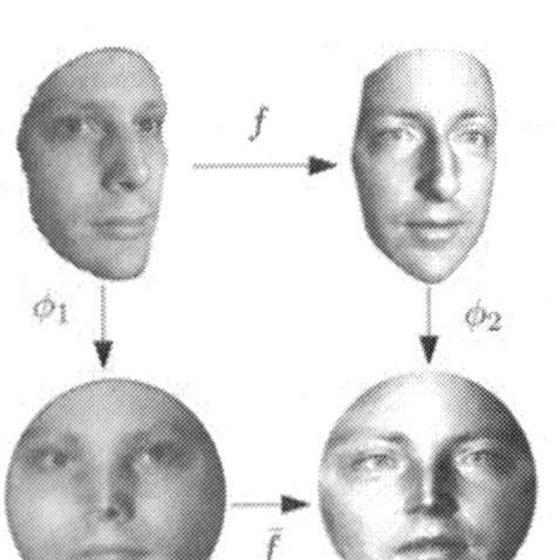

Fig. 10.13. Surface matching framework.

Therefore, the 3D surface matching problem $\mathbf{f}: S_1 \to S_2$ is converted to a 2D matching problem $\bar{\mathbf{f}}: D \to D$. We can apply the harmonic map technique to find $\bar{\mathbf{f}}$. It is also highly desirable to match feature points. Suppose $\{p_1, p_2, \cdots, p_n\} \subset S_1$ are feature points on S_1, $\{q_1, q_2, \cdots, q_n\} \subset S_2$ are corresponding feature points on S_2, such that

$$\mathbf{f}(p_k) = q_k, \quad 1 \leqslant k \leqslant n.$$

Then we can enforce the correspondences by adding the following constraints to the $\bar{f}$:

$$\bar{\mathbf{f}} \circ \phi_1(p_k) = \phi_2(q_k), \quad 1 \leqslant k \leqslant n.$$

Furthermore $\bar{\mathbf{f}}$ is harmonic and satisfies the Laplace equation

$$\Delta \bar{\mathbf{f}}(v) = 0, \quad v \notin \{\phi_1(p_1), \phi_1(p_2), \cdots, \phi_1(p_n)\}.$$

If the surfaces are topological disks, then the canonical domain is the unit planar disk. If the surfaces are topological spheres, then the canonical domain is the unit sphere. For high genus closed surfaces, the canonical domain should be the fundamental polygon (Fig. 10.14). In the following, we introduce the matching algorithm for high genus surfaces.

Assume the surface is of genus g, we first compute a canonical basis of homotopy group, denoted as

$$\{a_1, b_1, a_2, b_2, \cdots, a_g, b_g\}$$

such that all base curves only intersect at a unique base point p. Then we slice the surface along the base curves to get a fundamental domain D, the boundary of D is

$$\partial D = a_1 b_1 a_1^{-1} b_1^{-1} a_2 b_2 a_2^{-1} b_2^{-1} \cdots a_g b_g a_g^{-1} b_g^{-1}.$$

We then fix the boundary to a convex polygon with $4g$ sides, and compute a harmonic map $\phi_1 : \bar{S}_1 \to D$ with the fixed boundary condition.

The following algorithm explains the details to compute a map between two genus g closed surfaces (Fig. 10.15 and 10.16).

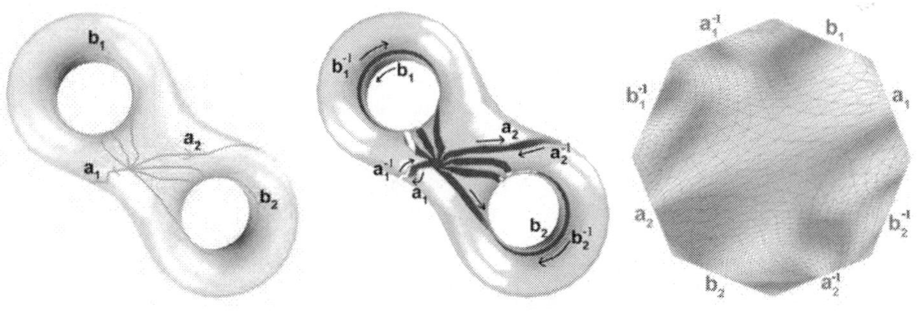

Fig. 10.14. Construct a fundamental polygon.

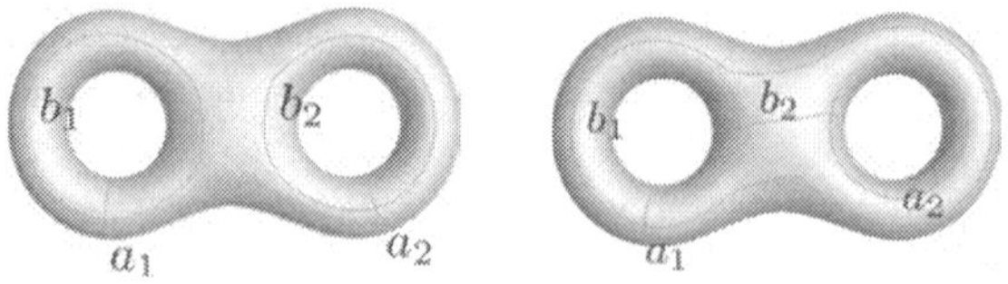

Fig. 10.15. Different sets of canonical homology basis of a two hole torus surface.

Algorithm 20: Map between two genus g closed surfaces

input : Two genus g closed meshes M_1, M_2, corresponding feature points $\{p_1, p_2, \cdots, p_n\} \subset M_1$ and $\{q_1, q_2, \cdots, q_n\} \subset M_2$, p_i, q_j's are vertices
output: A one-to-one continuous map $f : M_1 \to M_2$, $f(p_k) = q_k, 1 \leqslant k \leqslant n$.
forall M_k, $k = 1, 2$ **do**
Compute a canonical set of homotopy group basis of M_k;
Slice M_k along the homotopy basis to get $\bar{M}_k$;
Map the boundary of $\bar{M}_k$ to a planar polygon with $4g$ sides $g_k : \partial M_k \to \partial D$;
Compute a harmonic map with fixed boundary condition

$$\phi_k : \bar{M}_k \to D, \quad \Delta\phi_k \equiv 0, \quad \phi_k|_{\partial M_k} = g_k.$$

end
Compute $\bar{\mathbf{f}} : D \to D$, such that $\bar{\mathbf{f}}$ satisfies the feature correspondence constraints,

$$\bar{\mathbf{f}}(\phi_1(p_k)) = \phi_2(q_k),$$

and is harmonic

$$\Delta\bar{\mathbf{f}}(\phi_1(v)) = 0, \quad \forall v \notin \{p_1, p_2, p_3, \cdots, p_n\}.$$

Then desired map is

$$\mathbf{f} \leftarrow \boldsymbol{\phi}_2^{-1} \circ \bar{\mathbf{f}} \circ \boldsymbol{\phi}_1.$$

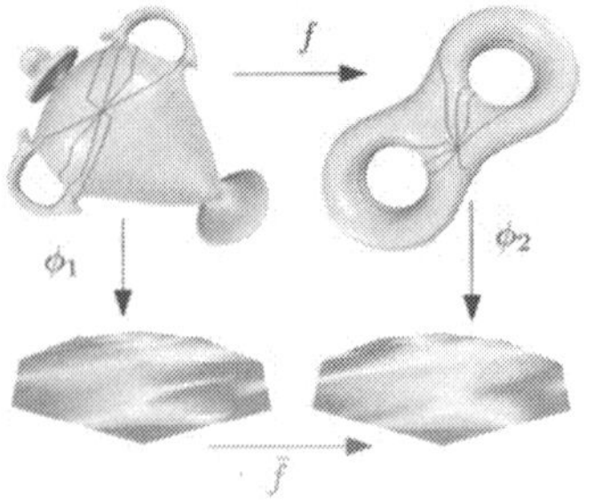

Fig. 10.16. Matching two genus two surfaces.

As we discussed before, all the homeomorphisms between two surfaces can be classified by homotopy. Two maps $\mathbf{f_1}, \mathbf{f_2} : M_1 \to M_2$ are homotopic if there exists a homotopy

$$\mathbf{F} : M_1 \times [0,1] \to M_2,$$

such that $\mathbf{F}$ is continuous, $\mathbf{F}(\cdot,0) = \mathbf{f_1}$, $\mathbf{F}(\cdot,1) = \mathbf{f_2}$. The homotopy type of the mapping is controlled by the choice of canonical homology basis. Given a high genus surface, there are infinite canonical homology bases. Fixing the homology basis on M_1, by choosing different canonical homology basis on M_2, we can construct maps with different homotopy types. Fig. 10.17 shows two maps between a vase surface and a two holed torus model, which are with different homotopy types.

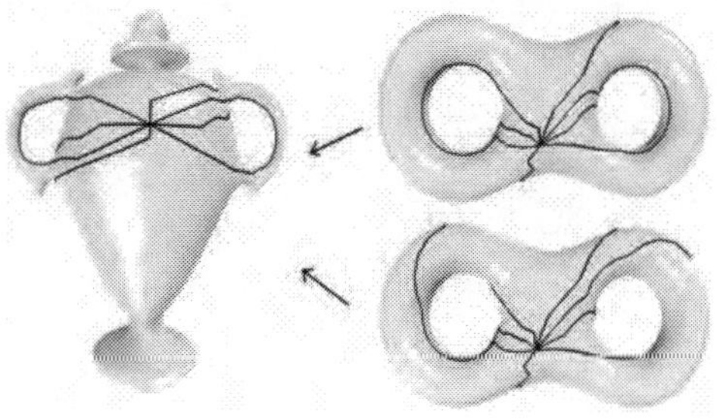

Fig. 10.17. Homeomorphisms between a vase model and a two hole torus, the maps are of different homotopy types.

Once two surfaces are matched, we can construct a geometric morphing between them by linearly interpolating the positions of corresponding vertices. Fig. 10.18 demonstrates such a morphing between the vase model and the two hole torus.

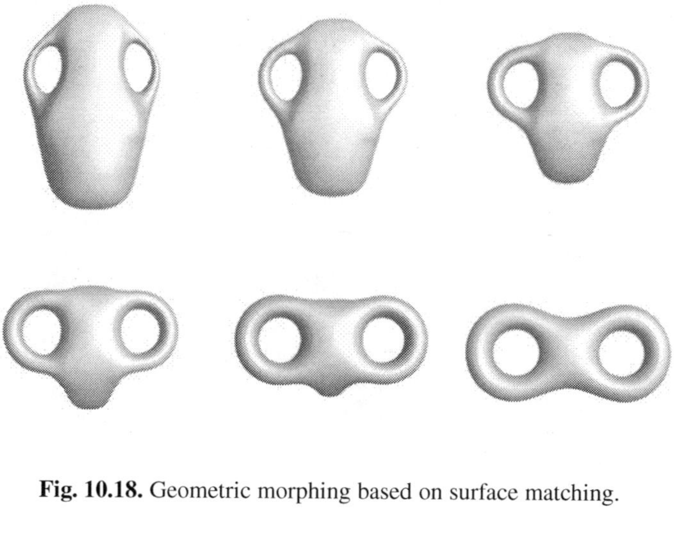

Fig. 10.18. Geometric morphing based on surface matching.

Further Readings

Finite element approximation of harmonic energy and Laplace operator are discussed in [67] for the purpose of computing discrete minimal surface. Harmonic maps in the mesh setting was discussed in [68] and applied in parameterization in [69]. Harmonic maps of topological disks are introduced in [70], [71]. Different approaches for harmonic maps of topological spheres are introduced in [72], [73], and [74]. Non-linear heat diffusion method is introduced in [22] and [23]. Other methods for conformal mappings will be introduced in the later chapters. More thorough references can be found in the following survey works [41] and [42].

Problems

10.1. Conjugate Gradient
In order to solve discrete Laplace equation, we can use conjugate gradient algorithm to solve the linear system

$$Q\mathbf{x} - \mathbf{b} = 0,$$

or equivalently the quadratic minimization problem

$$\min\left(\frac{1}{2}\mathbf{x}^T Q\mathbf{x} - \mathbf{b}^T\mathbf{x}\right),$$

where Q is a given $n \times n$ symmetric positive definite matrix and $\mathbf{b}$ is a given n vector.

The algorithm requires n iterations, starting from an arbitrary initial guess $\mathbf{x}_0$. Let i represent the iteration number, $\mathbf{x}_i$ the solution approximation, $\mathbf{d}_i$ the search direction, $\mathbf{r}_i$ the residual, defined as $\mathbf{b} - Q\mathbf{x}_i$.

Algorithm 21: Conjugate Gradient Algorithm

input : A symmetric positive definite matrix Q, a vector $\mathbf{b}$
output: Solution to $Q\mathbf{x} - \mathbf{b} = 0$

Initialization. Let $\mathbf{x}_0 \leftarrow \mathbf{0}$, $\mathbf{r}_0 \leftarrow -Q\mathbf{x}_0 + \mathbf{b}$, $\mathbf{d}_0 \leftarrow \mathbf{r}_0$

for $i = 0$ *to* $n-1$ **do**

$$\begin{aligned} \alpha &\leftarrow \frac{\mathbf{r}_i^T \mathbf{d}_i}{\mathbf{d}_i^T Q\mathbf{d}_i} \\ \mathbf{x}_{i+1} &\leftarrow \mathbf{x}_i + \alpha\mathbf{d}_i \\ \mathbf{r}_{i+1} &\leftarrow \mathbf{r}_i - \alpha Q\mathbf{d}_i \end{aligned}$$

if $||\mathbf{r}_i|| < \varepsilon$ **then**
stop, the solution has been found.
else

$$\begin{aligned} \beta &\leftarrow \frac{\mathbf{r}_{i+1}^T \mathbf{d}_i}{\mathbf{d}_i^T Q\mathbf{d}_i} \\ \mathbf{d}_{i+1} &\leftarrow -\mathbf{r}_{i+1} + \beta\mathbf{d}_i \end{aligned}$$

end
end

(a) Implement the sparse matrix data structure.

(b) Implement the conjugate gradient algorithm for solving a linear system.

10.2. Harmonic Map for Topological Disks
(a) Implement the harmonic map algorithm for topological disks. Set the edge weight to be one and use checker board texture mapping to visualize the conformality of the map.
(b) Use the cotangent formula for the edge weights, visualize the conformality and compare with the result in (a).
(c) If the images of the boundary vertices are set to be on a concave planar curve, test where there are some flipped faces in the harmonic map.

10.3. Harmonic Map for Topological Spheres
(a) Implement the spherical harmonic map algorithm.
(b) Double cover a topological disk and map the doubled mesh to the sphere using the spherical harmonic map algorithm.
(c) Implement the stereo-graphic projection.
(d) Implement the Möbius transformations, and map the original mesh to the unit disk.
(e) Visualize the conformality and compare with the result using the harmonic map algorithm with Dirichlet boundary condition.

10.4. Harmonic Map with Constraints
Add the feature point constraint to the harmonic map. Design the algorithm, formulate the linear system, and implement.

10.5. Volumetric Harmonic Map
Suppose we want to compute a volumetric harmonic map:
(a) Deduce the volumetric edge weight formula, which is the counter part of the cotangent formula for the 2D mesh.
(b) Design an algorithm for harmonic maps from a topological solid sphere (3 disk) to the unit solid sphere with Dirichlet boundary condition.
(c) Implement the algorithm.
(d) Verify if the map induces any flipped tetrahedron.

11

Harmonic Forms and Holomorphic Forms

Fig. 11.1. Holomorphic one-form.

The topological concepts of chain and homotopy are intuitive and easy to be visualized. But in practice, chains on meshes are with complicated combinatorial structures, and difficult to manipulate. The concepts of cochains, forms, and cohomology are abstract and often invisible, because they are functional instead of geometric. But the calculation of forms is very direct and simple on computers. Most differential geometric problems can be tackled using differential forms. This chapter focus on the computation of cohomology, harmonic forms, and holomorphic forms (Fig. 11.1).

11.1 Characteristic Forms

Suppose M is a mesh represented by the half-edge data structure. A chain is represented as a list of consecutive half-edges

$$\gamma = \{h_1, h_2, \cdots, h_n\},$$

where $h_i = [v_{i_1}, v_{i_2}]$. A simplicial 1-form is represented as a linear function defined on the half-edges

$$\omega : E \to \mathbb{R}^1, \quad \omega([v_i, v_j]) = -\omega([v_j, v_i]).$$

The integration of a 1-form ω along a 1-chain γ is defined as the following:

$$\int_\gamma \omega = \langle \omega, \gamma \rangle = \sum_{[v_i, v_j] \in \gamma} \omega([v_i, v_j]).$$

The following algorithm computes the integration,

Algorithm 22: Integration of a 1-form along a 1-chain

input : A 1-form ω, a 1-chain γ
output: The integration of ω along γ, $\int_\gamma \omega$

$$s \leftarrow 0$$

forall *half-edge* $[v_i, v_j] \in \gamma$ **do**
  $s \leftarrow \omega([v_i, v_j]) + s$
end
return s;

Suppose M is a simply connected mesh, ω is a closed 1-form. Then ω is an exact 1-form. Therefore, there exists a 0-form $f : M \to \mathbb{R}$, such that $\omega = df$. The 0-form f can be computed by integrating ω.

First, we select a root vertex v_0, and use width first search to traverse all the vertices. Suppose the path from the room v_0 to the vertex v_i is γ_i, then we define the value of $f(v_i)$ as

$$f(v_i) = \int_{\gamma_i} \omega.$$

If we choose a different root vertex or use a different way to traverse the mesh, then we will get another 0-form f'. Because the mesh is simply connected, ω is exact and $f - f'$ is a constant function. The following algorithm describes the details of finding f from ω.

Algorithm 23: Integrate a closed 1-form on a topological disk

input : A simply connected mesh M, a closed 1-form ω, a root vertex v_0
output: A function $f: M \to \mathbb{R}$, such that $\omega = df$, $f(v_0) = 0$.
Initialize a queue

$$Q \leftarrow v_0,$$
$$f(v_0) \leftarrow 0.$$

while *the queue Q is not empty* **do**
    $v \leftarrow pop\ Q$;
    forall $[v, w] \in M$ **do**
        if *w has not been accessed* **then**
            $f(w) \leftarrow f(v) + \omega([v, w])$;
            push w to the queue
        end
    end
end

11.2 Wedge Product

A closed 1-form has the property of $d\omega = 0$. Suppose $[v_0, v_1, v_2]$ is an arbitrary face. Then

$$d\omega([v_0, v_1, v_2]) = \omega\partial[v_0, v_1, v_2] = \omega[v_0, v_1] + \omega[v_1, v_2] + \omega[v_2, v_0] = 0.$$

If we consider the mesh M is embedded in $\mathbb{R}^3$, then there exists a unique vector $\mathbf{w}$ on the plane spanned by vertices $\mathbf{v}_0, \mathbf{v}_1, \mathbf{v}_2$, such that

$$\begin{aligned} \langle \mathbf{w}, \mathbf{v}_1 - \mathbf{v}_0 \rangle &= \omega([v_0, v_1]), \\ \langle \mathbf{w}, \mathbf{v}_2 - \mathbf{v}_1 \rangle &= \omega([v_1, v_2]), \\ \langle \mathbf{w}, \mathbf{v}_0 - \mathbf{v}_2 \rangle &= \omega([v_2, v_0]). \end{aligned}$$

The vector $\mathbf{w}$ can be explicitly calculated as:

$$\mathbf{w} = -\omega[v_0, v_1]\mathbf{v}_2 \times \mathbf{n} - \omega[v_2, v_0]\mathbf{v}_1 \times \mathbf{n} - \omega[v_1, v_2]\mathbf{v}_0 \times \mathbf{n}.$$

The formulae can be easily proved using the formulae in the proof of Lemma 10.9. Therefore, any closed 1-form ω corresponds to a vector valued 2-form

$$\mathbf{w} : F \to \mathbb{R}^3.$$

The following algorithm converts a closed 1-form to a vector valued 2-form.

Algorithm 24: Convert a closed 1-form to a vector valued 2-form

input : A closed 1-form ω
output: A vector valued 2-form $\mathbf{w}$
forall *face* $[v_i, v_j, v_k]$ **do**

$$\mathbf{w}([v_i, v_j, v_k]) \leftarrow -\omega[v_i, v_j]\mathbf{v}_k \times \mathbf{n} - \omega[v_k, v_i]\mathbf{v}_j \times \mathbf{n} - \omega[v_j, v_k]\mathbf{v}_i \times \mathbf{n}$$

end

The wedge product of two closed 1-forms is a 2-form, which can be constructed explicitly. Let ω_1 and ω_2 be the closed 1-forms. We first convert them to vector valued 2-forms, $\mathbf{w}_1$ and $\mathbf{w}_2$, then for a face $[v_i, v_j, v_k]$,

$$\omega_1 \wedge \omega_2([v_i, v_j, v_k]) = \mathbf{w}_1 \times \mathbf{w}_2 \cdot \mathbf{n}A,$$

where A is the area of the face and $\mathbf{n}$ is the normal of the face. The following algorithm computes the wedge product of two closed 1-forms.

Algorithm 25: Wedge product

input : Two closed 1-forms ω_1, ω_2 on a mesh M
output: A 2-form $\omega_1 \wedge \omega_2$
Convert ω_1 to a vector valued 2-form $\mathbf{w}_1$;
Convert ω_2 to a vector valued 2-form $\mathbf{w}_2$;
forall *face* $[v_i, v_j, v_k]$ **do**
    $\mathbf{n} \leftarrow (\mathbf{v}_j - \mathbf{v}_i) \times (\mathbf{v}_k - \mathbf{v}_i)$;
    $\omega_1 \wedge \omega_2[v_i, v_j, v_k] \leftarrow \mathbf{w}_1([v_i, v_j, v_k]) \times \mathbf{w}_2([v_i, v_j, v_k]) \cdot \mathbf{n}$;
end

11.3 Characteristic 1-Form

Suppose γ is a closed 1-chain. The characteristic 1-form ω of γ is a special closed 1-form, which satisfies the following condition: for any closed 1-form τ, the integration of $\omega \wedge \tau$ on M equals to the integration of τ on γ, namely,

$$\langle \gamma, \tau \rangle = < M, \omega \wedge \tau >, \int_\gamma \tau = \int_M \omega \wedge \tau,$$

where

$$\int_M \omega \wedge \tau = \sum_{[v_i, v_j, v_k] \in M} \omega \wedge \tau([v_i, v_j, v_k]).$$

In the following, we describe the procedure to compute the characteristic 1-form of a closed loop.

Suppose we are given a closed mesh M and a closed loop γ. We slice M along γ to get an open mesh $\bar{M}$. $\bar{M}$ has two boundaries, the orientation of one boundary is consistent with that of γ, denoted as γ^+; the orientation of the other boundary is opposite to that of γ, denoted as γ^-,

$$\partial \bar{M} = \gamma^+ + \gamma^-.$$

We can find a harmonic function $f : \bar{M} \to \mathbb{R}$, such that

$$\forall v \in \gamma^+, \quad f(v) = 1, \quad \forall v \in \gamma^-, \quad f(v) = 0,$$

and for all interior vertices,

$$\forall v \notin \partial \bar{M}, \quad \Delta f(v) = \sum_{[v,w] \in \bar{M}} k_{v,w}(f(v) - f(w)) = 0.$$

Then the gradient of f, df is an exact one form on $\bar{M}$. Any half-edge h on M corresponds uniquely to a half-edge $\bar{h}$ on $\bar{M}$. We can define a closed 1-form on M by

$$\omega(h) = df(\bar{h}), h \in M, \bar{h} \in \bar{M}.$$

Then ω is the characteristic 1-form of γ. The following algorithm describes the details of computing the characteristic 1-form:

Algorithm 26: Compute the Characteristic 1-form of a closed 1-chain

input : A closed 1-chain γ on a closed mesh M
output: The characteristic 1-form ω of γ
Slice M along γ to get an open mesh $\bar{M}$,

$$\partial \bar{M} = \gamma^+ \cup \gamma^-.$$

Record the correspondence between the half-edges of M and those of $\bar{M}$,

$$M \in h \rightarrow \bar{h} \in \bar{M}.$$

Compute a harmonic function $f : \bar{M} \rightarrow \mathbb{R}$, with Dirichlet boundary conditions

$$\begin{cases} \Delta f(v) = 0, & \forall v \notin \partial M \\ f_{\gamma^+} & = 1, \\ f_{\gamma^-} & = 0. \end{cases}$$

Set the closed 1-form on M,

$$\omega(h) \leftarrow df(\bar{h}).$$

Suppose γ_1 and γ_2 are homologous, then their corresponding characteristic 1-forms ω_1 and ω_2 are cohomologous. If both ω_1 and ω_2 are harmonic 1-forms, then according to Hodge theorem, they are equal to each other.

11.4 Computing Cohomology Basis

Suppose M is a closed mesh of genus g. We want to compute a set of closed 1-forms, which form the basis of its cohomology group $H^1(M,\mathbb{R})$. First, we compute a basis of the homology group $H_1(M,\mathbb{R})$; then, for each closed 1-chain in the basis, we compute its characteristic 1-form. All the characteristic 1-forms form a basis of the cohomology group $H^1(M,\mathbb{R})$.

Algorithm 27: Cohomology basis $H^1(M,\mathbb{R})$

input : A closed mesh of genus g
output: A set of basis of $H^1(M,\mathbb{R})$

$$\Omega = \{\omega_1, \omega_2, \cdots, \omega_{2g}\}$$

Compute a set of basis of the homology group $H_1(M,\mathbb{Z})$:

$$\Gamma = \{\gamma_1, \gamma_2, \cdots, \gamma_{2g}\}.$$

forall $\gamma_i \in \Gamma$ **do**
    Compute the characteristic 1-form ω_i of γ_i
end
The set of 1-forms

$$\Omega \leftarrow \{\omega_1, \omega_2, \cdots, \omega_{2g}\}$$

is the cohomology group basis $H^1(M,\mathbb{R})$.

Alternatively, we can use the following method, which is much simpler and more efficient. We compute a cut graph G of the mesh M and compute a spanning tree T of G. Suppose

$$G - T = \{e_1, e_2, \cdots, e_{2g}\}.$$

We slice M along G to obtain a topological disk $\bar{M}$. Each half-edge $\bar{h} \in \bar{M}$ has a unique corresponding half-edge $h \in M$. Suppose

$$\partial \bar{M} = \{v_0, v_1, \cdots, v_k, v_{k+1}, \cdots, v_{n-1}\},$$

where all of the vertices are sorted in the cyclic order. Suppose half-edges h_i^+ and h_i^- attach the edge e_i on M. Then on the open mesh

$$\bar{h}_i^+ = [v_k, v_{k+1}], \bar{h}_i^- = [v_s, v_{s+1}].$$

We construct a function $f_i : \bar{M} \to \mathbb{R}$ in the following way

$$f_i(v_j) = \begin{cases} 0 \ \forall v_j \notin \partial \bar{M} \\ 1 \ \forall v_j \in \partial \bar{M}, k < j \leqslant s \\ 0 \ \forall v_j \in \partial \bar{M}, s < j \leqslant k \end{cases}$$

Then we define ω_1 on M,

$$\omega_i(h) = df_i(\bar{h}).$$

Then

$$\Omega = \{\omega_1, \omega_2, \cdots, \omega_{2g}\}$$

form the cohomology basis of $H^1(M, \mathbb{R})$.

Algorithm 28: Cohomology basis $H^1(M, \mathbb{R})$

input : A closed mesh of genus g
output: A set of basis of $H^1(M, \mathbb{R})$

$$\Omega = \{\omega_1, \omega_2, \cdots, \omega_{2g}\}$$

Compute a cut graph G of M;
Compute a spanning tree T of G;

$$G - T = \{e_1, e_2, \cdots, e_{2g}\}.$$

Slice M along G to get a topological disk $\bar{M}$;
The boundary of $\bar{M}$ is a list of vertices

$$\partial \bar{m} = \{v_0, v_1, \cdots, v_{n-1}\}$$

sorted in a cyclic order;
forall $e_i \in G - T$ **do**
  Find half-edges attaching to e_i, h_i^+ and h_i^-;
  Suppose $\bar{h}_i^+ = [v_k, v_{k+1}]$ and $\bar{h}_i^- = [v_s, v_{s+1}]$;
  Construct a function $f_i : \bar{M} \to \mathbb{R}$, such that

$$f(v_j) = \begin{cases} 1, & v_j \in \partial \bar{M},\ k < j \leqslant s, \\ 0, & otherwise. \end{cases}$$

  Define 1-form on M by

$$\omega_i(h) \leftarrow df_i(\bar{h});$$

end

11.5 Harmonic 1-Form

In each cohomologous class, there are infinite closed forms. In practice, it is highly desirable to choose a unique representative for each cohomologous class. According to Hodge theory, each cohomologous class has a unique harmonic form. In this section, we explain the algorithm to compute harmonic 1-forms.

Suppose ω is a closed 1-form. Then locally ω is the gradient of some function $f: M \to \mathbb{R}$ (0-form). If f is harmonic, then ω is a harmonic 1-form. Namely, a harmonic 1-form satisfies the following equation:

$$\Delta f(v) = \sum_{[v,w]\in M} k_{v,w} df([v,w]) = \sum_{[v,w]\in M} k_{v,w}\omega([v,w]) = 0. \tag{11.1}$$

Given a closed 1-form ω, we can add an exact 1-form df, such that $\omega + df$ is the unique harmonic 1-form cohomologous to ω. Then $f: M \to \mathbb{R}$ should satisfy the following condition:

$$\sum_{[v,w]\in M} k_{v,w}(\omega([v,w]) + f(w) - f(v)) = 0. \tag{11.2}$$

This equation is a discrete Poisson equation, which can be easily solved using linear algebraic methods.

The group of all harmonic 1-forms is isomorphic to the cohomology group $H^1(M,\mathbb{R})$. We first compute a set of basis of the cohomology group; then for each base 1-form ω_i, we find a unique exact 1-form df_i, such that $\omega_i + df_i$ is harmonic. The following algorithm computes the basis of the group formed by all harmonic 1-forms (Fig. 11.2). The algorithm can be generalized to compute harmonic n-forms, where n is greater than one.

Algorithm 29: Harmonic 1-form basis

input : A closed genus g mesh
output: Compute a set of basis of harmonic 1-form group
Compute a set of basis of cohomology group

$$\Omega = \{\omega_1, \omega_2, \cdots, \omega_{2g}\}.$$

forall $\omega_i \in \Omega$ **do**
  Find $f_i : M \to \mathbb{R}$ by solving equation

$$\sum_{[v,w]\in M} k_{v,w}(\omega_i([v,w]) + f_i(w) - f_i(v)) = 0, \quad \forall v \in M,$$

  $\omega_i \leftarrow \omega_i + df_i.$
end
Output Ω.

Harmonic 1-forms can be easily visualized by texture mapping. We compute a fundamental domain $\bar{M}$ of the mesh M and integrate ω on the fundamental domain to obtain a harmonic function f, such that $\omega = df$ on $\bar{M}$. We use the function value on each vertex

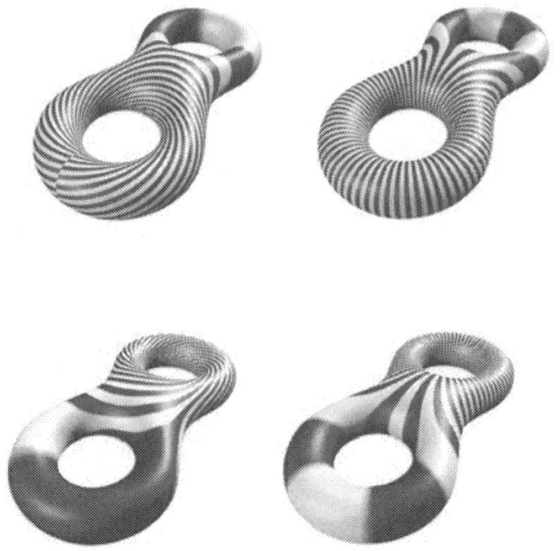

Fig. 11.2. The basis of the group of harmonic 1-forms.

as one dimensional texture coordinates and visualize the 1-form by texture mapping a regular stripe. In the following discussion, we use half-edges to represent corners.

Algorithm 30: Visualization of a closed 1-form ω

input : A closed mesh of genus g, a closed 1-form ω
output: Texture coordinates u of corners $u(h)$, where h is a half-edge
Compute the fundamental domain $\bar{M}$ of M using algorithm 5;
Integrate ω on $\bar{M}$ to get texture coordinates $u(v)$ for each vertex, using algorithm 24;
forall *Corner h in M* **do**
  Locate the unique corresponding half-edge $\bar{h}$ in $\bar{M}$;
  Get the target vertex v of $\bar{h}$;

$$u(h) \leftarrow u(v).$$

end

11.6 Hodge Star Operator

Suppose S is a smooth surface, with an atlas $\{(U_\alpha, \phi_\alpha)\}$. Then a differential 1-form has local representation

$$\omega = f_\alpha dx_\alpha + g_\alpha dy_\alpha,$$

where (x_α, y_α) are local coordinates, f_α and g_α are smooth functions. The Hodge star operator is defined as

$$^*\omega = f_\alpha dy_\alpha - g_\alpha dx_\alpha.$$

We say $^*\omega$ is conjugate to ω (Fig. 11.3).

Fig. 11.3. Harmonic 1-forms conjugate to those in figure 11.2.

A 1-form is a linear functional on the tangential vector fields. Suppose the surface has a Riemannian metric $\mathbf{g}$. Then we can associate each 1-form ω with a vector field $\mathbf{w}$, such that for any vector field $\mathbf{v}$,

$$\omega(\mathbf{v}) = \langle \mathbf{w}, \mathbf{v} \rangle_{\mathbf{g}},$$

where $\langle , \rangle_{\mathbf{g}}$ represents the inner product induced by the Riemannian metric $\mathbf{g}$. If ω is closed, then $\mathbf{w}$ is curl-free. If ω is harmonic, then $\mathbf{w}$ is both curl-free and divergence-free. Assume $^*\mathbf{w}$ is the vector field corresponding to $^*\omega$. Then at each point,

$$^*\mathbf{w} = \mathbf{n} \times \mathbf{w}.$$

Namely, we rotate $\mathbf{w}$ at each point by 90 degree about the normal, then we get the vector field corresponding to the conjugate 1-form. In the above discussion, we use a Riemannian metric. However in fact, the Hodge star operator only depends on the conformal structure of S.

Suppose ω is a closed 1-form on a mesh M. We can convert it to the vector valued 2-form $\mathbf{w}$. Suppose two faces $[v_i, v_j, v_k]$ and $[v_j, v_i, v_l]$ are adjacent to the edge $\{v_i, v_j\}$, then the following consistency holds:

$$\omega[v_i, v_j] = \langle \mathbf{w}([v_i, v_j, v_k]), v_j - v_i \rangle = \langle \mathbf{w}([v_j, v_i, v_l]), v_j - v_i \rangle \tag{11.3}$$

Then we rotate the vector on each face by 90 degree about the normal to get $^*\mathbf{w}$. But in this case, we cannot convert $^*\mathbf{w}$ to a closed 1-form directly, because the consistency relation 11.3 doesn't hold any more.

$$\langle {}^*\mathbf{w}([v_i, v_j, v_k]), v_j - v_i \rangle \neq \langle {}^*\mathbf{w}([v_j, v_i, v_l]), v_j - v_i \rangle \tag{11.4}$$

This is the mainly caused by the discrete approximation. In order to mitigate the problem, we can $^*\omega[v_i,v_j]$ by taking the average:

$$^*\omega([v_i,v_j]) = \frac{1}{2}(\langle {}^*\mathbf{w}([v_i,v_j,v_k]),\mathbf{v}_j-\mathbf{v}_i\rangle + \langle {}^*\mathbf{w}([v_j,v_i,v_l]),\mathbf{v}_j-\mathbf{v}_i\rangle).$$

Algorithm 31: Approximation of Hodge star

input : A closed 1-form ω
output: An approximation of Hodge star of ω, $^*\omega$
Convert ω to the vector valued 2-form $\mathbf{w}$;
forall *a face* $f \in M$ **do**

$$^*\mathbf{w}(f) \leftarrow \mathbf{n}\times\mathbf{w}$$

end
forall *half-edge* $[v_i,v_j] \in M$ **do**
Locate the faces f_0, f_1 attaching to the edge;

$$^*\omega([v_1,v_2]) = \frac{1}{2}(\langle {}^*\mathbf{w}(f_0),\mathbf{v}_2-\mathbf{v}_1\rangle + \langle {}^*\mathbf{w}(f_1),\mathbf{v}_2-\mathbf{v}_1\rangle).$$

end

A more sophisticated and accurate algorithm based on the wedge product for computing the Hodge star operator on harmonic 1-forms will be introduced in the next section.

11.7 Holomorphic 1-Form

Conjugate harmonic 1-form

Suppose S is a Riemann surface. Then all of the holomorphic 1-forms on S form a group (Fig. 11.4). Each holomorphic 1-form ζ can be decomposed to a real part and an imaginary part, both parts are real harmonic 1-forms,

$$\zeta = \omega_1 + \omega_2\sqrt{-1}.$$

Furthermore the imaginary part harmonic 1-form is conjugate to the real part harmonic 1-form, $\omega_2 = {}^*\omega_1$, namely

$$\zeta = \omega + {}^*\omega\sqrt{-1}.$$

Therefore, we can approximate a holomorphic 1-form by computing harmonic 1-form ω and its conjugate $^*\omega$. Because of the approximation error of the Hodge star operator, we adapt the following more accurate method.

Suppose M is a closed mesh with genus g. We first compute a set of basis of the harmonic 1-form group,

$$\Omega = \{\omega_1,\omega_2,\cdots,\omega_{2g}\}.$$

Suppose ω is a harmonic 1-form. Then ω can be represented as the linear combination of ω_k's:

$$\omega = \sum_{k=1}^{2g} \lambda_k \omega_k.$$

the 1-form conjugate to ω, ${}^*\omega$ is also harmonic,

$${}^*\omega = \sum_{k=1}^{2g} \mu_k \omega_k.$$

Then the linear coefficients $\{\mu_k\}$ can be computed using the following method:

$$\int_M \omega_i \wedge {}^*\omega = \sum_k \mu_k \int_M \omega_i \wedge \omega_k, \quad i = 1, 2, \cdots, 2g.$$

This can form a linear system

$$\begin{pmatrix} \int_M \omega_1 \wedge \omega_1 & \int_M \omega_1 \wedge \omega_2 & \cdots & \int_M \omega_1 \wedge \omega_{2g} \\ \int_M \omega_2 \wedge \omega_1 & \int_M \omega_2 \wedge \omega_2 & \cdots & \int_M \omega_2 \wedge \omega_{2g} \\ \vdots & \vdots & & \vdots \\ \int_M \omega_{2g} \wedge \omega_1 & \int_M \omega_{2g} \wedge \omega_2 & \cdots & \int_M \omega_{2g} \wedge \omega_{2g} \end{pmatrix} \begin{pmatrix} \mu_1 \\ \mu_2 \\ \vdots \\ \mu_{2g} \end{pmatrix} = \begin{pmatrix} \int_M \omega_1 \wedge {}^*\omega \\ \int_M \omega_2 \wedge {}^*\omega \\ \vdots \\ \int_M \omega_{2g} \wedge {}^*\omega \end{pmatrix}.$$

It is easy to see that $\omega_k \wedge \omega_k = 0$, $\omega_i \wedge \omega_j = -\omega_j \wedge \omega_i$. The matrix on the left-hand side can be easily computed. The right-hand side needs approximation. For example, we want to compute $\omega_k \wedge {}^*\omega$. First, we convert ω_k and ω to the vector valued 2-forms, $\mathbf{w}_k$ and $\mathbf{w}$; second, we approximate ${}^*\mathbf{w}$ by $\mathbf{n} \times \mathbf{w}$; third, $\omega_k \wedge {}^*\omega$ is approximated by

$$\omega_k \wedge {}^*\omega(f) = \mathbf{w}(f) \times {}^*\mathbf{w}(f) \cdot \mathbf{n}_f A_f,$$

where f is a face in M, $\mathbf{n}_f$ is the normal to the face, A_f is the area of the face. We obtain

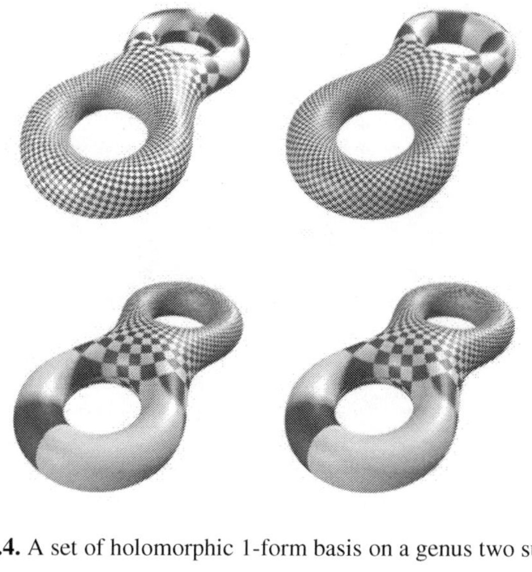

Fig. 11.4. A set of holomorphic 1-form basis on a genus two surface.

$$\int_M \omega_k \wedge {}^*\omega = \sum_{f \in M} \omega_k \wedge {}^*\omega(f) = \mathbf{w}(f) \times {}^*\mathbf{w}(f) \cdot \mathbf{n}_f A_f.$$

Therefore, the linear equation can be solved directly to obtain μ_k's. The following algorithm summarizes the procedure to compute the conjugate harmonic 1-form.

Algorithm 32: Conjugate harmonic 1-form

input : A harmonic 1-form ω on a closed mesh M
output: The conjugate harmonic 1-form ${}^*\omega$

Compute a basis of harmonic 1-forms on M,

$$\Omega = \{\omega_1, \omega_2, \cdots, \omega_{2g}\}.$$

forall $\omega_k \in \Omega$ **do**
  Convert ω_k to a vector valued 2-form $\mathbf{w}_k$
end
Convert ω to a vector valued 2-form $\mathbf{w}$;
forall *Face* $f \in M$ **do**

$${}^*\mathbf{w}(f) \leftarrow \mathbf{n}_f \times \mathbf{w}(f)$$

end
forall $\omega_k \in \Omega$ **do**
  $s_k \leftarrow 0$;
  forall *face* $f \in M$ **do**

$$s_k \leftarrow s_k + \mathbf{w}_k(f) \times {}^*\mathbf{w}(f) \cdot \mathbf{n}_f A_f.$$

  end
end
forall $\omega_i, \omega_j \in \Omega, i < j$ **do**
  Compute $\omega_i \wedge \omega_j$ using algorithm 26;
  $a_{ij} \leftarrow \int_M \omega_i \wedge \omega_j$;
  $a_{ji} \leftarrow -a_{ij}$
end
Solve the linear equation

$$(a_{ij})(\mu_j) = (s_i),$$

Set

$${}^*\omega \leftarrow \sum_{k=1}^{2g} \mu_k \omega_k.$$

Return ${}^*\omega$.

Basis for holomorphic 1-forms

The computation of the basis of the holomorphic 1-form group is straightforward. After obtaining the basis of harmonic 1-forms, we compute their conjugate harmonic 1-forms, and pair each harmonic 1-form with its conjugate to form a holomorphic 1-form.

Algorithm 33: Compute a basis of the real vector space of all holomorphic 1-forms

input : A closed mesh M of genus g
output: Holomorphic 1-form basis $\{\zeta_1, \zeta_2, \cdots, \zeta_{2g}\}$
Compute a harmonic 1-form basis of M using algorithm 30

$$\Omega = \{\omega_1, \omega_2, \cdots, \omega_{2g}\},$$

forall $\omega_k \in \Omega$ **do**
  Compute the conjugate of ω_k, ${}^*\omega_k$ using the algorithm 34;
  $\zeta_k \leftarrow \omega_k + \sqrt{-1}^*\omega_k$.
end
$\{\zeta_1, \zeta_2, \cdots, \zeta_{2g}\}$ form a basis for the real vector space of all holomorphic 1-forms.

The vector space of all holomorphic 1-forms is of $2g$ real dimensions, but g complex dimensions. The following algorithm computes the g complex base holomorphic 1-forms. The key idea is to use the canonical homology basis.

Algorithm 34: Compute a basis of the complex vector space of all holomorphic 1-forms

input : A closed mesh M of genus g
output: Holomorphic 1-form basis $\{\zeta_1, \zeta_2, \cdots, \zeta_g\}$

Compute a canonical homology group basis,

$$\Gamma = \{a_1, b_1, a_2, b_2, \cdots, a_g, b_g\},$$

using algorithm 6;
forall $a_k \in \Gamma, b_k \in \Gamma$ **do**
  Compute the characteristic 1-form ω_k for a_k, η_k for b_k, using algorithm 27;
  Convert ω_k to harmonic 1-form using algorithm 30;
  Convert η_k to harmonic 1-form using algorithm 30;
end
forall ω_k **do**
  Compute the conjugate of ω_k as ${}^*\omega_k$, using algorithm 34;

$$\zeta_k \leftarrow \omega_k + \sqrt{-1}^*\omega_k.$$

end
Then

$$\{\zeta_1, \zeta_2, \cdots, \zeta_g\}$$

form a basis for the group of holomorphic 1-forms.

Visualization of holomorphic 1-forms

Holomorphic 1-forms can be easily visualized by texture mapping. Basically, we compute the fundamental domain of the mesh, then the holomorphic 1-form becomes an exact form

df. The corresponding complex valued function $f : M \to \mathbb{C}$ assigns the texture coordinates for each vertex. f can be computed by integrating the holomorphic 1-form on the fundamental domain. The following algorithm describes the details.

Algorithm 35: Visualize holomorphic 1-form

```
input  : A holomorphic 1-form
output: A complex valued function f : M → ℂ
Compute a fundamental domain M̄ of M, using the algorithm 5;
Choose a root vertex v0 ∈ M̄;
Initialize a vertex queue Q ← v0;
while Q is not empty do
    v ← pop Q;
    forall [v,w] ∈ M do
        if w has not been accessed then
            f(w) ← f(v) + ω([v,w]);
            Insert w to Q.
        end
    end
end
```

Fig. 11.4 shows a set of holomorphic 1-form basis on a genus two surface.

11.8 Inner Product Among 1-Forms

We can introduce the inner product in the linear space of all closed 1-forms, therefore, we can construct orthonormal basis of 1-forms.

Suppose ω_1, ω_2 are two closed 1-forms. Then the inner product of them is defined as

$$\langle \omega_1, \omega_2 \rangle = \int_M \omega_1 \wedge {}^*\omega_2.$$

We first convert ω_k to a vector valued 2-form $\mathbf{w}_k$. Then

$$\langle \omega_1, \omega_2 \rangle = \sum_{f \in M} \langle \mathbf{w}_1(f), \mathbf{w}_2(f) \rangle A(f),$$

where f is a face on the mesh and $A(f)$ is the area of the face.

Suppose M is of genus g, $\{\omega_1, \omega_2, \cdots, \omega_{2g}\}$ is an orthonormal basis,

$$\langle \omega_k, \omega_k \rangle = 1, \quad \langle \omega_i, \omega_j \rangle = 0, \quad i \neq j.$$

Suppose ω is a closed 1-form. Then ω has a unique representation

$$\omega = \sum_{k=1}^{2g} \lambda_k \omega_k,$$

the linear coefficients λ_k's can be computed directly as

$$\lambda_k = \langle \omega, \omega_k \rangle.$$

The following algorithm describes how to convert a basis of closed 1-forms to an orthonormal basis.

Algorithm 36: Schmidt orthonormalization

input : A genus g closed mesh M, a cohomology basis

$$\Omega = \{\omega_1, \omega_2, \cdots, \omega_{2g}\}$$

output: A set of orthonormal basis
Normalize ω_1,

$$\omega_1 \leftarrow \frac{\omega_1}{\sqrt{\langle \omega_1, \omega_1 \rangle}}$$

forall $\omega_k \in \Omega,\ k > 1$ **do**
  $\omega_k \leftarrow \omega_k - \sum_{i<k} \langle \omega_i, \omega_k \rangle \omega_i;$
  $\omega_k \leftarrow \frac{\omega_k}{\sqrt{\langle \omega_k, \omega_k \rangle}}$
end

11.9 Holomorphic Forms on Surfaces with Boundaries

Double Covered Surface

Suppose M is a compact surface of genus g with b boundaries, in order to compute the holomorphic 1-forms on it, we can use the double covering technique introduced in algorithm 7.

First, we construct two copies of M, with opposite orientations, denoted as M^+ and M^-. For any vertex $v^+ \in M^+$, it has a unique vertex $v^- \in M^-$. We glue M^- and M^+ along the corresponding boundaries, namely,

$$\bar{M} = M^+ \cup M^- / \sim,$$

where the equivalence relation is defined as

$$v^- \sim v^+, \quad v^- \in \partial M^-, \quad v^+ \in \partial M^+.$$

Then we define a reflection automorphism of $\bar{M}$,

$$\phi : \bar{M} \to \bar{M}, \quad \phi(v^+) \to \phi(v^-), \quad \forall v^+ \notin \partial M^+,$$

and

$$\phi(v) \to v, \quad \forall v \in \partial M^+.$$

Then $\phi^2 = id$.

Symmetric and Anti-Symmetric Hamronic 1-forms

Suppose ω is a 1-form on M, such that

$$\omega([v_i,v_j]) = 0, \quad \forall [v_i,v_j] \in \partial M.$$

Then we can define a 1-form $\bar{\omega}$ on $\bar{M}$ in the following way:

$$\begin{aligned} \bar{\omega}([v_i^+,v_j^+]) &\leftarrow \omega([v_i,v_j]),\\ \bar{\omega}([v_i^-,v_j^-]) &\leftarrow \omega([v_i,v_j]). \end{aligned}$$

Then $\bar{\omega}$ has the following symmetry:

$$\phi^*\bar{\omega} = \bar{\omega},$$

where $\phi^*\bar{\omega}$ is the pull back of $\bar{\omega}$, namely,

$$\phi^*\bar{\omega}([v_i,v_j]) = \bar{\omega}([\phi(v_i),\phi(v_j)]).$$

We can also define an anti-symmetric 1-form $\bar{\omega}$ on $\bar{M}$:

$$\begin{aligned} \bar{\omega}([v_i^+,v_j^+]) &\leftarrow \omega([v_i,v_j]),\\ \bar{\omega}([v_i^-,v_j^-]) &\leftarrow -\omega([v_i,v_j]). \end{aligned}$$

Then $\bar{\omega}$ is anti-symmetric:

$$\phi^*\bar{\omega} = -\bar{\omega}.$$

Suppose $\bar{\omega}$ is a 1-form on $\bar{M}$. Then it can be decomposed to two parts,

$$\bar{\omega} = \frac{1}{2}(\bar{\omega} + \phi^*\bar{\omega}) + \frac{1}{2}(\bar{\omega} - \phi^*\bar{\omega}),$$

where $\frac{1}{2}(\bar{\omega} + \phi^*\bar{\omega})$ is symmetric, because

$$\phi^*[\frac{1}{2}(\bar{\omega} + \phi^*\bar{\omega})] = \frac{1}{2}[\phi^*\bar{\omega} + (\phi^2)^*\bar{\omega}] = \frac{1}{2}(\phi^*\bar{\omega} + \bar{\omega}).$$

Similarly $\frac{1}{2}(\bar{\omega} + \phi^*\bar{\omega})$ is anti-symmetric

Suppose ω is a harmonic 1-form on M, such that $\omega([v_i,v_j]) = 0$ for all $[v_i,v_j] \in \partial M$. Then ω corresponds to a unique symmetric harmonic 1-form and a unique anti-symmetric harmonic 1-form on $\bar{M}$. By computing symmetric and anti-symmetric harmonic 1-forms on $\bar{M}$, we can find all such harmonic 1-forms on M.

Therefore, we can compute the basis of all harmonic 1-forms on $\bar{M}$ and get the symmetric components and anti-symmetric components, which correspond to harmonic 1-forms on M. We can select a linearly independent subset of those harmonic 1-forms on M, which form the harmonic 1-form basis for M. The following algorithm explains the computational details.

Algorithm 37: Compute harmonic 1-form basis for compact surface with boundaries

input : A genus g mesh with b boundaries M
output: Harmonic 1-form basis of M
Compute the double covering of M, denoted as $\bar{M}$ using the algorithm 7. Then $\bar{M}$ is of genus $2g+b-1$;
Compute the harmonic 1-form basis of $\bar{M}$;

$$\bar{\Omega} = \{\bar{\omega}_1, \bar{\omega}_2, \cdots, \bar{\omega}_{4g+2b-2}\}$$

forall $\bar{\omega}_k \in \bar{\Omega}$ **do**
    Compute $\phi^* \bar{\omega}$;

$$\begin{aligned} \bar{\tau}_{2k} &\leftarrow \bar{\omega}_k + \phi * \bar{\omega}_k \\ \bar{\tau}_{2k+1} &\leftarrow \bar{\omega}_k - \phi * \bar{\omega}_k, \end{aligned}$$

end
forall $\bar{\tau}_k$ **do**
    forall $[v_i, v_j] \in M$ **do**

$$\tau_k([v_i, v_j]) \leftarrow \bar{\tau}_k([v_i^+, v_j^+]).$$

    end
end
Schmidt orthogonalize $\{\tau_1, \tau_2, \cdots, \tau_{4g+2b-2}\}$, choose the maximal linearly independent set.

Fig. 11.5 to Fig. 11.8 illustrates the holomorphic 1-form on a compact surface with boundaries. We slice the surface open at the bottom of the feet, the tips of horns, the mouth and the tip of the tail. Then we compute a holomorphic 1-form and visualize it using texture mapping checkerboard.

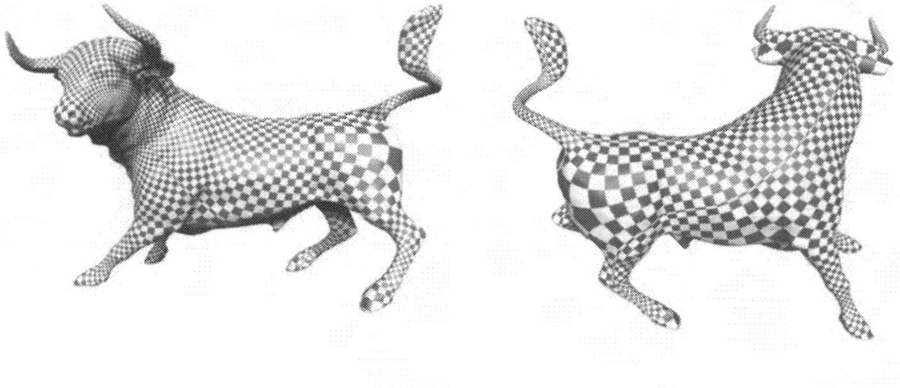

Fig. 11.5. Holomorphic 1-form on a surface with boundaries.

Fig. 11.6. Holomorphic 1-form on a genus one surface with boundaries.

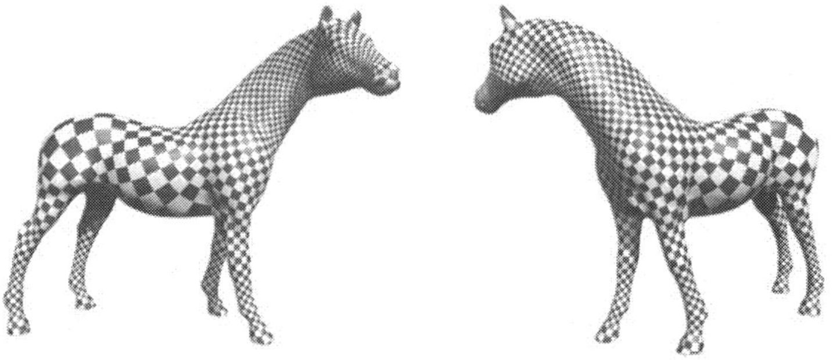

Fig. 11.7. Holomorphic 1-form on a genus zero surface with five boundaries, the cuts are on the mouth and the toes.

11.10 Zero Points and Critical Trajectories

All holomorphic 1-forms form a linear space. Any holomorphic 1-form can be represented as the linear combination of the basis:

$$\zeta = \sum_{k=1}^{g} \lambda_k \zeta_k, \quad \lambda \in \mathbb{C}.$$

According to Hopf theorem, every holomorphic 1-form has zero points,the total zero point index equal to $2-2g$. In general, one can select special local coordinates compatible with the conformal structure, such that the holomorphic 1-form at the zero point has local representation

$$\omega = zdz.$$

Therefore the zero point has index -1. In general, there are $2g-2$ zero points.

The positions of zero points are determined by the conformal structure of the surface and the cohomologous class of the holomorphic 1-form (Fig. 11.9).

Suppose $\mathbf{f} : M \to \mathbb{C}$ are the texture coordinates induced by a holomorphic 1-form ζ.

Fig. 11.8. Holomorphic one-forms on teapot surfaces.

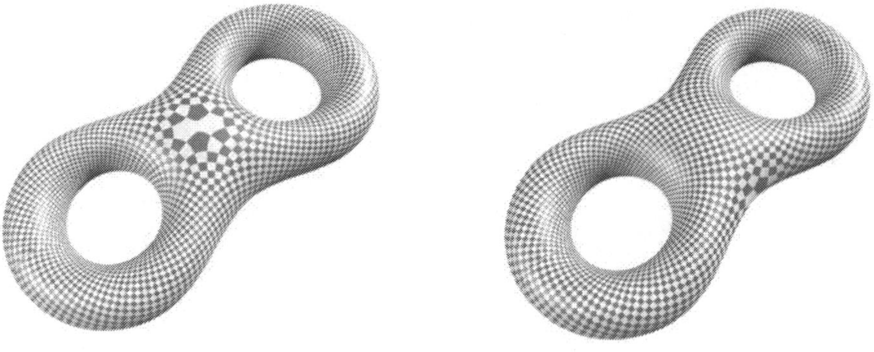

Fig. 11.9. Different holomorphic one-forms with zero points.

Definition 11.1 (Trajectory). *Suppose γ is a curve on the surface. If $\mathbf{f}(\gamma)$ is a straight line parallel to the real axis, then γ is called a* horizontal trajectory *of ζ. If $\mathbf{f}(\gamma)$ is a straight line parallel to the imaginary axis, then γ is called a* vertical trajectory. *The trajectories through zero points are called the* critical trajectories.

In general, the length of a trajectory on a surface could be infinite. If all the critical horizontal trajectories are of finite length, then all horizontal trajectories are finite. The surfaces can be segmented by critical trajectories. Each segment can be mapped to a rectangle on the texture domain.

In Fig. 11.10, the zero point is in the center of the third frame. The horizontal and vertical trajectories are illustrated explicitly. In the neighborhood of the zero point, the horizontal trajectories can be represented as the level sets

$$xy = const,$$

the vertical trajectories can be represented as

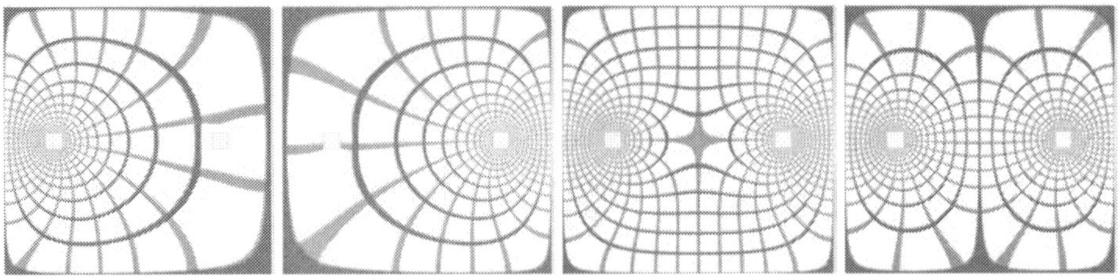

(a) ω_1 (b) ω_2 (c) $\omega_1 + \omega_2$ (d) $\omega_1 - \omega_2$

Fig. 11.10. Holomorphic 1-forms on a two-holed disk. ω_1 and ω_2 are two base forms, their different linear combinations induce different holomorphic 1-forms.

$$x^2 - y^2 = const,$$

both of them are hyperbolic curves. Fig. 11.11 demonstrates the segmentation induced by the critical horizontal trajectories. Each segment is mapped onto a rectangle in the texture domain.

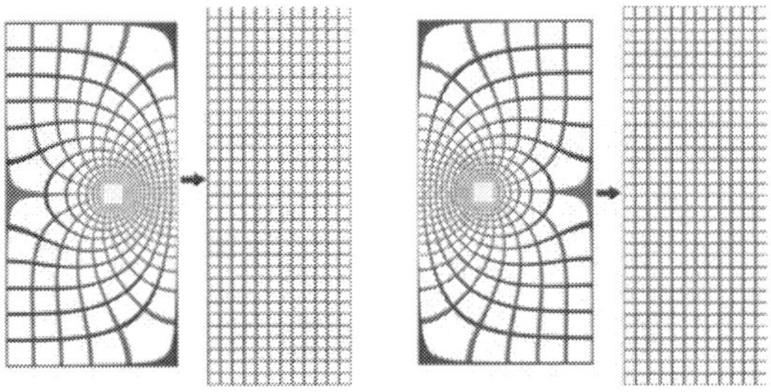

Fig. 11.11. Segmentation induced by the critical horizontal trajectories.

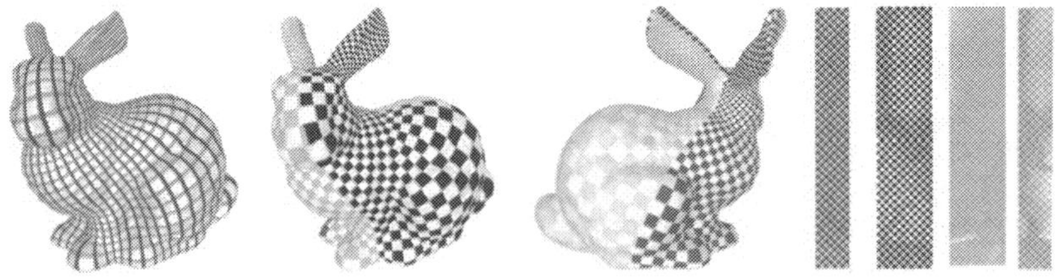

Fig. 11.12. Segmentation by critical horizontal trajectories of the bunny surface.

Fig. 11.12—11.14 show several examples of surface segmentation induced by the horizontal critical trajectories of holomorphic 1-forms.

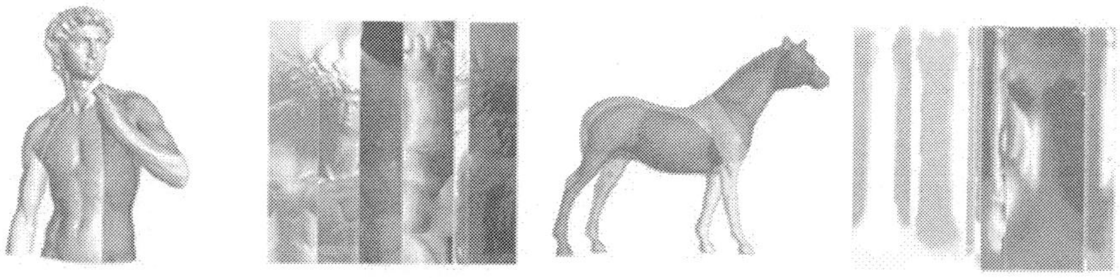

Fig. 11.13. Surface segmentation induced by the critical trajectories of a holomorphic 1-form.

Fig. 11.14. Surface segmentation induced by the critical trajectories of a holomorphic 1-form.

11.11 Flat Metric Induced by Holomorphic 1-Forms

Suppose ω is a holomorphic 1-form on a Riemann surface. Then we can define a Riemannian metric

$$ds^2_\omega = \omega\bar{\omega},$$

where $\bar{\omega}$ is the complex conjugate of ω. In the discrete case, this means we can define the length of a half-edge $[v_i, v_j]$ as $\sqrt{\omega([v_i, v_j])\bar{\omega}([v_i, v_j])}$.

It is easy to verify that the Gaussian curvature induced by ds^2_ω is zero everywhere, except at the zero points of ω. If the index of zero points are of -1, then the cone angles are -2π. Namely, the metric induced by a holomorphic 1-form is flat with cone singularities.

Let S be a Riemann surface with nontrivial topology, ω be a holomorphic 1-form, ds^2_ω be the induced flat metric. Suppose $(\bar{S}, \pi)$ is the universal covering space of S, π is the projection. Then the pull back metric $\pi^* ds^2_\omega$ is a flat metric with cone singularities on $\bar{S}$. Suppose $\phi : \bar{S} \to \bar{S}$ is a deck transformation, $\pi \circ \phi = \pi$. Then ϕ is an isometric transformation with respect to $\pi^* ds^2_\omega$, therefore ϕ is a rigid motion on the plane $\mathbb{R}^2$. We can see that ϕ must be a translation.

Theorem 11.2. *A deck transformation of $(\bar{S}, \pi^* ds^2_\omega)$ is a translation. The holonomy group of $(\bar{S}, \pi^* ds^2_\omega)$ is rotation free.*

Proof. Let ϕ be a deck transformation. $p, q \in S$, $\bar{p}_k, \bar{q}_k \in \bar{S}$ are pre-images of p and q respectively,

$$\pi(\bar{p}_k) = p, \quad \pi(\bar{q}_k) = q.$$

Suppose

$$\phi(\bar{p}_0) = \phi(\bar{p}_1), \quad \phi(\bar{q}_0) = \phi(\bar{q}_1),$$

$\bar{\gamma} : [0,1] \to \bar{S}$ is a path, such that

$$\bar{\gamma}(0) = \bar{p}_0, \quad \bar{\gamma}(1) = \bar{p}_1.$$

Then the projection of $\bar{\gamma}$ is a closed curve γ on S. Then

$$\bar{p}_1 - \bar{p}_0 = \int_\gamma \omega.$$

Similarly, we can show $\bar{q}_1 - \bar{q}_0 = \int_{\gamma'} \omega$, where γ and γ' are homotopic to each other. Because ω is closed, therefore $\int_\gamma \omega = \int_{\gamma'} \omega$. So

$$\bar{q}_1 - \bar{q}_0 = \bar{p}_1 - \bar{p}_0.$$

Because q is arbitrarily chosen, therefore, ϕ is a translation. □

In practice, it is highly desirable to choose the positions of the zero points, because the zero points are singularities, where the mapping from the surface to the texture domain is not one-to-one and cause technical problems. Suppose $\{\omega_1, \omega_2, \cdots, \omega_g\}$ is a set of holomorphic 1-form basis on the surface S, $\{p_1, p_2, \cdots, p_g\}$ is a set of points on the surface arbitrarily chosen, which are the desired positions of zero points. Then we want to solve the following equations:

$$\sum_{k=1}^{g} \lambda_k \omega_k(p_i) = 0, \quad i = 1, 2, \cdots, g.$$

In the discrete case, $\omega(v)$ can be approximated using the following method:

$$\omega(v) = \sum_{[v,w] \in M} \frac{\omega([v,w])}{|\mathbf{w} - \mathbf{v}|}.$$

Fig. 11.15 shows the results of locating zero points on a two holed torus model.

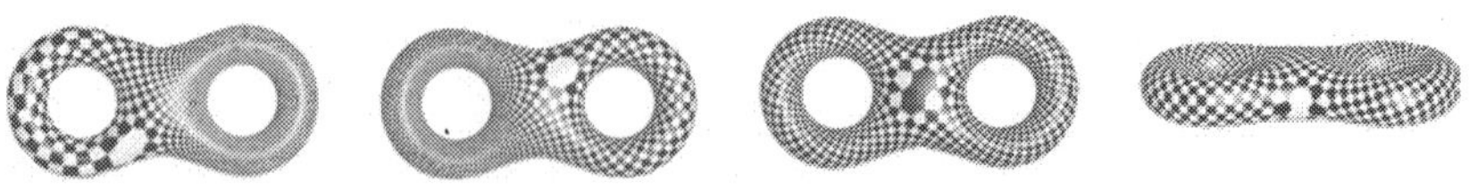

Fig. 11.15. Holomorphic 1-forms on the two hole torus model. The zero points are located at different positions.

In engineering applications, we can require the induced metric ds^2_ω to have various properties. For example, the area distortion induced by the metric should be as uniform as possible, or the new metric allocates more area for a special region of interests on the surface (Fig. 11.16). Therefore, we can formulate different properties as different energy

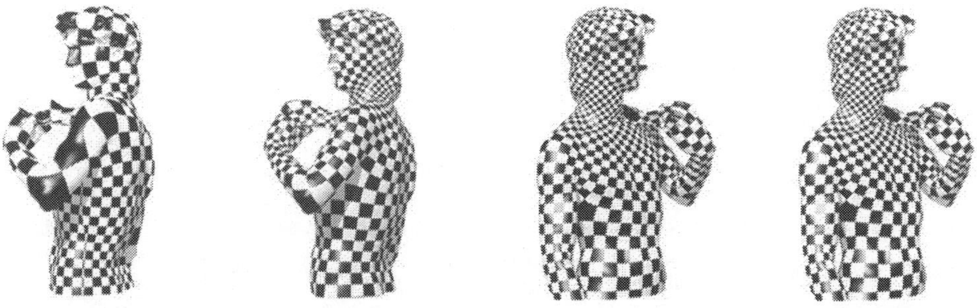

(a) (b) (c) (d)

Fig. 11.16. Zero point allocation. (a) Zero point is originally at left shoulder. (b) Put zero point at left armpit. (c) Zero point is originally at right shoulder. (d) Put zero point at right armpit, most uniform result.

forms of ω; by optimizing the energies, we can compute the desired linear coefficients λ_k's. In the following Fig. 11.17, we show the optimization results for the Stanford bunny surface. The original ω induces a very non-uniform flat metric, as shown in the first column. The sizes of the checkers are quite different. After optimization, the uniformity of the flat metric has been improved, and the sizes of the checkers are similar.

(a) Least uniform (b) Most uniform

Fig. 11.17. Uniform global conformal parameterization. Least uniform conformal parameterization is shown in (a). Most uniform conformal parameterization is shown in (b).

The next Fig. 11.18 shows different regions on surfaces emphasized. The first image emphasizes the left ear, the second one emphasizes the right ear.

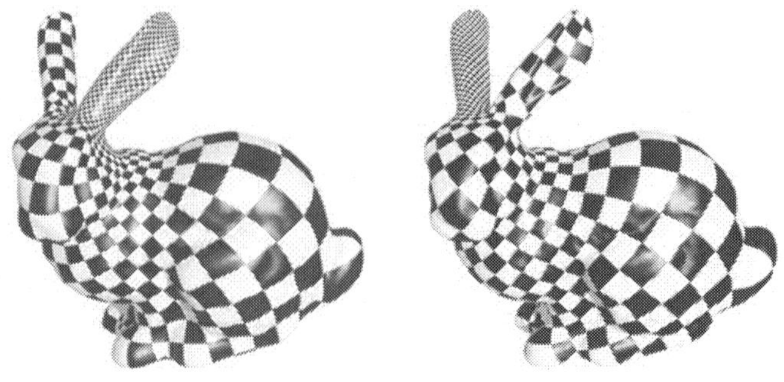

(a) Emphasize left ear (b) Emphasize right ear

Fig. 11.18. Emphasized conformal parameterization. (a) Maximize the parameter area of the left ear. (b) Maximize the parameter area of the right ear.

11.12 Conformal Invariants

All surfaces can be classified by the conformal equivalence relation. If there exists a conformal map from S_1 to S_2, and the inverse map is also conformal, then they are conformally equivalent. In this section, we discuss the algorithms to compute conformal invariants. Two surfaces sharing the same conformal invariants are conformally equivalent.

Topological Triangles

Suppose S_1, S_2 are topological disks, and $p_0, p_1, p_2 \in \partial S_1$, $q_0, q_1, q_2 \in S_2$ are points on the boundaries. Then we construct a conformal map $\phi_1 : S_1 \rightarrow D$ from S_1 to the unit disk D, such that $\phi_1(p_0) = 0, \phi_1(p_1) = 1, \phi_1(p_2) = i$. Similarly, we can construct a conformal map $\phi_2 : S_2 \rightarrow D$ from S_2 to D, such that $\phi_2(q_0) = 0, \phi_2(q_1) = 1, \phi_2(q_2) = i$. Then we can construct a conformal map

$$\phi = \phi_2^{-1} \circ \phi_1 : S_1 \rightarrow S_2,$$

such that

$$\phi(p_0) = q_0, \quad \phi(p_1) = q_1, \quad \phi(p_2) = q_2.$$

Therefore all topological triangles are conformally equivalent.

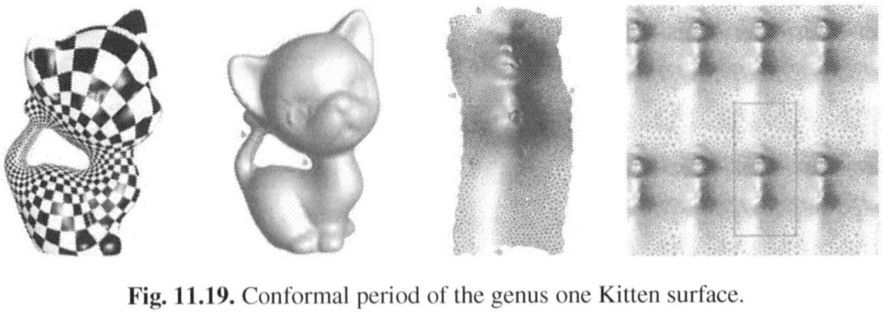

Fig. 11.19. Conformal period of the genus one Kitten surface.

Topological Quadrilateral

We consider topological quadrilateral S with four boundary corners p_0, p_1, p_2, p_3. Then it can be conformally mapped to a planar rectangle R, such that all corners of S are mapped to the corners of R. The ratio between the height and the width of R is called the *conformal module* of S. If two topological quadrilaterals share the same module, then there exists a conformal map, which maps the corresponding corners to each other. The algorithm to compute the module of a topological quadrilateral is as follows.

Algorithm 38: Conformal the conformal module of a topological quadrilateral

input : A topological quadrilateral M, with four corner vertices $\{v_0, v_1, v_2, v_3\}$ sorted counter-clock-wisely
output: The conformal module of M
Extract the boundary vertices as a cyclic list

$$\partial M = \{v_0, \cdots, v_1, \cdots, v_2, \cdots, v_3, \cdots\}$$

Make two copies of M, M^+, and M^-;
Reverse the orientation of M^-;
Glue M^+ and M^- along the boundary segments

$$\{v_1, \cdots, v_2\}, \quad \{v_3, \cdots, v_0\}$$

to form $\bar{M}$, which is a topological annulus;
Compute the holomorphic 1-form basis of $\bar{M}$ using algorithm 34, denote it as ω;
Integrate ω on M to map M to a planar rectangle R using algorithm 23;
The module of S is the ratio between the height and the width of R.

Therefore, all conformal equivalence classes of topological quadrilaterals form a one-dimensional space.

Topological Annulus

Suppose M is a topological annulus. Then it can be conformally mapped to a planar annulus. We fix the outer radius to be one. Then the inner radius is called the conformal module. The following algorithm computes the conformal module.

Algorithm 39: Conformal module of a topological annulus

input : A topological annulus M
output: The conformal module of M
Compute the homology basis of M as γ using algorithm 6;
Compute the holomorphic 1-form basis of M as ω using algorithm 34;
Let $\int_\gamma \omega = a + bi$. Then the module is $e^{-\frac{2\pi a}{b}}$.

Topological Torus

For a topological torus, we need to specify the homotopy type of the maps. One way to specify the homotopy type of a homeomorphism is to introduce the concept of a *marked surface*. A marked surface is a surface with a set of canonical homology basis. Suppose S_1, S_2 are two genus one closed surfaces, $\Gamma_1 = \{a^1, b^1\}$ is a set of canonical homology basis of S_1, $\Gamma_2 = \{a^2, b^2\}$ is a set of canonical homology basis of S_2. $\phi : S_1 \to S_2$ is a homeomorphism, such that

$$[a^2] = [\phi(a^1)], \quad [b^2] = [\phi(b^1)].$$

By changing the canonical homology basis on S_2, we can traverse all homotopy types of homeomorphisms between S_1 and S_2.

The conformal invariants of a marked topological torus S can be specified in the following way. Suppose ω is the basis of holomorphic 1-forms on S. Then ω is also a holomorphic 1-form on the universal covering space $\bar{S}$ of S. We select four vertices $\{\bar{v}_0, \bar{v}_1, \bar{v}_2, \bar{v}_3\} \subset \bar{S}$, such that $\pi(\bar{v}_k) = v \in S$. Furthermore, the path connecting $\bar{v}_0$ to $\bar{v}_k$ is $\bar{\gamma}_k$, $k = 1,2,3$. Then $[\pi(\bar{\gamma}_1)] = a$, $[\pi(\bar{\gamma}_2)] = a + b$, $[\pi(\bar{\gamma}_3)] = b$. We integrate ω on $\bar{S}$. Then $\bar{v}_0, \bar{v}_1, \bar{v}_2, \bar{v}_3$ is a parallelogram, the shape of the parallelogram is the conformal module of S. In details, we define $\phi : \bar{S} \to \mathbb{C}$ in the following way:

$$\phi(\bar{v}) = \int_{\bar{v}_0}^{\bar{v}} \omega,$$

where the integration is along an arbitrary path from $\bar{v}_0$ to $\bar{v}$ in $\bar{S}$. Then the coordinates of the corners of the parallelogram are $\phi(\bar{v}_0), \phi(\bar{v}_1), \phi(\bar{v}_2), \phi(\bar{v}_3)$, and the conformal module of the torus is

$$Mod(S) = \frac{\phi(\bar{v}_3)}{\phi(\bar{v}_1)}.$$

For two marked topological tori, there exists a conformal map between them if and only if they share the same conformal module. Therefore, all the conformal classes of topological tori is a two-dimensional space. (Fig. 11.19)

High Genus Surface

For high genus surfaces, the conformal invariant is described by the so-called *period matrix*. Suppose S is of genus g, with marks $\{a_1, b_1, a_2, b_2, \cdots, a_g, b_g\}$. $\{\omega_1, \omega_2, \cdots, \omega_g\}$ is a set of holomorphic 1-form basis. Then the complex matrix is

$$\left(\int_{a_i} \omega_j \right) = A + \sqrt{-1}B,$$

then the period matrix is defined as $A^{-1}B$. Two marked high genus surfaces with the same period matrices are conformally equivalent. The following algorithm describes the computational details of the period matrix.

Algorithm 40: Period matrix

input : A marked genus g closed surface M with a set of canonical homology basis Γ
output: Period matrix of (M,Γ)
Compute a holomorphic 1-form basis $\{\omega_1,\omega_2,\cdots,\omega_g\}$ using algorithm 34;
Compute

$$\left(\int_{a_i}\omega_j\right)=A+\sqrt{-1}B.$$

return $A^{-1}B$.

11.13 Conformal Mappings for Multi-Holed Annuli

Suppose the input mesh is a multi-holed annulus M, such that all the boundaries are

$$\partial M=\gamma_0-\gamma_1-\gamma_2-\cdots-\gamma_n.$$

We choose γ_0 as the outer boundary as shown in Fig. 11.20.

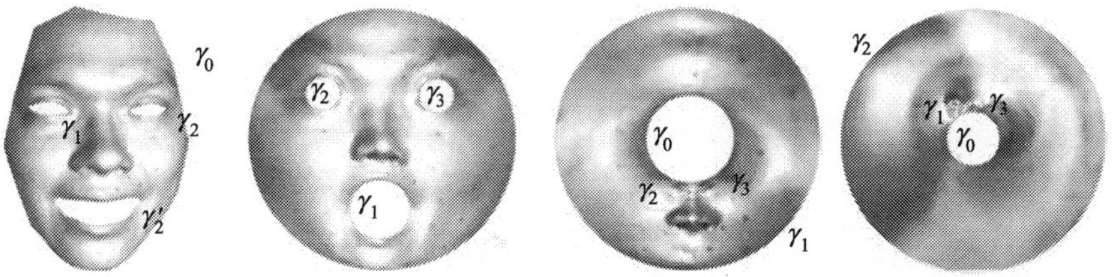

Fig. 11.20. Conformal mapping a multi-holed annulus to canonical domains.

We first compute the *harmonic measures* of M. A harmonic measure of γ_k with respect to the surface M is the solution to the Dirichlet problem,

$$\Delta f_i\equiv 0,\quad f_i|_{\gamma_j}=\delta_{ij}=\begin{cases}1, & i=j,\\ 0, & i\neq j.\end{cases}$$

Let η_k be df_k, $k=1,2,\cdots,n$. Then η_k's are exact harmonic 1-forms on M.

Then we compute the shortest paths $\tau_1,\tau_2,\cdots,\tau_n$, τ_k connects γ_k to γ_0. We slice M along τ_k to get M_k and compute harmonic functions $g_k: M_k\to\mathbb{R}$, such that

$$g_k|_{\tau_k^+}=1,\quad g_k|_{\tau_k^-}=0.$$

where τ_k^+ and τ_k^- are the two boundary segments corresponding to the cut τ_k on M_k. g_k could be arbitrary on the other vertices of M_k. Then we map dg_k back to M, which is a closed 1-form, denoted as ω_k. Then the set of closed 1-forms

$$\{\omega_1, \omega_2, \cdots, \omega_n\}$$

forms the basis of the first cohomology group of M, $H^1(M, \mathbb{R})$. We then diffuse ω_k to be harmonic by finding a function $h_k : M \to \mathbb{R}$ using the harmonic 1-form algorithm 30, such that

$$\Delta(\omega_k + dh_k) \equiv 0.$$

Then $\{\eta_1, \eta_2, \cdots, \eta_n\}$ and $\{\omega_1, \cdots, \omega_n\}$ form a harmonic 1-form basis of M. We compute the Hodge star (conjugate) of η_k's,

$${}^*\eta_k = \sum_{i=1}^{n} \mu_i \omega_i + \sum_{j=1}^{n} \iota_j \tau_j,$$

$\{\mu_i, \iota_j\}$ can be computed using the conjugate harmonic 1-form algorithm 34 directly. Then the holomorphic 1-forms

$$\eta_k + \sqrt{-1}^* \eta_k$$

form a basis of the holomorphic 1-form group on M. Let the period of the conjugate of η_k be

$$\alpha_{kj} = \int_{\gamma_k} {}^*\eta_j.$$

We can construct a holomorphic 1-form

$$\omega = \sum_{i=1}^{n} \lambda_i (\eta_i + \sqrt{-1}^* \eta_i),$$

such that

$$\begin{pmatrix} \alpha_{01} & \alpha_{02} & \cdots & \alpha_{0n} \\ \alpha_{11} & \alpha_{12} & \cdots & \alpha_{1n} \\ \vdots & \vdots & & \vdots \\ \alpha_{n-1,1} & \alpha_{n-1,2} & \cdots & \alpha_{n-1,n} \end{pmatrix} \begin{pmatrix} \lambda_0 \\ \lambda_1 \\ \vdots \\ \lambda_{n-1} \end{pmatrix} = \begin{pmatrix} 2\pi \\ 0 \\ \vdots \\ 0 \end{pmatrix}.$$

The above linear equation system has a unique solution, it also induces the following result:

$$\lambda_1 \alpha_{n,1} + \lambda_2 \alpha_{n,2} + \cdots + \lambda_n \alpha_{n,n} = -2\pi.$$

Then we can define a complex single-valued function $f : M \to \mathbb{C}$,

$$f(q) = e^{\int_p^q \omega},$$

where p is a base point, arbitrarily chosen. The following theorem claims that f maps M to the annulus, γ_0 and γ_n are mapped to the outer and inner circle, and all the other boundaries are mapped to concentric circular arcs.

Theorem 11.3 (Canonical Domains for Multi-holed Annuli). *The function f affects a one-to-one conformal mapping of M onto the annulus $1 < |z| < e^{\lambda_0}$ minus $n-2$ concentric arcs situated on the circles $|z| = e^{\lambda_i}$, $i = 1, 2, \cdots, n-2$.*

Fig. 11.21. Conformal mappings of annuli. The faces are sliced along the mouths and mapped onto planar annuli, using the harmonic measure and its conjugate.

Fig. 11.20 shows the conformal mappings of a three-holed human face surface. The planar domains are annuli with concentric arcs. The second frame shows the conformal mapping computed using the Euclidean Ricci flow method, which is a non-linear process and maps all the boundaries to the circular holes. The mapping in the third frame transforms γ_0 to the inner circle and γ_1 to the outer circle. The mapping in the third frame transforms γ_0 to the inner circle and γ_2 to the outer circle.

Fig. 11.21 illustrates the conformal mapping results for human faces with cuts along the mouths. We compute their harmonic measures, and compute the conjugates, which give us the desired conformal mappings.

Further Readings

Conformal mappings based on the Cauchy-Riemann equation using least square method was introduced in [75]. Holomorphic forms were first introduced by the authors in the first author's thesis [6] and applied for global conformal parameterizations for surfaces with arbitrary topologies in [5]. Discrete forms are studied in [76]. Holomorphic 1-forms have a broad range of applications. Manifold spline theory is established in [16]; the fundamental theoretic result claims that if a spline scheme can be defined on a manifold, then the manifold admits an affine structure, and vice versa. Affine structures are constructed using holomorphic forms in [16], and for manifold triangular B-Splines [20], manifold T-Splines [17], etc. Optimal conformal parameterization based on holomorphic 1-forms is introduced in [9]. In the medical imaging field, holomorphic 1-forms are applied for conformal brain mapping in [25] and conformal virtual colon flattening [26]. Holomorphic 1-forms are also applied in computer vision for shape matching, stitching [15], and shape analysis [14]. It has been applied for vector field design in [77] and [78]. A more flexible parameterization method is introduced in [79]. Other methods for conformal mappings will be introduced in the later chapters. More thorough references can be found in the survey works [41] and [42].

Problems

11.1. Harmonic 1-Forms
Suppose the input mesh is a closed mesh.
(a) Implement the algorithm for computing the cut graph.
(b) Implement the algorithm for constructing cohomology $H^1(M,\mathbb{R})$ basis generators.
(c) Implement the algorithm for computing harmonic 1-forms by diffusing the cohomology group generators.

11.2. Holomorphic 1-Forms
Suppose the input mesh is a closed mesh.
(a) Implement the algorithm for wedge product of two closed 1-forms.
(b) Implement the algorithm for Hodge star operator.
(c) Implement the algorithm for computing holomorphic 1-form generators.

11.3. Trajectories
(a) Design an algorithm to locate zero points of a holomorphic 1-form, implement it.
(b) Design an algorithm to trace critical trajectories on the mesh.

11.4. Holomorphic 1-Forms on Open Meshes
Suppose the input mesh is with boundaries and handles.
(a) Design an algorithm to compute its holomorphic 1-form basis without double covering.
(b) Implement the algorithm and compare with the results obtained by double covering.

11.5. Re-Meshing
(a) If the input mesh is a planar triangulation, given a point on the plane, design an algorithm to locate the face which contains the point. (Point location algorithm)
(b) Design an algorithm for re-meshing a triangular mesh to quad-mesh using holomorphic 1-forms.

11.6. Conformal Module
(a) Implement the algorithm to compute the conformal module for an annulus.
(b) Implement the algorithm to compute the conformal module for a topological rectangle (a topological disk with four corner vertices.)

11.7. Period Matrix
Implement the algorithm for computing period matrices for closed surfaces with high genus.

11.8. Manifold Spline
(a) Design an algorithm to construct a Euclidean atlas for a high genus closed mesh using holomorphic 1-form with zero points removed.
(b) Implement conventional B-Spline surfaces on each chart.
(c) How to make the B-Spline patches defined on different charts glued together coherently?
(d) How to fill the holes caused by the zero points?

12

Discrete Ricci Flow

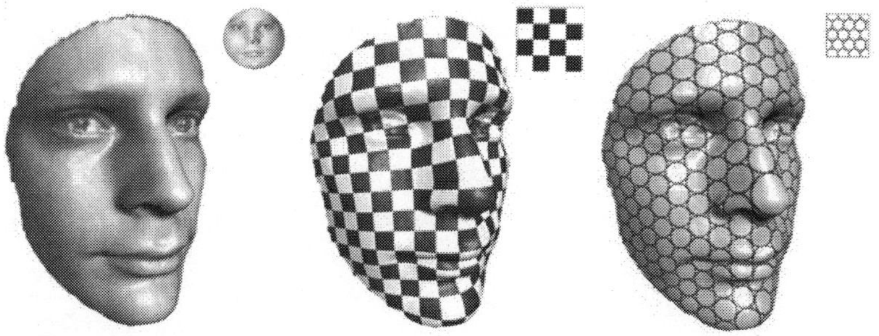

Fig. 12.1. Conformal mappings transform infinitesimal circles to infinitesimal circles and preserve the intersection angles among the circles. Here, the infinitesimal circles are approximated by circles with finite radii, and all the intersection angles among the circles are zeros, i.e. adjacent circles are tangent to each other.

This chapter introduces some practical algorithms for simulating smooth surface Ricci flow introduced in Chapter 7.2. The major application is to design a Riemannian metric conformal to the original metric by user-defined Gaussian curvature.

We first introduce the theories of discrete Ricci flow, then describe the algorithms. The main idea is very intuitive. Basically, we can imagine that the surface is triangulated, and each vertex is covered by a small disk.We can change the radii freely but preserve all the intersection angles. This is an approximation of conformal metric deformation. The metric deformation will change the Gaussian curvature at the vertices. Bing given a prescribed target curvature, we can find the correct radii. In fact, we can construct special energy forms, which involve both the initial Gaussian curvature, the target Gaussian curvature, and also the circle radii. The desired configuration of the circle radii is the energy minimizer. Furthermore, the energy is convex, therefore, the optima is global and the solution is unique.

The elegant theory leads to powerful algorithms. The discrete Ricci flow method is very versatile, it can handle most tasks hardly handled by other methods. The cost is that the discrete Ricci flow method is highly non-linear.

12.1 Circle Packing Metric

Conformal mappings have a special property that they transform infinitesimal circles to infinitesimal circles and keep the intersection angles among the infinitesimal circles. Fig. 12.1 visualizes this property. A human face surface is conformally mapped to the planar unit disk by a Riemann mapping as shown in the first column. We texture map a checkerboard onto the surface as shown in the third column. All the corner angles are well-preserved. If we replace the checkerboard by a *circle packing pattern* as shown in the second column, we can see that circles on the texture are mapped to circles on the surface and their tangency relations are well preserved.

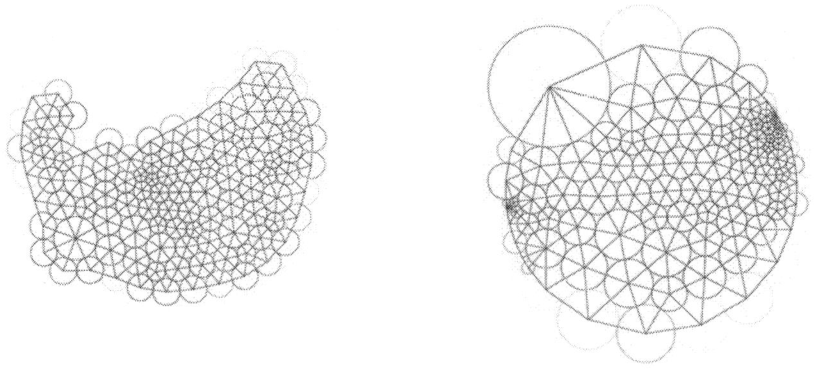

Fig. 12.2. Discrete conformal mapping approximated by circle packing.

Fig. 12.2 explains the concept of circle packing. Suppose T is a planar triangulation, each vertex v_i is associated with a circle $c(v_i, \gamma_i)$, which represents a circle centered at v_i with radius γ_i. Suppose $[v_i, v_j]$ is an edge of T. Then two circles $c(v_i, \gamma_i)$ and $c(v_j, \gamma_j)$ intersect each other or are tangent to each other. Suppose the vertex set of T is

$$V = \{v_1, v_2, \cdots, v_n\}.$$

Then

$$\Gamma : V \to \mathbb{R}^+, \quad \Gamma(v_i) = \gamma_i,$$

represents the circle radii of $c(v_i, \gamma_i)$'s. The edge length of $[v_i, v_j]$ is determined by

$$l_{ij} = \gamma_i + \gamma_j.$$

As shown in Fig. 12.3, planar circle packing can be generalized to surface circle packing in the following way.

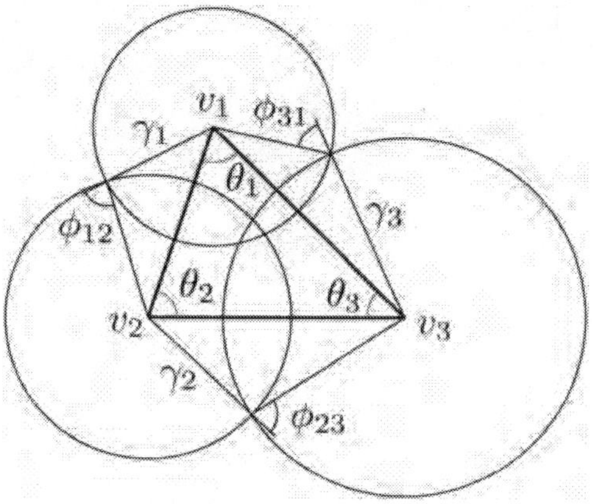

Fig. 12.3. Circle packing metric.

1. The planar triangulation T is a triangulation of a surface of arbitrary topology.
2. Each circle is generalized to a cone centered at the vertex.
3. Suppose $[v_i, v_j]$ is an edge on the triangulation, the circle $c(v_i, \gamma_i)$ and $c(v_j, \gamma_j)$ intersect at an acute angle. We represent the circle intersection angle by $\Phi : E \to [0, 2\pi)$,

   $$\Phi([v_i, v_j]) = \phi_{ij},$$

   where ϕ_{ij} is the intersection angle between $c(v_i, \gamma_i)$ and $c(v_j, \gamma_j)$. If ϕ_{ij} is zero, then $c(v_i, \gamma_i)$ and $c(v_j, \gamma_j)$ are tangent to each other.
4. The edge length is determined by the following Euclidean cosine law

   $$l_{ij} = \sqrt{\gamma_i^2 + \gamma_j^2 + 2\gamma_i\gamma_j \cos\phi_{ij}}.$$

In the following, we formally define the circle packing metric for a triangular mesh.

Definition 12.1 (Discrete Metric). *Suppose M is a triangular mesh. A* discrete metric *is a function defined on non-oriented edges of M,*

$$l : E \to \mathbb{R}^+,$$

such that for each triangle $[v_i, v_j, v_k]$, *the triangle inequality holds*:

$$l([v_i, v_j]) + l([v_j, v_k]) > l([v_k, v_i]).$$

If the mesh is assigned with a circle packing, then it has a natural circle packing metric.

Definition 12.2 (Circle Packing Metric). *Suppose* M *is a triangular mesh, with a circle packing* (M, Γ, Φ). *Then the* circle packing metric *is defined as*

$$l([v_i, v_j]) = \sqrt{\gamma_i^2 + \gamma_j^2 + 2\cos\phi_{ij}\gamma_i\gamma_j}.$$

If all the intersection angles are acute, then the edge lengths induced by a circle packing satisfy triangle inequality. Fig. 12.4 shows a circle packing on the Stanford bunny surface.

Fig. 12.4. Circle packing on the Stanford bunny surface.

If S is a smooth surface with a Riemannian metric $\mathbf{g} = (g_{ij})$, then a conformal metric is $e^{2u}\mathbf{g}$, where $u : S \to \mathbb{R}$ is a function on the surface. In the discrete case, we can define conformal circle packing metric analogically.

Definition 12.3 (Conformal Circle Packing Metric). *Suppose* M *is a triangular mesh with two different circle packing metrics* (M, Γ_1, Φ_1) *and* (M, Γ_2, Φ_2). *If*

$$\Phi_1 \equiv \Phi_2,$$

then two circle packing metrics are conformally equivalent.

Namely, two circle packings with the same intersection angles but different circle radii induce conformal circle packing metrics.

Definition 12.4 (Weighted Mesh). *Suppose M is a triangular mesh, $\Phi : E \to [0, \frac{\pi}{2}]$ is the intersection angle function, called the edge weight. Then (M, Φ) is called a weighted mesh.*

A weighted mesh represents a family of conformal equivalent circle packing metrics. Therefore, each Φ represents a conformal structure of the mesh M.

Fig. 12.2 demonstrates two conformal circle packing metrics of the same combinatorial triangulation T. We can define a quasi-conformal map induced by the two circle packing metrics,

$$\phi : (M, \Gamma_1, \Phi_1) \to (M, \Gamma_2, \Phi_1).$$

Intuitively, the identity map of the combinatorial triangulation with two metrics gives us the quasi-conformal map. On each face, the map is linear; globally, the map is piecewise linear. Fig. 12.2 demonstrates one such quasi-conformal map. We can see that ϕ maps an irregular planar domain $|M|$ to the unit disk. ϕ is a discrete approximation of the Riemann mapping from the domain $|M|$ to the unit disk.

Proposition 12.5. *Let K be a combinatorial closed disk (simply connected, finite and with nonempty boundary). Then there exists a univalent circle packing $P \subset \mathbb{D}$ for K, where $\mathbb{D}$ is the planar unit disk, such that for any edge $[v_i, v_j]$, the circles $c(v_i, \gamma_i)$ and $c(v_j, \gamma_j)$ are tangent to each other; every boundary circle $c(v_k, \gamma_k), v_k \in \partial K$ is tangent to the unit circle. Two such circle packings differ by a Möbius transformation.*

Such kind of circle packing is called the maximal circle packing of K. The detailed proof can be found in [37].

Given a planar domain D, we can construct a sequence of circle packings $P_i = (T_i, \Gamma_i, \phi_i)$, such that T_i is the regular planar tessellation by equilateral triangles, the radii Γ_i are constant and go to zero, $\lim_{k\to\infty} \gamma \to 0$. T_i exhausts D from the interior eventually. For each circle packing $P_i = (T_i, \Gamma_i, \phi_i)$, we compute its maximal circle packing $\tilde{P}_i = (T_i, \tilde{\Gamma}_i, \phi_i)$. The discrete conformal map induced by the two circle packings is

$$\phi_i : P_i \to \tilde{P}_i.$$

Furthermore, suppose $p, q \in D$, we use a Möbius transformation to ensure $\phi_i(p)$ is the origin, $\phi_i(q)$ is on the real axis.

Thurston [36] conjectured that such discrete conformal maps $\{\phi_i\}$ converge to the real Riemann mapping $\phi_\infty : D \to \mathbb{D}$. Rodin and Sullivan proved Thurston's conjecture in [80]. Later, He and Schramm [81] used circle packing to prove the classical Riemann mapping theorem using elementary geometry. The whole literature of conventional complex analysis can be based on circle packing [37]. Conformal mapping between surfaces can also be approximated by circle packing as shown in Fig 12.5.

In the following discussion, we generalize the circle packing method from domains on the complex plane to general Riemann surfaces.

12.2 Discrete Gaussian Curvature

In the smooth case, Gaussian curvature is defined by means of a Gauss map. Suppose $G : S \to \mathbb{S}^2$, $p \to \mathbf{n}(p)$ is the Gauss map of the surface, which maps a point $p \in S$ to its

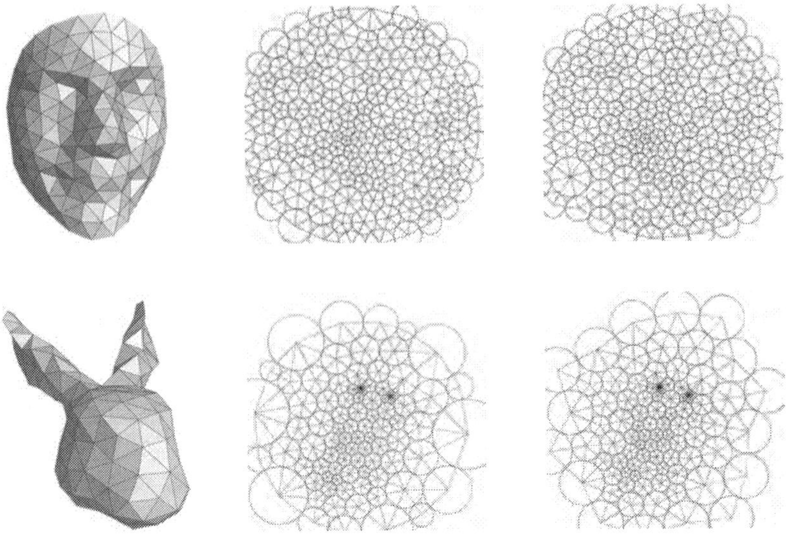

Fig. 12.5. Approximating conformal mappings by circle packing. The circle radii are changed while tangency relations are preserved. The second column shows a circular boundary condition; the third column shows a free boundary condition. For the free boundary condition, all circle radii of boundary vertices are equal.

normal $\mathbf{n}(p)$. We choose a neighborhood of a point p, denoted as D. $G(D)$ is the Gauss image of D on the unit sphere $\mathbb{S}^2$. Then the Gauss curvature at p is the limit of the area ratio between $G(D)$ and D, when D shrinks to p.

The discrete Gauss map of a mesh can be defined using the face normal.

Definition 12.6 (Discrete Gauss Map). *Suppose M is a triangular mesh, embedded in $\mathbb{R}^3$. Then each face f has a unique normal $\mathbf{n}(f)$. For each point $p \in f$, and $p \notin \partial f$, the Gauss map is defined as*

$$G(p) = \mathbf{n}(f).$$

For points on edges, the Gauss map is not defined.

Fig. 12.6 Illustrates the definition of discrete Gaussian curvature.

Definition 12.7 (Discrete Gaussian Curvature). *Suppose v is an interior vertex of M, $v_0, v_1, \cdots, v_n$ are all neighboring vertices of v, sorted counter-clock-wisely. The normal of face $[v, v_k, v_{k+1}]$ is denoted as $\mathbf{n}_k$. Then we consider the spherical polygon*

$$G(v) = [\mathbf{n}_0, \mathbf{n}_1, \cdots, \mathbf{n}_n].$$

All the edges are great arcs of the sphere. The order of the $\mathbf{n}_i$'s might be counter-clock wisely or clock-wisely, determined by the local geometry of v. Then we define the discrete Gauss curvature of v as the area of the spherical polygon, denoted as $K(v)$.

If the vertices of $G(v)$ are sorted counter-clock wisely, then the area is positive. Otherwise, the area is negative. In the following, we show that the discrete Gauss curvature can be computed using the corner angles on faces.

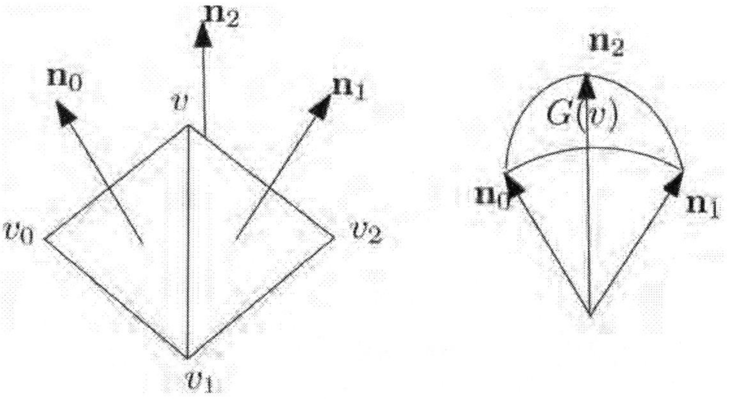

Fig. 12.6. Discrete Gaussian curvature

Proposition 12.8. *Suppose all the corner angles surrounding an interior vertex v are* $\alpha_0, \alpha_1, \cdots, \alpha_n$. *Then*

$$K(v) = 2\pi - \sum_{k=0}^{n} \alpha_k.$$

Proof. As shown in Fig. 12.6, the normals $\mathbf{n}_k$ and $\mathbf{n}_{k+1}$ and the great arc connecting them $\overline{\mathbf{n}_k\mathbf{n}_{k+1}}$ on the unit sphere form a plane π_k, the edge $[v, v_{k+1}]$ is orthogonal to both $\mathbf{n}_k$ and $\mathbf{n}_{k+1}$, therefore, $[v, v_{k+1}]$ is orthogonal to the plane π_k. Therefore, $[v, v_{k+1}]$ is orthogonal to the tangent vector at $\mathbf{n}_{k+1}$ to the arc $\overline{\mathbf{n}_k\mathbf{n}_{k+1}}$. Similarly, edge $[v, v_{k+2}]$ is orthogonal to the tangent vector at $\mathbf{n}_{k+1}$ to the arc $\overline{\mathbf{n}_{k+1}\mathbf{n}_{k+2}}$.

Therefore the edges of the angle $\alpha_{k+1} = \angle v_{k+1} v v_{k+2}$ are perpendicular to the $k+1$-th inner angle of the spherical polygon $G(v)$ at $\mathbf{n}_{k+1}$, therefore, α_{k+1} equals π minus the $(k+1)$-th inner angle of $G(v)$, hence α_{k+1} equals the $(k+1)$-th exterior angle of $G(v)$.

According to the Gauss-Bonnet theorem,

$$\int_{G(v)} dA + \sum \alpha_i = 2\pi,$$

we have $K(v) = \int_{G(v)} dA = 2\pi - \sum \alpha_i$. □

The Gauss-Bonnet theorem bridges differential geometry and topology. It also holds for discrete meshes as formulated in the discrete Gauss-Bonnet theorem.

Theorem 12.9 (Discrete Gauss-Bonnet Theorem). *Suppose M is a closed triangular mesh with a discrete metric. Then*

$$\sum_{v \in M} K(v) = 2\pi\chi(M),$$

where $\chi(M)$ *is the Euler number of the mesh,* $\chi(M) = V + F - E$.

Proof. The summation of three angles of each face equals to π. The total corner angles of a mesh equals to

$$\sum \alpha = F\pi.$$

The total discrete Gauss curvature is

$$\sum_{v \in M} K(v) = \sum_{v \in M} 2\pi - \sum \alpha = 2V\pi - F\pi.$$

Each face has three edges, and each edge is shared by two faces,

$$3F = 2E.$$

Therefore

$$\sum_{v \in M} K(v) = 2\pi(V + F - E).$$

□

Now, we generalize the above results to meshes with boundaries. First, we define Gauss curvature of boundary vertices.

Definition 12.10 (Discrete Gaussian Curvature of Boundary Vertex). *Suppose M is a mesh with boundaries, $v \in \partial M$, all the corner angles surrounding v are $\alpha_0, \alpha_1, \cdots, \alpha_n$ then the discrete Gauss curvature is*

$$K(v) = \pi - \sum_{k=0}^{n} \alpha_k.$$

Then discrete Gauss-Bonnet theorem still holds for meshes with boundaries. Let $\bar{M}$ be the double covering of M. Then the following relations hold:

$$\sum_{v \in M} K(v) = \frac{1}{2} \sum_{\bar{v} \in \bar{M}} K(\bar{v}),$$

also

$$\chi(M) = \frac{1}{2}\chi(\bar{M}).$$

Gauss-Bonnet theorem shows that the total Gauss curvature is independent of the choice of the metric, it is solely a topological invariant.

12.3 Discrete Surface Ricci Flow

Suppose S is a smooth surface with a Riemannian metric $\mathbf{g} = (g_{ij})$. The Ricci flow deforms the metric $\mathbf{g}(t)$ according to its induced Gaussian curvature $K(t)$, where t is the time parameter:

$$\frac{dg_{ij}(t)}{dt} = -2K(t)g_{ij}(t), \tag{12.1}$$

with the constraint that the total surface area is preserved. If we represent the Riemannian metric in the following form:

$$\mathbf{g}(t) = e^{2u(t)}\mathbf{g}(0),$$

then the surface Ricci flow can be written as

$$\frac{du(t)}{dt} = -K(t). \tag{12.2}$$

In a physical sense, the curvature evolution induced by the Ricci flow is exactly the same as the heat diffusion on the surface:

$$\frac{dK(t)}{dt} = -\Delta_{\mathbf{g}(t)} K(t),$$

where $\Delta_{g(t)}$ is the Laplace-Beltrami operator induced by the metric $\mathbf{g}(t)$.

The Ricci flow can be easily modified to compute a metric with a prescribed curvature $\bar{K}$, and the flow becomes

$$\frac{du(t)}{dt} = 2(\bar{K} - K)g_{ij}(t), \tag{12.3}$$

with the area preserving constraint.

The discrete surface Ricci flow is the exact analogy of the smooth surface Ricci flow as described in equation (12.3). The profound relation between circle packing and Ricci flow was discovered by Chow and Luo in [82].

Suppose that we are given a mesh M with a circle packing metric (M, Γ, Φ). The circle centered at the vertex v_i is with radius γ_i. The Gaussian curvature of v_i is K_i, the target Gaussian curvature is $\bar{K}_i$. Then the discrete Ricci flow is defined as

$$\frac{du_i(t)}{dt} = (\bar{K}_i - K_i(t)). \tag{12.4}$$

with the constraint that $\sum_i u_i = constant$.

Definition 12.11 (Convergence). *A solution to equation* (12.4) exists *and is* convergent, *if*

1. $\lim_{t\to\infty} K_i(t) = \bar{K}_i, \forall i,$
2. $\lim_{t\to\infty} \gamma_i(t) = \bar{\gamma}_i \in \mathbb{R}^+, \forall i.$

A convergent solution converges exponentially *if there are constants c_1, c_2, so that for all time $t \geqslant 0$,*

$$|K_i(t) - \bar{K}_i| \leqslant c_1 e^{-c_2 t},$$

and

$$|\gamma_i(t) - \bar{\gamma}_i| \leqslant c_1 e^{-c_2 t}.$$

Chow and Luo proved that the discrete Ricci flow is exponentially convergent.

Theorem 12.12. *Suppose (M, Φ) is a closed weighted mesh. Given any initial circle packing metric based on the weighted mesh, the solution to the discrete Ricci flow* (12.4) *in the Euclidean geometry with the given initial value exists for all time and converges exponentially fast.*

For a given weighted mesh (M, Φ), we say a Gaussian curvature $K : V \to \mathbb{R}$ is *admissible*, if there exists a circle packing metric (M, Φ, Γ), which induces a Gaussian curvature that equals to K. The following theorem was proven by Chow and Luo, which describes the space of all the admissible Gaussian curvatures.

Theorem 12.13. *Suppose* (M,Φ) *is a weighted mesh,* $\Phi : E \to [0, \frac{\pi}{2})$*,* I *is a subset of vertices* V*. Then*

$$\sum_{i\in I} K_i > -\sum_{(e,v)\in Lk(I)} (\pi - \Phi(e)) + 2\pi\chi(F_I),$$

where F_I *is the subcomplex consisting of cells whose vertices are in* I *and*

$$Lk(I) = \{(e,v) | e \text{ is an edge so that } e \cap I = \emptyset \text{ and the vertex } v \in I \text{ so that } e,v \text{ form a triangle}\}.$$

The following algorithm 42 describes the process to compute the discrete surface Ricci flow in details.

In practice, if the desired metric is required to be conformal to the initial induced Euclidean metric, then the initial circle packing metric needs to be carefully chosen, such that the circle packing metric is as close to the original induced Euclidean metric as possible. If there are too many skinny triangles or obtuse angles in the mesh, it is difficult to find a good initial circle packing metric. Therefore, in this situation, we can re-mesh the surface first to improve the quality of triangulations.

During the process, we add one normalization step to ensure

$$\sum_{v_i\in M} \log \gamma_i = 0.$$

This is equivalent to the area preserving constraint in the smooth surface Ricci flow. The step length δ also needs to be carefully selected. Because the whole process is highly non-linear, if δ is too big, the process is unstable; if δ is too small, the process converges too slowly.

Algorithm 41: Discrete surface Ricci flow

input : A mesh M embedded in $\mathbb{R}^3$, target curvature $\bar{K}$, curvature error threshold ε
output: A circle packing metric (M,Γ,Φ) which induces $\bar{K}$

Compute the initial circle packing metric (M,Γ_0,ϕ);
while $\max|K_i - \bar{K}_i| > \varepsilon$ **do**
  forall *edge* $e = [v_i, v_j] \in M$ **do**
    Compute the edge length

$$l_{ij} \leftarrow \sqrt{\gamma_i^2 + \gamma_j^2 + 2\gamma_i\gamma_j \cos\phi_{ij}}$$

  end
  forall *Corner in mesh* $c_i \in [v_i, v_j, v_k]$ **do**
    Compute the corner angle c_i

$$c_i \leftarrow \cos^{-1}\frac{l_{ij}^2 + l_{ki}^2 - l_{jk}^2}{2l_{ij}l_{ik}}.$$

  end
end

forall *Vertex* $v_i \in M$ **do**
Compute the Gaussian curvature of v_i;

$$K_i \leftarrow 2\pi.$$

forall *Corners c attached to* v_i **do**

$$K_i \leftarrow K_i - c.$$

end
end
forall *Vertex* $v_i \in M$ **do**

$$\gamma_i \leftarrow \gamma_i + \delta\gamma_i \times (\bar{K}_i - K_i)$$

where δ is the step length.
end

$$s \leftarrow 0,$$

forall *Vertex* $v_i \in M$ **do**

$$s \leftarrow s + \log\gamma_i,$$

end
forall *Vertex* $v_i \in M$ **do**

$$\gamma_i \leftarrow \gamma_i \times \exp -\frac{s}{|V|}.$$

end

12.4 Newton's Method

The discrete surface Ricci flow in equation (12.4), is in fact the negative gradient flow of some convex energy. The desired circle packing metric is the optimum of the energy. In this section, we explicitly formulate the energy and its Hessian matrix, therefore the optimum can be computed using Newton's method with much higher efficiency.

Fig. 12.7 illustrates the circle packing metric of a triangle $[v_1, v_2, v_3]$. Three circles $c(v_i, \gamma_i)$ are centered at each vertex, each pair has a common chord. Three common chords intersect at a point O, called the *center* of the triangle. The center can be obtained in another way. There exists a unique circle orthogonal to the three circles, the center of the circle is also at O. The distance from O to the edge $[v_i, v_j]$ is h_{ij}.

Let $u_i = \log\gamma_i$. By direct calculation, it can be verified that

$$\frac{\partial\theta_j}{\partial u_i} = \frac{\partial\theta_i}{\partial u_j} = h_{ij}.$$

Because of $\theta_1 + \theta_2 + \theta_3 = \pi$,

$$\frac{\partial\theta_i}{\partial u_i} = -\frac{\partial\theta_j}{\partial u_i} - \frac{\partial\theta_k}{\partial u_i} = -\frac{\partial\theta_i}{\partial u_j} - \frac{\partial\theta_i}{\partial u_k}. \tag{12.5}$$

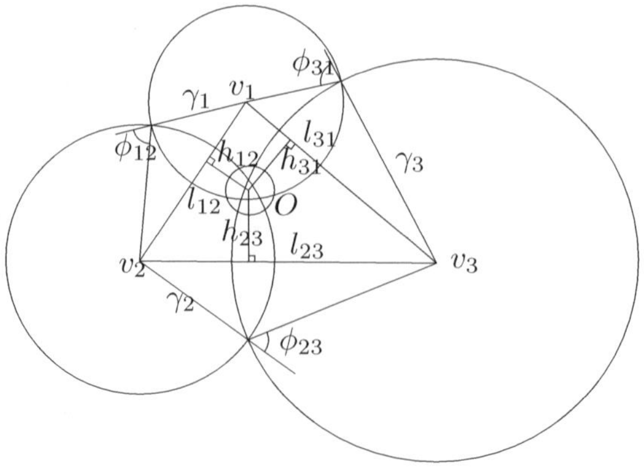

Fig. 12.7. Circle packing metric on a triangle.

Now we define the edge weight w_{ij} in the following way: suppose two faces $[v_i, v_j, v_k]$ and $[v_j, v_i, v_l]$ share an edge $[v_i, v_j]$, O_k is the center in $[v_i, v_j, v_k]$, O_l is the center in $[v_j, v_k, v_l]$, h^k_{ij} is the distance from O_k to the edge $[v_i, v_j]$, h^l_{ji} is the distance from O_l to the edge $[v_j, v_i]$. Then

$$w_{ij} = h^k_{ij} + h^l_{ji}.$$

If $[v_i, v_j]$ is a boundary edge and only attaches to one face $[v_i, v_j, v_k]$, then

$$w_{ij} = h^k_{ij}.$$

From equation (12.5) it is easy to show that

$$\frac{\partial K_i}{\partial u_j} = \frac{\partial K_j}{\partial u_i}. \tag{12.6}$$

Now let (M, Φ) be a weighted mesh. We consider all possible circle packing metrics, namely, all the possible radii functions $\Gamma : V \to \mathbb{R}^+$. Suppose the vertex set of M is $\{v_1, v_2, \cdots, v_n\}$. Then the corresponding radii is

$$\gamma = (\gamma_1, \gamma_2, \cdots, \gamma_n).$$

For the convenience of discussion, we take the logarithm of the radii,

$$\mathbf{u} = (u_1, u_2, \cdots, u_n),$$

where $u_i = \log \gamma_i$. All possible $\mathbf{u}$'s form the n-dimensional space homeomorphic to $\mathbb{R}^n$. We call it the *u-space*. We add one normalization constraint,

$$\sum_{k=1}^{n} u_k = 0.$$

Then all normalized $\mathbf{u}$'s form an $(n-1)$-dimensional space homeomorphic to $\mathbb{R}^{n-1}$. We call it *the normalized u-space*.

We define a differential 1-form ω in the u-space,

$$\omega = \sum_{i=1}^{n} K_i du_i.$$

Then

$$d\omega = \sum_{[v_i,v_j]\in M} \left(\frac{\partial K_i}{\partial u_j} - \frac{\partial K_j}{\partial u_i} \right) du_j \wedge du_i.$$

Because of equation (12.6), $d\omega$ is zero. Therefore, ω is a closed 1-form. Because u-space is simply connected, therefore, ω is also exact. We choose a special metric

$$\mathbf{u}_0 = (0,0,\cdots,0),$$

and define the following energy function:

$$E(\mathbf{u}) = \int_{\mathbf{u}_0}^{\mathbf{u}} \sum K_i du_i.$$

Then the energy is independent of the choice of the integration path. The gradient of the energy is

$$\nabla E = \left(\frac{\partial E}{\partial u_1}, \frac{\partial E}{\partial u_2}, \cdots, \frac{\partial E}{\partial u_n} \right) = (K_1, K_2, \cdots, K_n).$$

Therefore, discrete surface Ricci flow in equation (12.4) is exactly the negative gradient flow of the energy.

We can further compute the Hessian matrix of E with respect to $\mathbf{u}$. If $[v_i, v_j]$ is an edge on the mesh M, then

$$\frac{\partial^2 E(\mathbf{u})}{\partial u_i \partial u_j} = \frac{\partial K_i}{\partial u_j} = -w_{ij}.$$

Otherwise $\frac{\partial^2 E(\mathbf{u})}{\partial u_i \partial u_j} = 0$. Furthermore,

$$\frac{\partial^2 E(\mathbf{u})}{\partial u_i^2} = \frac{\partial K_i}{\partial u_i} = \sum_{[v_i,v_j]\in M} w_{ij}.$$

In the normalized u-space, the Hessian matrix is diagonal dominant, therefore it is positive definite. Hence, the energy $E(\mathbf{u})$ is convex and has a unique global minimum.

We denote the Hessian matrix as $\Delta(\mathbf{u})$. Then from the above calculation, we get

$$d\mathbf{k} = \Delta(\mathbf{u}) d\mathbf{u}.$$

In the discrete Ricci flow $d\mathbf{u} = -\mathbf{k}dt$, we get

$$\frac{d\mathbf{k}}{dt} = -\Delta(\mathbf{u})\mathbf{k}.$$

Therefore, the discrete Gaussian curvature evolves like a heat diffusion process.

In order to compute a metric satisfying the prescribed curvature $\bar{K}$, the energy can be reformulated as

$$E(\mathbf{u}) = \int_{\mathbf{u}_0}^{\mathbf{u}} \sum_{i=1}^{n} (\bar{K}_i - K_i) du_i, \tag{12.7}$$

the desired metric is the global optimum of the energy. By using Newton's method, the energy can be minimized very efficiently. Fig. 12.8 demonstrates the conformal parameterization results using Newton's method. The parameterization quality highly depends on the choice of the target curvature.

Algorithm 42: Newton's method of discrete Ricci flow

input : A mesh M embedded in $\mathbb{R}^3$, target curvature $\bar{K}$, curvature error threshold ε
output: A circle packing metric (M, Γ, Φ) which induces $\bar{K}$

Compute the initial circle packing metric (M, Γ_0, ϕ);
Compute the initial curvature K.

$$\mathbf{u} \leftarrow 0.$$

while $\max|K_i - \bar{K}_i| > \varepsilon$ **do**
  forall *edge* $e = [v_i, v_j] \in M$ **do**
    Compute the edge weight $w_{ij}(\mathbf{u})$ to form the Hessian matrix.
  end

$$\begin{aligned} d\mathbf{u} &\leftarrow \Delta^{-1}(\bar{K} - K) \\ \mathbf{u} &\leftarrow \mathbf{u} + d\mathbf{u} \\ \mathbf{k} &\leftarrow K(\mathbf{u}). \end{aligned}$$

end
$\bar{\mathbf{u}} \leftarrow \mathbf{u}$

Fig. 12.8. Conformal parameterization using Newton's method for optimizing the discrete Ricci energy.

12.5 Isometric Planar Embedding

Once the flat metric is computed, the mesh can be isometrically embedded onto the plane $\mathbb{R}^2$. If the mesh has complicated topology, we need to embed a finite portion of its universal covering space . In theory, this step is very trivial, but in practice, because of accumulation of the numerical error, accurate planar embedding is very challenging to compute.

Instead, we formulate planar embedding as a variational problem, and solve it using the optimization method. This approach is more robust and accurate than the direct approach.

Suppose M is a topological disk (or a finite portion of the universal covering space of a mesh), with a flat metric $l : E \to \mathbb{R}^+$, we want to flatten M onto the plane $\mathbb{R}^2$. We denote the map as

$$\phi : M \to \mathbb{R}^2, \quad \phi(v_i) = (x_i, y_i),$$

such that

$$|\phi(v_i) - \phi(v_j)| = l([v_i, v_j]), \quad \forall [v_i, v_j] \in M.$$

Suppose $f = [v_i, v_j, v_k]$ is a face of the mesh. Then we can embed it in a canonical way in the xy-plane, such that

$$(x_i, y_i) = (0, 0), \quad (x_j, y_j) = (l_{ij}, 0), \quad (x_k, y_k) = (l_{ki} \cos\theta_i, l_{ki} \sin\theta_i),$$

where

$$\theta_i = \cos^{-1} \frac{l_{ij}^2 + l_{ki}^2 - l_{jk}^2}{2 l_{ij} l_{ki}}.$$

We treat (x_i, y_i) as local coordinates of $[v_i, v_j, v_k]$. Let

$$\begin{aligned}
\mathbf{s}_i &= [(x_k, y_k) - (x_j, y_j)] \times \mathbf{n}, \\
\mathbf{s}_j &= [(x_i, y_i) - (x_k, y_k)] \times \mathbf{n}, \\
\mathbf{s}_k &= [(x_j, y_j) - (x_i, y_i)] \times \mathbf{n},
\end{aligned}$$

where $\mathbf{n} = (0, 0, 1)$ is the face normal.

Then $\phi : M \to \mathbb{R}^2$ has two components (ϕ_1, ϕ_2), $\phi_k : M \to \mathbb{R}$, ϕ_k's are piecewise linear functions on the mesh. Then

$$\nabla\phi_1(f) = u_i \mathbf{s}_i + u_j \mathbf{s}_j + u_k \mathbf{s}_k, \quad \nabla\phi_2(f) = v_i \mathbf{s}_i + v_j \mathbf{s}_j + v_k \mathbf{s}_k.$$

If ϕ is an isometric map, then

$$\nabla\phi_2(f) = \mathbf{n} \times \nabla\phi_1(f).$$

In practice, we fix the image of one face, and minimize the following energy to obtain ϕ:

$$E(\phi) = \sum_{f \in M} |\nabla\phi_2(f) - \mathbf{n} \times \nabla\phi_1(f)|^2,$$

and

$$\phi = \min_{\tau} E(\tau).$$

The variables of the above energy $E(\phi)$ are the planar images of all vertices $\{\phi(v_i) = (x_i, y_i)\}$ and $E(\phi)$ is a quadratic form. So optimizing $E(\phi)$ is a linear problem.

Fig. 12.9 demonstrates the isometric embedding computed using this method. The target curvature is set to be zero everywhere on the genus one Kitten model. The conformal flat metric is computed using Newton's method. In the middle frame, a fundamental domain is flattened onto the plane isometrically. In the right frame, a finite portion of the universal covering space is isometrically flattened on the plane.

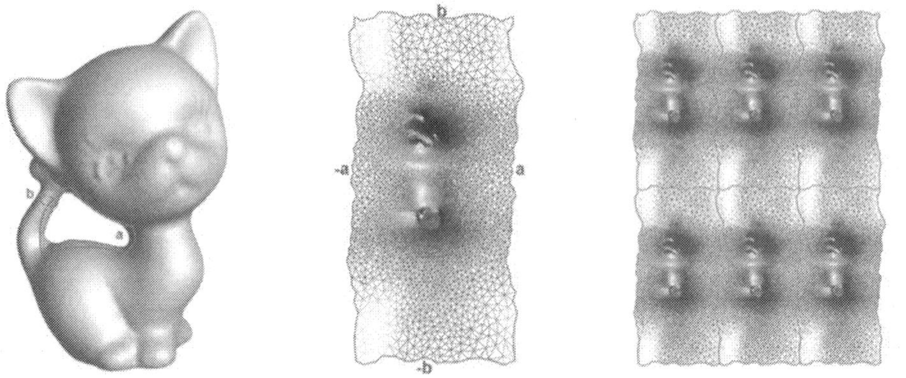

Fig. 12.9. Isometric embedding of the Kitten model with a flat metric.

12.6 Surfaces with Boundaries

Genus zero surfaces with several boundaries can be conformally mapped onto a planar multi-holed annulus. Each boundary is mapped to a circle, the centers and the radii of the boundary circles are determined automatically by the geometry of the surface. Therefore, the centers and the radii are the conformal invariants.

Topological Disk

We first consider a topological disk case. The goal is to find a conformal mapping to map it onto the unit disk. The target curvatures for interior vertices are zeros. The main difficulty is how to determine the target curvatures for boundary vertices. Suppose $\partial M = \{v_1, v_2, \cdots, v_n\}$. Then if we set the boundary vertex curvatures evenly, $k_i = \frac{2\pi}{n}$, then we will not get a unit disk after running Ricci flow. The final planar shape will be a convex planar polygon. In fact, the target curvatures on boundary vertices should satisfy the following requirements:

$$\frac{k_i}{l_{i-1,i} + l_{i,i+1}} \equiv c, \quad \forall v_i \in \partial M, \tag{12.8}$$

and $\sum_{v_i \in \partial M} k_i = 2\pi$, where l_{ij} are the edge length under the target metric, k_i is the target vertex curvature. In the beginning of the algorithm, we do not know the edge length under the target metric yet, therefore, we cannot set the target curvature. In the process, we need to adjust the target curvature step by step.

Multi-holed Annulus

A similar idea can be applied for multi-holed annuli. We select one boundary as the exterior boundary, the other boundaries as the interior boundaries. The total curvature of exterior boundary vertices is 2π. For each interior boundary, the total curvature is -2π. The target curvature can be determined at the start. We adjust the target curvature according to equation 12.8, such that the relation holds under the current metric.

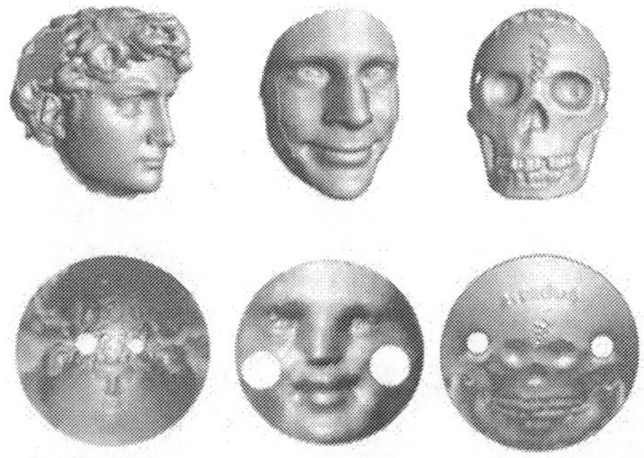

Fig. 12.10. Surfaces which are topologically equivalent but not conformally equivalent.

Algorithm 43: Newton's method of discrete Ricci flow for multi-holed annulus

input : A multi-holed annulus M
output: A flat circle packing metric (M, Γ, Φ) which maps the boundaries to circles

Compute the initial circle packing metric (M, Γ_0, ϕ);
Compute the initial curvature K.
Trace boundaries, store each boundary as a list of ordered vertices

$$\Gamma_1, \Gamma_2, \cdots, \Gamma_m.$$

Set Γ_1 as the outer boundary,

$$\theta_1 \leftarrow 2\pi, \quad \theta_2 \leftarrow -2\pi, \ \cdots, \ \theta_m \leftarrow -2\pi.$$

forall *Interior Vertex* $v \in M$ **do**

$$\bar{K}(v) \leftarrow 0.$$

end
forall *Boundaries* Γ_k **do**
  forall $v \in \Gamma_i$ **do**

$$\bar{k}(v) \leftarrow \frac{\theta_k}{|\Gamma_k|}$$

  end
end
repeat
  Compute target metric according to $\bar{K}$ using Newton's method in algorithm 43;
  forall *Boundaries* Γ_k **do**
    Compute the total length of Γ_k under current metric, s_k;
  end
  forall $v_i \in \Gamma_i$ **do**

$$\bar{k}(v_i) \leftarrow \theta_k \frac{l_{i-1,i} + l_{i,i+1}}{2s_k}$$

  end
until *The maximal difference of current target curvature and previous target curvature is less than* ε;

Figure R. 10 shows examples of conformal mappings of two holed annuli. The total target curvature for the outer boundary is 2π. The total curvature for each inner boundary is -2π. The target curvatures on boundary vertices are incrementally adjusted during the computation. The centers and radii of circles of inner boundaries on the parameter domain are conformal invariants. In the figure, we can see that the three surfaces in Fig. 12.10 are not conformally equivalent.

We have more flexibility in setting the boundary target curvatures using Ricci flow. We can choose two boundaries, and set their total target curvatures to be zeros, and the total target curvatures of other inner boundaries are still -2π. Then the Ricci flow method will map the surface to periodic strips with circular holes. Figure 12.11 shows such an example. Where two boundaries are mapped to parallel lines, the other boundaries are mapped to circles.

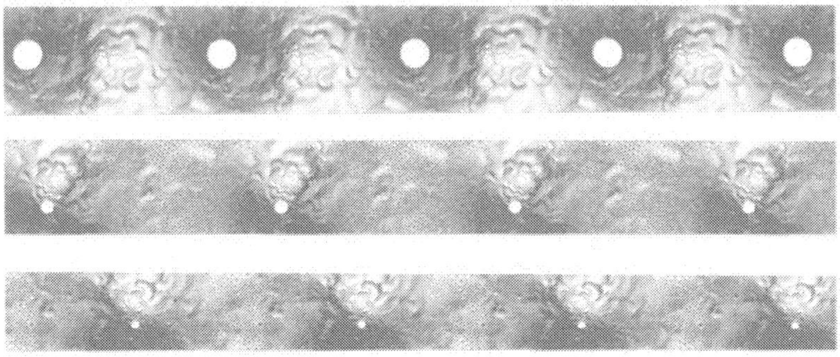

Fig. 12.11. Tiling of different global conformal parameterizations for the David Head surface with three boundaries using curvature control by discrete Euclidean ricci flow.

12.7 Optimal Parameterization Using Ricci Flow

Suppose M is a topological disk. Then for an arbitrary planar domain D, there will be conformal mappings from M to D. Although conformal mappings introduce no angle distortion, they introduce area distortions. In engineering practice, it is highly desirable to obtain conformal mappings with uniform area distortion. This section discusses one rigorous method to choose the target domain to achieve the most uniform conformal mappings.

Given a weighted mesh (M,Φ) with n vertices, all of the possible curvatures form an $(n-1)$-dimensional convex polytope determined by both the connectivity of M and the edge angle Φ, denoted as Ω_k. For any interior vertex, the target curvature will be zero. According to the discrete Gauss-Bonnet theorem, the summation of boundary vertex curvature equals to $2\pi\chi(M)$. All such kind of curvature form a convex polytope as defined in the following.

Definition 12.14 (Admissible Curvature Space). *Given a weighted mesh (M,Φ) with boundaries, the admissible curvature space is the intersection of hyper-planes with Ω_k,*

$$\Pi_k := \bigcap_{v_i \notin \partial M} \{k_i = 0\} \bigcap \{ \sum_{v_j \in \partial M} k_j = 2\pi\chi(M)\} \bigcap \Omega_k.$$

Suppose $\mathbf{u}$ is a circle packing metric of (M,Φ) in the normalized u-space. Then $\mathbf{u}$ uniquely determines a curvature $\mathbf{k}$. From previous discussion, we can compute the Jacobi matrix of the mapping

$$K : \mathbf{u} \to \mathbf{k}.$$

By explicit calculation, we can get

$$d\mathbf{k} = \Delta d\mathbf{u},$$

where Δ is the Hessian matrix of the discrete Ricci energy (12.7). In the normalized u-space, Δ is positive definite. Therefore it is invertible, $d\mathbf{u} = \Delta^{-1} d\mathbf{k}$.

The optimal parameterization problem is equivalent to optimizing some energy form $E(\mathbf{k})$, which measures the uniformity of the area distortion, in the admissible curvature space Π_k.

If the energy is differentiable, the problem can be solved using the *projected gradient* method. We compute the gradient of $E(\mathbf{k})$ with respect to $\mathbf{k}$, denoted as $\nabla_k E$. Then project the gradient to the affine subspace Π_k, and update $\mathbf{k}$ along the projected gradient direction, until the projected gradient equals zero. Namely, at a critical point of $E(\mathbf{k})$, $\nabla_k E$ is orthogonal to the admissible curvature space Π_k.

Let $\mathbf{e}_i$ denote the normal to the plane $\{k_i = 0\}$, $\mathbf{d}$ be the unit normal to the plane $\{\sum_{v_i \in \partial M} k_i = 2\pi\chi(M)\}$. The following algorithm optimizes $E(\mathbf{k})$ using the projected gradient method.

Algorithm 44: Optimal discrete conformal parameterization using projected gradient method

input : A mesh M with boundaries
output: A flat circle packing metric (M,Γ,Φ) which minimizes an energy $E(\mathbf{k})$

Compute the initial circle packing metric (M,Γ_0,ϕ);
Compute the initial curvature K;
repeat
  $\mathbf{u} \leftarrow K^{-1}(\mathbf{k})$;
  Compute the gradient of E with respect to $\mathbf{u}$, $\nabla_u E$;
  $\nabla_k E \leftarrow \Delta(\mathbf{u})^{-1} \nabla_u E$;
  forall $v_i \notin \partial M$ **do**

$$\nabla_k E \leftarrow \nabla_k E - \langle \nabla_k E, \mathbf{e}_i \rangle \mathbf{e}_i$$

  end
  $\nabla_k E \leftarrow \nabla_k E - \langle \nabla_k E, \mathbf{d} \rangle \mathbf{d}$;
  $\mathbf{k} \leftarrow \mathbf{k} - \lambda \nabla_k E$;
until $|\nabla_k E| < \varepsilon$;

Proposition 12.15. *Suppose* $\mathbf{k}$ *is an interior point of* Π_k, *also an optimum for an energy form* $E(\mathbf{u})$. *Then all the components of* $\nabla_k E$ *corresponding to the boundary vertices are equal.*

Proof. If $\mathbf{k}$ is an optima of $E(\mathbf{k})$, then $\nabla_k E \perp \Pi_k$. Therefore,

$$\nabla_k E = \sum_{v_i \notin \partial M} \lambda_i \mathbf{e}_i + \mu \mathbf{d}.$$

where $\mathbf{d}$'s components on boundary vertices are equal. □

We can design different energy forms $E(\mathbf{k})$ to measure the uniformity of the conformal parameterization. Because $\mathbf{u}$ is in the normalized u-space,

$$\sum_{v_i \in M} k_i = 0,$$

the variance

$$E(\mathbf{u}) = \sum_{v_i \in M} (u_i - u_i^0)^2$$

measure the uniformity of the parameterization, where $\mathbf{u}^0$ is the initial circle packing metric. This energy can be reformulated as

$$\sum_{v_i \in M} (u_i - u_i^0)^2 = \int_{\mathbf{u}_0}^{\mathbf{u}} (\mathbf{u} - \mathbf{u}^0) d\mathbf{u} = \int_{\mathbf{k}_0}^{\mathbf{k}} (\mathbf{u} - \mathbf{u}^0) \Delta^{-1}(\mathbf{k}) d\mathbf{k},$$

because Δ is positive definite, therefore Δ^{-1} is also positive definite. Then we can replace Δ^{-1} by another positive definite matrix, and expect the new energy form also measures the uniformity of the parameterization. If we use an identity matrix to replace Δ^{-1}, then we get the Legendre dual of the Ricci energy,

$$E(\mathbf{k}) = \int_{\mathbf{k}_0}^{\mathbf{k}} \sum_i (u_i - u_i^0) dk_i.$$

The Hessian matrix of $E(\mathbf{k})$ is $\Delta^{-1}(\mathbf{k})$, which is positive definite when restricted in the normalized u-space. Therefore the energy is convex, it has a unique global minimum, which can be achieved by the direct projected gradient method. It can also be minimized using the following curvature diffusion with free boundaries,

Algorithm 45: Curvature diffusion with free boundaries

input : A mesh M with boundaries
output: A most uniform circle packing metric

Compute the initial circle packing metric (M, Φ, Γ);
Compute the initial curvature $\mathbf{k}_0$;
while $\max|k_i| > \varepsilon, v_i \notin \partial M$ **do**
  forall $v_i \notin \partial M$ **do**
    $du_i \leftarrow -k_i$;
    $u_i \leftarrow u_i + \lambda du_i$;
  end
  $c \leftarrow \frac{\sum_{v_i \in M} u_i}{|V|}$ **forall** *vertex* $v_i \in M$ **do**
    $u_i \leftarrow u_i - c.$
  end
end

This algorithm can also lead to the global minimum of the uniformity energy, because the change of the radii on boundary vertices are same. Fig. 12.12 and 12.13 show two cases of optimal conformal parameterization using the projected gradient method. The initial parameterizations are highly non-uniform, the final results are greatly improved.

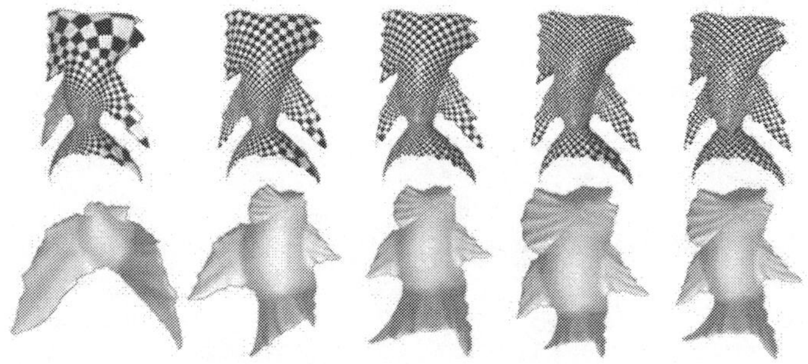

Fig. 12.12. Conformal parameterization optimization process. The initial parameterization is gradually optimized to reach the most uniform result.

Fig. 12.13. Optimized conformal parameterization of horse and camel models. The initial parameterizations on the left are gradually improved and reached the most uniform conformal mapping on the right.

12.8 Hyperbolic Ricci Flow

In previous discussions, we assumed the surface was embedded in the three-dimensional Euclidean space $\mathbb{R}^3$ and approximated by a piecewise-linear triangle mesh. Each face on the mesh is a Euclidean triangle. The corner angles and the edge lengths are governed by the Euclidean cosine law. In this case, we say the mesh is with *the Euclidean background geometry*.

We can consider a mesh with *the hyperbolic background geometry*. Given a mesh M, we treat each face as a hyperbolic triangle, each edge is a geodesic (a hyperbolic line), the corner angles and the edge lengths are governed by the hyperbolic cosine law. The discrete Ricci flow method can be generalized to the hyperbolic mesh.

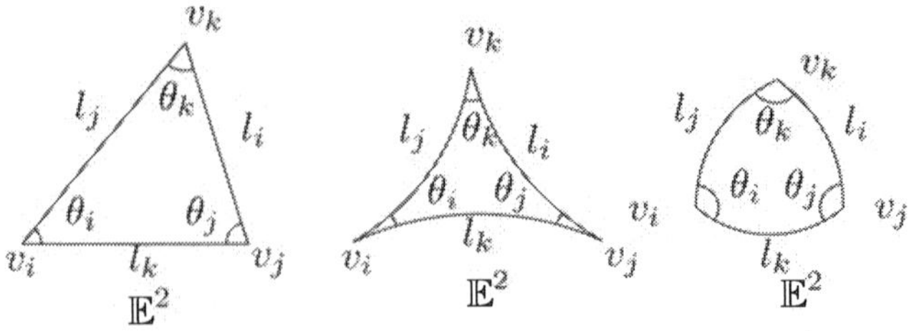

Fig. 12.14. Euclidean, hyperbolic and spherical triangles.

The cosine law for triangles with different background geometries in Fig. 12.14 are

$$\begin{array}{lll} l_k^2 & = l_i^2 + l_j^2 - 2l_i l_j \cos\theta_k, & \mathbb{E}^2, \\ \cosh l_k^2 & = \cosh l_i \cosh l_j + \sinh l_i \sinh l_j \cos\theta_k, & \mathbb{H}^2, \\ \cos l_k^2 & = \cos l_i \cos l_j - \sin l_i \sin l_j \cos\theta_k, & \mathbb{S}^2. \end{array}$$

If a mesh M is in the hyperbolic background geometry and with a circle packing metric, the edge length should be calculated using the hyperbolic cosine law. Suppose the vertices v_i and v_j are connected by an edge $[v_i, v_j]$, the circles centered at v_i and v_j have radii γ_i, γ_j respectively, and the intersection angle between the two circles is ϕ_{ij}. Then the edge length is

$$\cosh l_{ij}^2 = \cosh\gamma_i \cosh\gamma_j + \sinh\gamma_i \sinh\gamma_j \cos\phi_{ij}.$$

The discrete Gaussian curvature of each vertex is calculated in a similar way as the Euclidean Ricci flow:

$$K_i = \begin{cases} 2\pi - \sum_{[v_i,v_j,v_k]\in M} \theta_i^{jk}, & \text{interior vertex,} \\ \pi - \sum_{[v_i,v_j,v_k]\in M} \theta_i^{jk}, & \text{boundary vertex.} \end{cases}$$

where θ_i^{ij} is the corner angle at v_i in the face $[v_i, v_j, v_k]$. The discrete conformal factor u_i is defined differently for a hyperbolic or a spherical mesh M with a circle packing metric (M, Γ, Φ),

$$u_i = \begin{cases} \log\gamma_i, & \mathbb{E}^2, \\ \log\tanh\frac{\gamma_i}{2}, & \mathbb{H}^2, \\ \log\tan\frac{\gamma_i}{2}, & \mathbb{S}^2. \end{cases}$$

The discrete Ricci flow for a hyperbolic or spherical mesh is in the same form:

$$\frac{du_i(t)}{dt} = \bar{K}_i - K_i.$$

In practice, we always set the target curvature $\bar{K}_i$ to zeros for all vertices. The discrete Ricci flow is the negative gradient flow of the discrete Ricci energy, which has the same form as the discrete Euclidean Ricci flow. Suppose the vertex set is $\{v_1, v_2, \cdots, v_n\}$, the discrete conformal factor vector is $\mathbf{u} = \{u_1, u_2, \cdots, u_n\}$. Then the following symmetry holds for hyperbolic or spherical meshes:

$$\frac{\partial K_i}{\partial u_j} = \frac{\partial K_j}{\partial u_i},$$

therefore the discrete hyperbolic, Euclidean, or spherical energy is

$$E(\mathbf{u}) = \int_{\mathbf{u}_0}^{\mathbf{u}} \sum_{i=1}^{n} (\bar{K}_i - K_i) du_i.$$

The hyperbolic Ricci energy is convex if all the intersection angles ϕ_{ij} are acute. Therefore, it has a global minimum, which is the metric $\bar{\mathbf{u}}$. $\bar{\mathbf{u}}$ induces the target curvature $\bar{\mathbf{k}}$.

The spherical Ricci energy is not convex, but $\bar{\mathbf{u}}$ is a critical point of the energy. It can be reached using Newton's method.

In order to get a hyperbolic metric for a surface with a negative Euler number, we can simply set the target curvatures to zeros for all vertices and minimize the discrete hyperbolic Ricci energy. Because the energy is convex, Newton's method leads to the unique global minimum with arbitrary initial metric $\mathbf{u}$ stably. The elements of the Hessian matrix are computed using the following formulae:

$$\frac{\partial^2 E(\mathbf{u})}{\partial u_j \partial u_i} = -\frac{\partial K_i}{\partial u_j}.$$

Details can be deduced by direct computation as shown in [2].

12.9 Hyperbolic Embedding

Suppose M is a mesh with hyperbolic background geometry. Let $(\bar{M}, \pi)$ be the universal covering space of M, π be the projection. Suppose (Γ, Φ) is a hyperbolic circle packing metric on M. Then the pull back metric on $\bar{M}$ is also a hyperbolic circle packing metric, denoted as $(\bar{M}, \pi^*\Gamma, \pi^*\Phi)$. If all the discrete vertex curvatures of (M, Γ, Φ) are zeros, then all the discrete vertex curvatures of $(\bar{M}, \pi^*\Gamma, \pi^*\Phi)$ are also zeros. Therefore, $(\bar{M}, \pi^*\Gamma, \pi^*\Phi)$ can be isometrically embedded onto the hyperbolic space $\mathbb{H}^2$. The embedding process is more complicated than Euclidean case.

12.9.1 Poincaré Disk Model

The most popular model for the hyperbolic space $\mathbb{H}^2$ is the Poincaré disk model. The Poincaré disk model is the unit disk on the complex plane,

$$|z| < 1,$$

with the Riemannian metric:

$$ds^2 = \frac{4dzd\bar{z}}{(1-\bar{z}z)^2}, \quad z \in \mathbb{C}.$$

This metric is conformal to the Euclidean metric $dzd\bar{z}$. All the rigid motions are Möbius transformations:

$$z \to e^{i\theta} \frac{z - z_0}{1 - \bar{z}_0 z}, \quad z_0 \in \mathbb{C}, \quad \theta \in [0, 2\pi).$$

Suppose $p,q \in \mathbf{H}^2$ are two points in the Poincaré disk, the hyperbolic geodesic (hyperbolic line) is the circular arc through p and q and orthogonal to the unit circle $|z| = 1$. The hyperbolic line intersects the unit circle at a and b, as shown in Fig. 12.15. The hyperbolic distance between p and q is

$$d(pq) = |\log \frac{|p-a|/|p-b|}{|q-a|/|q-b|}|.$$

The hyperbolic lines through the origin are also Euclidean lines.

12.9.2 Embedding the Fundamental Domain

Suppose a hyperbolic triangle $[v_i, v_j, v_k]$ as shown in the middle of the Fig. 12.14 is given, the edge lengths are l_i, l_j, l_k, where l_i is the edge length for the edge against v_i, l_j for the edge against v_j and l_k for the edge against v_k. Then we can embed the face onto $\mathbb{H}^2$ simply by

$$v_i = 0, \quad v_j = \frac{e^{l_k} - 1}{e^{l_k} + 1}, \quad v_k = \frac{e^{l_j} - 1}{e^{l_j} + 1} e^{i\theta_i}. \tag{12.9}$$

After embedding the first face $[v_i, v_j, v_k]$, we can embed the faces sharing one edge with it. Suppose $[v_j, v_i, v_l]$ is adjacent to the first face, v_i, v_j have been set. We need to determine the position of v_l. We use $c(v, \gamma)$ to denote a hyperbolic circle centered at v with radius γ. Then v_l is one of the two intersection points of $c(v_i, l_i)$ and $c(v_j, l_j)$, which makes the orientation of the triangle $[v_i, v_j, v_l]$ counter-clock-wise.

In the Poincaré disk, the locus of a hyperbolic circle $c(v, \gamma)$ is also a Euclidean circle $C(V, R)$, where

$$V = \frac{2 - 2\mu^2}{1 - \mu^2|v|^2} v, \quad R^2 = |V|^2 - \frac{|v|^2 - \mu^2}{1 - \mu^2|v|^2},$$

and

$$\mu = \frac{e^r - 1}{e^r + 1}.$$

Therefore, the center of a hyperbolic circle seems not to be in the center of the circle in the Euclidean sense. It is possible that a hyperbolic circle is with an infinite radius. In this case, the center is on the circle, which is the tangent point between the hyperbolic circle and the unit circle as shown in the right frame of Fig. 12.15. The left hyperbolic circle is with the infinite radius.

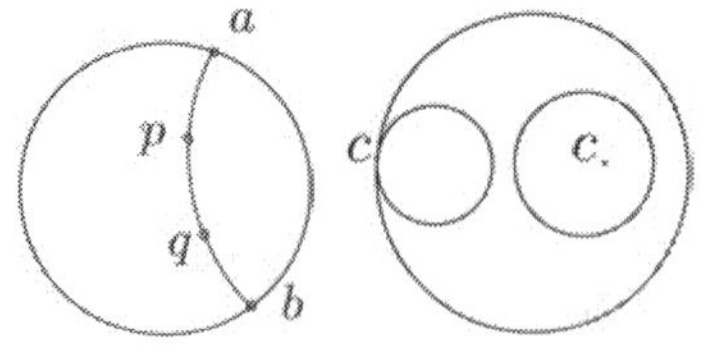

Fig. 12.15. Hyperbolic line and hyperbolic circle.

The following algorithm describes how to embed a hyperbolic mesh, which is a topological disk, onto the Poincaré disk.

Algorithm 46: Hyperbolic embedding of a topological disk
input : A simply connected mesh M with single boundary and hyperbolic metric
output: Hyperbolic embedding of M onto the Poincaré disk

Select a root face f, embed f using formula (12.9);
Put all faces sharing an edge with f in a queue Q;
while *Q is not empty* **do**
  Pop the queue, $f \leftarrow$ *pop* Q, suppose $f = [v_i, v_j, v_k]$, where v_i, v_j have been embedded;
  Convert the hyperbolic circles $c(v_i, l_i)$ and $c(v_j, l_j)$ to Euclidean circles;
  Compute the intersections of the two Euclidean circles;
  Choose the intersection which makes the orientation of the face counter-clock-wise, which is the coordinates of v_k;
  Put all the neighboring faces of f to the queue, which have not been embedded;
end

12.9.3 Hyperbolic Embedding of the Universal Covering Space

Suppose M is a high genus surface, $(\bar{M}, \pi)$ is its universal covering space. Suppose $\bar{M}$ has a hyperbolic metric, therefore it can be isometrically embedded onto the Poincaré disk. Fig. 12.16 explains the process to compute the embedding.

A deck transformation $\rho : \bar{M} \to \bar{M}$ satisfies the following property:

$$\pi \circ \rho = \pi.$$

Furthermore, a deck transformation is isometric. Therefore a deck transformation is a rigid motion of the Poincaré disk, namely, a Möbius transformation. All the deck transformations form the deck transformation group, which is isomorphic to the fundamental group of M.

Suppose D is a fundamental domain of M, $\phi : D \to \mathbb{H}^2$ is the embedding of D onto the Poincaré disk. $Deck(M)$ is the deck transformation group, $\rho \in Deck(M)$ is a Möbius transformation. Then

$$\bigcup_{\rho \in Deck(M)} \rho \circ \phi(D) = \mathbb{H}^2,$$

all the copies of the embedding of a fundamental domain covers the whole Poincaré disk, namely, the embedding of the universal covering space covers the whole Poincaré disk.

The key to the embedding of the whole universal covering space is to find the deck group generators. This can be achieved in the following manner. Suppose M is a closed genus g mesh. First we choose a base point $v \in M$. Then we choose a set of canonical fundamental group generators through v,

$$\{a_1, b_1, a_2, b_2, \cdots, a_g, b_g\}.$$

We slice the mesh along the fundamental group generators to form a fundamental domain D. The boundary of D is

$$\partial D = a_1 b_1 a_1^{-1} b_1^{-1} a_2 b_2 a_2^{-1} b_2^{-1} \cdots a_g b_g a_g^{-1} b_g^{-1}.$$

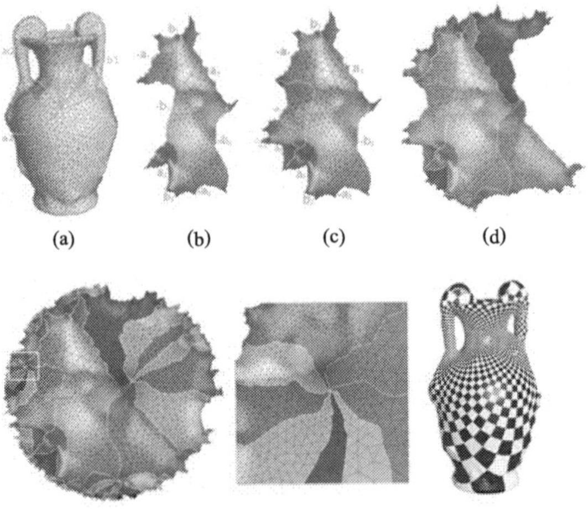

(a) (b) (c) (d)

(e) (f) (g)

Fig. 12.16. From left to right: (a) Genus two vase model (b) One canonical fundamental domain embedded in Poincaré disk (c) $2g$ Fuchsian group generators act on vase model, which are rigid motions in the hyperbolic space. Different color indicates different periods (fundamental domains). The generators maps the central period to the colored ones respectively.

Then we embed D onto the Poincaré disk using the method described in the last subsection. $\phi : D \to \mathbb{H}^2$ is an embedding.

For simplicity, we omit the embedding ϕ in the following discussion, and assume D has an embedding in $\mathbb{H}^2$ already. Then a_i, b_j are curves in $\mathbb{H}^2$. There exists a unique Möbius transformation ρ_k, which maps a_k to a_k^{-1}, namely, for any pair $p \in a_k$ and $q \in a_k^{-1}$, if $\pi(p) = \pi(q)$ then

$$\rho(p) = q.$$

Therefore $\rho(a_k) = a_k^{-1}$. Similarly, there exists a unique τ_k which maps b_k to b_k^{-1}. Then

$$\{\rho_1, \tau_1, \rho_2, \tau_2, \cdots, \rho_g, \tau_g\}$$

are the generators of the Deck transformation group of M (Fuchsian group generator).

Now, let us focus on the computation of ρ_1, the other Fuchsian group generators can be obtained in the same way. Suppose the starting and ending points of a_1 are s,t, $\partial a_1 = t - s$. The starting and the ending points of a_1^{-1} are t_1, s_1, $\partial a_1^{-1} = s_1 - t_1$. It is obvious that all of them are the pre-images of the base point, $\pi : \{s,t,s_1,t_1\} \to v$. We want to find a Möbius transformation ρ_1, such that

$$\rho_1(s) = s_1, \quad \rho_1(t) = t_1.$$

Suppose the hyperbolic line through s and t is $\overline{st}$ and the hyperbolic line through s_1 and t_1 is $\overline{s_1 t_1}$. Then ρ_1 maps $\overline{st}$ to $\overline{s_1 t_1}$. We first construct a Möbius transformation to map s to the origin,

$$\phi_1 : z \to \frac{z - s}{1 - \bar{s}z}.$$

Then ϕ_1 maps $\overline{st}$ to a radial Euclidean line. Suppose the angle from the real axis to $\phi(\overline{st})$ is θ, and

$$\phi_2 : z \longrightarrow e^{-i\theta} z.$$

Then $\phi_2 \circ \phi_1$ maps $\overline{st}$ to the real axis, s to the origin, t to the point $\frac{e^d-1}{e^d+1}$, where d is the hyperbolic distance between s and t. We can use a similar method to construct a Möbius transformation $\bar{\phi}_2 \circ \bar{\phi}_1$ to map s_1 to the origin, t_1 to $\frac{e^d-1}{e^d+1}$. Then

$$\rho_1 = \bar{\phi}_1^{-1} \circ \bar{\phi}_2^{-1} \phi_2 \circ \phi_1$$

maps a_1 to a_1^{-1}.

Once we have the generators of the Fuchsian group, we can transform the embedded fundamental domains and glue them coherently. Suppose D is the fundamental domain, $p \in D$ is an interior point, D_k is another fundamental domain, p_k is an interior point, and $\pi(p) = \pi(p_k)$. Suppose $\gamma \subset \bar{M}$ is a path connecting p to p_k. Then $\pi(\gamma)$ is a loop in M. Suppose the homotopy class of $\pi(\gamma)$ is

$$[\gamma] = s_1 s_2 \cdots s_k,$$

where the symbol

$$s_i \in \{a_1, b_1, a_1^{-1}, b_1^{-1}, \cdots, a_g, b_g, a_g^{-1}, b_g^{-1}\}, \quad 1 \leqslant i \leqslant k.$$

The corresponding Möbius transformation can be computed by replacing the generators of the fundamental group by the generators in Fuchsian group:

1. Replace a_i by τ_i^{-1}, a_i^{-1} by τ_i,
2. Replace b_i by ρ_i, b_i^{-1} by ρ_i^{-1}.

Then we get a unique Möbius transformation, which maps D to D_k. The following algorithm explains the computation details.

Algorithm 47: Hyperbolic embedding of the universal covering space of a high genus mesh

input : A high genus closed mesh M with the hyperbolic metric
output: Hyperbolic embedding of the finite portion of the universal covering space of M onto the Poincaré disk

Compute a canonical fundamental group generators;

$$\{a_1, b_1, a_2, b_2, \cdots, a_g, b_g\}.$$

Slice M along the generators to form a fundamental domain D;
Embed D onto the Poincaré disk using algorithm 47;
Compute the Fuchsian group generators:

$$\{\rho_1, \tau_1, \rho_2, \tau_2, \cdots, \rho_g, \tau_g\}$$

the Möbius transformation ρ_k maps a_k to a_k^{-1}, τ_k maps b_k to b_k^{-1};
For any element in the fundamental group γ_k, compute the corresponding deck transformation by replacing the generators of the fundamental group by the generators of the Fuchsian group;
For each element τ in Fuchsian group, map D to $\tau(D)$;
Stich $\tau(D)$'s coherently along their boundaries.

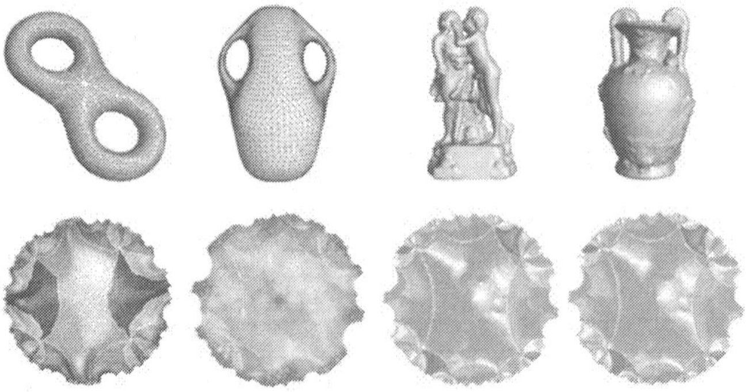

Fig. 12.17. Hyperbolic embedding of the universal covering spaces of high genus surfaces.

Once the universal covering space is embedded onto the Poincaré disk, we can use hyperbolic lines to separate the fundamental domains. Each fundamental domain becomes a hyperbolic polygon, which is called the fundamental polygon. Fig. 12.17 demonstrates the hyperbolic embedding of several high genus surfaces.

12.10 Hyperbolic Ricci Flow for Surfaces with Boundaries

Suppose M is a compact surface with boundaries. The Euler number $\chi(M)$ of M is negative. Then we can compute its hyperbolic metric by double covering. We denote the double covering of M as $\bar{M}$. Then $\bar{M}$ is a symmetric closed surface. $\chi(\bar{M}) = 2\chi(M)$ is still negative. We compute the hyperbolic metric of $\bar{M}$. Then all of the boundaries become hyperbolic lines. We copy the hyperbolic metric of $\bar{M}$ to M, which is the hyperbolic metric of M.

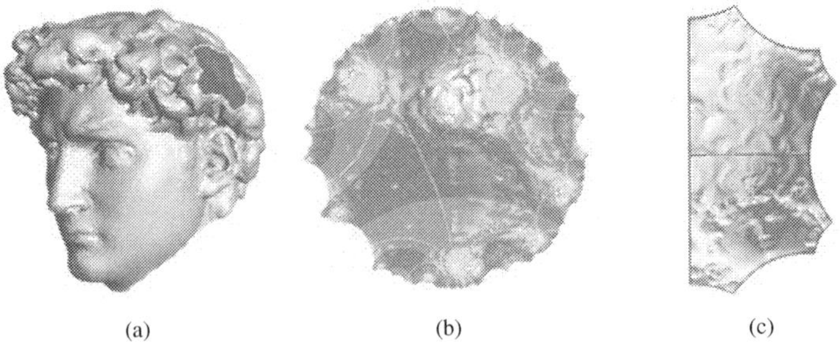

(a) (b) (c)

Fig. 12.18. (a) David head model, which is a two holed topological annulus. (b) Hyperbolic embedding of the universal covering space of the doubled surface. (c) Hyperbolic fundamental polygon.

Fig. 12.18 illustrates the process. The David head model is a two-holed annulus, its double covering is a genus two surface. Different copies are illustrated using different shading. We compute its hyperbolic metric by optimizing the hyperbolic Ricci energy and embed a finite portion of its universal covering space onto the Poincaré disk. We can see that the boundaries of the original surface are mapped to hyperbolic lines.

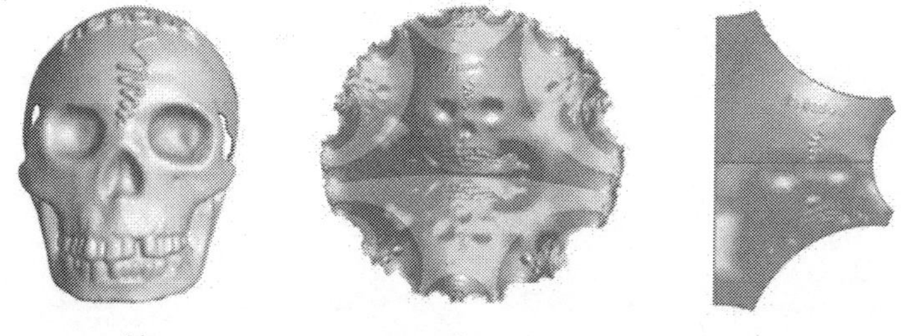

(a) (b) (c)

Fig. 12.19. (a) Skull model, which is a two holed topological disk. (b) Hyperbolic embedding of the universal covering space of the double covering. (c) Hyperbolic fundamental polygon.

The hyperbolic space $\mathbb{H}^2$ can also be modeled using upper half plane model $z : \mathrm{im}\, z > 0$. The Riemannian metric is defined as

$$ds^2 = \frac{dzd\bar{z}}{y}.$$

The geodesics are circular arcs perpendicular to the real axis. The following conformal map converts the Poincaré disk to the upper half plane,

$$z \to i\frac{1+z}{1-z}. \tag{12.10}$$

In Fig. 12.18 and 12.19, we compute the fundamental polygon in the Poincaré disk first, then convert the $\mathbb{H}^2$ model from the Poincaré disk to the upper half plane. The fundamental polygons in the upper half plane are shown in the third frames of each Figure respectively.

We also call a two holed annulus a pair of pants. Given a pair of pants with Riemannian metric, it has a unique hyperbolic metric ,which is conformal to the original metric, such that three boundaries become geodesics under the hyperbolic metric. Suppose the boundaries are $\gamma_1, \gamma_2, \gamma_3$. The hyperbolic lengths of them are (l_1, l_2, l_3), then the triple (l_1, l_2, l_3) represents the conformal class of the surface. The triple is the Teichmüller space coordinates of the pair of pants. In detail, take two surfaces M_1 and M_2, which are two pairs of pants and with hyperbolic metrics. Suppose $\partial M_1 = \{\gamma_1^1, \gamma_2^1, \gamma_3^1\}$, $\partial M_2 = \{\gamma_1^2, \gamma_2^2, \gamma_3^2\}$. If the length of γ_1^k equals to that of $\gamma_2^k, k = 1, 2, 3$, then M_1 and M_2 are conformally equivalent. The conformal map can be easily constructed. The fundamental polygons of M_1 and M_2 are congruent. By a Möbius transformation, we can transform the fundamental polygon of M_1 to that of M_2, this gives us the conformal map between them.

Suppose M is a pair of pants with hyperbolic metric, all boundaries $\{\gamma_1, \gamma_2, \gamma_3\}$ are geodesics. Then there exists a unique geodesic $\{\tau_1, \tau_2, \tau_3\}$, such that

$$\tau_i \perp \gamma_j, \quad \tau_i \perp \gamma_k, \quad i \neq j,\ i \neq k.$$

These geodesics bisect the boundaries and separate the pair of pants into two isometric hyperbolic hexagons, whose inner angles are right angles. Fig. 12.20 shows the two hyperbolic hexagons of the fundamental polygon of the David head model in Fig. 12.18. The center circle is the boundary of the Poincaré disk. $\{\gamma_1, \gamma_2, \gamma_3\}$ are the boundaries of the surface, which are hyperbolic lines, and $\{\tau_1, \tau_2, \tau_3\}$ are the hyperbolic lines orthogonal to the boundaries. The surface is partitioned into two identical hyperbolic hexagons with right corner angles.

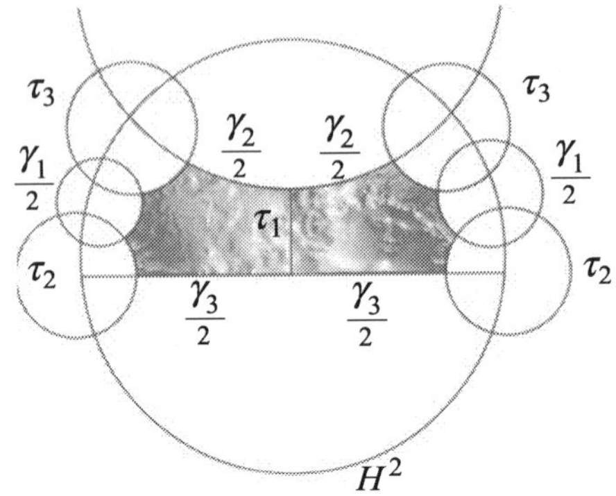

Fig. 12.20. Hyperbolic embedding of a pair of pants.

Further Readings

The thorough introduction to Ricci flow theories can be found in [38]. The surface Ricci flow theory was invented by Hamilton in [51]. The connection between the surface Ricci flow and the circle packing method is analyzed in [82]. The angle based flattening method was proposed by Sheffer in [83]. A more efficient and robust angle based flattening method was introduced in [84]. Circle packing was proposed by Thurston in [85]. An excellent text book on circle packing is by Stephenson [37]. Circle pattern was proposed in [86] and [87]. It is applied for surface parameterization in [88]. Discrete hyperbolic Ricci flow for computing general geometric structures is first introduced in [2]. Discrete Ricci flow for computing geodesic spectrum as the coordinates in the shape space is introduced in [4]. Discrete Euclidean Ricci flow was used to construct manifold splines with single singularity in [18]. Discrete Euclidean Ricci flow has been applied for shape analysis in computer vision [89]. More references on general surface parameterization algorithms can be found in the surveys [41] and [42].

Problems

12.1. Euclidean Ricci Flow Formulae Deduction

(a) In Euclidean Ricci flow, prove

$$\frac{\partial \theta_i}{\partial u_j} = \frac{\partial \theta_j}{\partial u_i}.$$

Therefore, the 1-form $\omega = \sum K_i du_i$ is a closed 1-form.
(b) Prove the Ricci energy

$$E(\mathbf{u}) = \int_{\mathbf{u}_0}^{\mathbf{u}} \omega$$

is convex in the sublinear space $\sum u_i = 0$. This is equivalent to showing that its Hessian matrix is positive definite.

12.2. Hyperbolic Ricci Flow Formulae Deduction
(a) In hyperbolic Ricci flow, prove

$$\frac{\partial \theta_i}{\partial u_j} = \frac{\partial \theta_j}{\partial u_i}.$$

Therefore, the 1-form $\omega = \sum K_i du_i$ is a closed 1-form.
(b) Prove the hyperbolic Ricci energy

$$E(\mathbf{u}) = \int_{\mathbf{u}_0}^{\mathbf{u}} \omega$$

is convex in the sublinear space $\sum u_i = 0$. This is equivalent to showing that its Hessian matrix is positive definite. Explicitly compute the Hessian matrix.

12.3. Spherical Ricci Flow Formulae Deduction
In spherical Ricci flow, prove

$$\frac{\partial \theta_i}{\partial u_j} = \frac{\partial \theta_j}{\partial u_i}.$$

Therefore, the 1-form $\omega = \sum K_i du_i$ is a closed 1-form.

12.4. Inverse Curvature Map
Give a mesh with a circle packing metric.
(a) Represent Gaussian curvature by the metric $\mathbf{u}$.

$$K : \mathbf{u} \to \mathbf{k}.$$

(b) Compute the Jacobi matrix of the curvature map in (a).
(c) Design an algorithm to compute the inverse curvature map K^{-1}.
(d) Compare the efficiency of your inverse curvature map method and of Newton's method for optimizing the Ricci energy.

12.5. Topological Pants
Suppose the input mesh is a pair of topological pants.
(a) Implement the algorithm to conformally map it to the unit disk with circular holes.
(b) Implement the algorithm to compute its hyperbolic metric.
(c) Implement the algorithm to embed its universal covering space on the Poincaré disk.
(d) Compute the Teichmüller coordinates of the mesh. Compare the coordinates of different meshes.

12.6. Topological Disk
Suppose the input mesh is a topological disk.
(a) Compute a flat metric, such that the boundary vertex curvature is zero, and all interior vertex curvatures are zeros except 2 vertices, where the curvature are π's.
(b) Compute a flat metric, such that the boundary vertex curvature is zero, and all interior vertex curvatures are zeros except 4 vertices, where the curvature are $\frac{1}{2}\pi$'s.
(c) Flatten the meshes with different flat metrics onto the plane.

12.7. Real Projective Structure
Besides the Poincaré disk and the upper half plane models, there is another model for hyperbolic space $\mathbb{H}^2$, the Beltrami-Klein model, which is also the unit disk. But all the geodesics are chords of the boundary of the unit disk, and all the rigid transformations are real projective transformations. The map from the Poincaré model to the Klein model is β :

$$\beta : z \to \frac{2z}{1+\bar{z}z}, \quad \beta^{-1} : z \to \frac{1-\sqrt{1-\bar{z}z}}{\bar{z}z}.$$

Prove that if ϕ is a Möbius transformation with the format

$$\phi(z) = e^{i\theta}\frac{z-z_0}{1-\bar{z}_0 z}, \quad z_0 = x_0 + iy_0$$

in the Poincaré model, then the corresponding rigid motion in the Klein model

$$\beta \circ \phi \circ \beta^{-1}$$

has the format

$$\frac{1}{\lambda}OT,$$

where rotation matrix O is

$$O = \begin{pmatrix} \cos\theta & -\sin\theta & 0 \\ \sin\theta & \cos\theta & 0 \\ 0 & 0 & 1 \end{pmatrix}$$

and T is

$$T = \begin{pmatrix} 1+x_0^2+y_0^2 & 2x_0y_0 & -2x_0 \\ 2x_0y_0 & 1-x_0^2+y_0^2 & -2y_0 \\ 2x_0 & 2y_0 & -1-x_0^2-y_0^2 \end{pmatrix}.$$

12.8. Real Projective Structure
(a) Design algorithm to compute the real projective structure of a surface with a negative Euler number.
(b) Implement the algorithm.

12.9. Concentrate All Curvature on One Point
By using the Euclidean Ricci flow, we can compute a flat metric conformal to the original one and all the curvatures are concentrated on one vertex, which we call the singularity.

Fig. 12.21 shows such kind of metric and the singularity. The canonical homotopy group generators are computed, the base point is the singularity. The fundamental domain is flattened onto the plane isometrically. The choice of the singularity affects the area distortion of the mapping. This metric is valuable for converting mesh to splines with only one extraordinary point.
(a) Design algorithm to compute such kind of metric.
(b) Implement the algorithm.
(c) Try to locate the singularity at different positions and compare the uniformity of the area distortion.

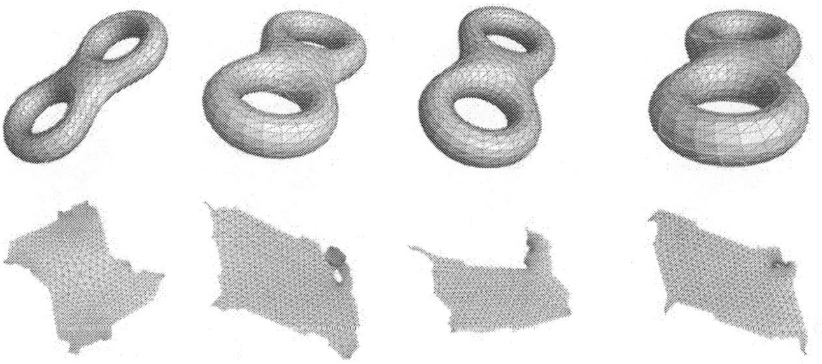

Fig. 12.21. Concentrating curvatures on one vertex. The first row gives the different positions of singularities on the same model. The second row shows their corresponding flat metric. From left to right, the area distortion increases.

12.10. Pants Decomposition
As shown in Fig. 12.22, suppose the input mesh M is with negative Euler number $\chi(M)$.
(a) Design an algorithm to decompose the mesh into several pairs of pants.
(b) Prove that the number of pairs of pants is $-\chi(M)$.
(c) Two pants decompositions are equivalent if their corresponding cuts are homotopic. If M is a genus 2 closed mesh, how many ways are there to decompose it to pairs of pants?

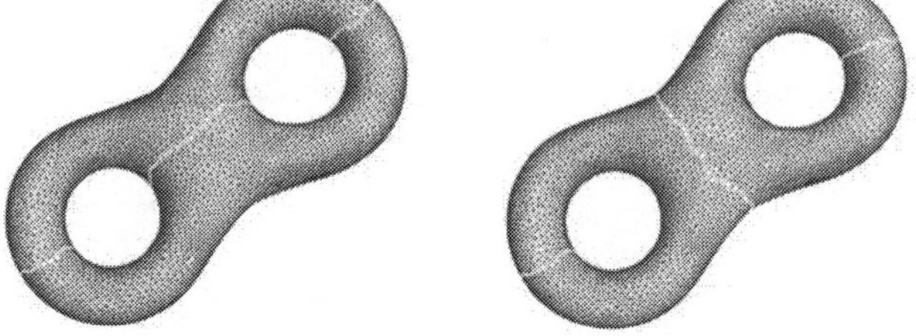

Fig. 12.22. Different pants decomposition for a genus two surface.

A

Major Algorithms

The followings are the list of major algorithms.

- Create a mesh from a vertex list and a face list 1.
- Slice a mesh along a curve 2.
- Compute a cut graph of a mesh 3.
- Prune the cut graph 4.
- Compute a fundamental domain of a mesh 5.
- Compute homology group basis 6.
- Double covering an open mesh 7.
- Compute a finite portion of the universal covering space 8.
- Compute a lift of a curve on the universal covering space 9.
- Homotopy detection 10.
- Shortest loop in a given homotopy class 12.
- Mesh subdivision 13.
- Compute a set of Canonical homotopy group generator 15.
- Harmonic map for a topological disk 16.
- Spherical conformal mapping 17.
- Riemann mapping 19.
- Map between two genus g closed surfaces 20.
- Conjugate gradient method 21.
- Integration of a 1-form a long a 1-chain 22.
- Integrate a closed 1-form on a topological disk 23.
- Wedge product 25.
- Cohomology basis $H^1(M,\mathbb{R})$ 27.
- Compute a set of harmonic 1-form basis 29.
- Conjugate harmonic 1-form 32.
- Compute a basis of the real vector space of holomorphic 1-form group 33.
- Compute harmonic 1-form basis for compact surface with boundaries 37.
- Compute the conformal module of a topological quadrilateral 38.
- Conformal module of a topological annulus 39.
- Period matrix 40.
- Discrete surface Ricci flow 41.

- Newton's method for minimizing discrete Ricci energy 42.
- Newton's method for minimizing discrete Ricci energy for multi-holed annulus 43.
- Hyperbolic embedding of the universal covering space of a high genus mesh 47.

B

Acknowledgement

The first author also appreciates all the students in the Center of Visual Computer at Stony Brook. Especially the following students, who learned the theory with great curiosity and implemented the major algorithms in the books with passion: Miao Jin, global conformal parameterization, discrete Ricci flow; Ying He, manifold spline, non-photorealistic rendering; Xiaohu Guo, point based geometry, simulation; Yang Wang, face tracking, surface stitching and matching; Xin Li, shape matching and search; Xiaotian Yin, characteristic class, computational topology; Hongyu Wang, manifold spline; Wei Zeng, re-meshing, shape comparison; Sen Wang, shape analysis and surface registration; Junfei Dai, geometric approximation theory, efficient rendering. Wei Hong, virtual colonoscopy; Shengying Li, efficient rendering using conformal geometry; Joe Marino, medical imaging; Chris Cartner, topological matching.

The first author thanks the National Science Foundation for the great supports.

References

[1] Xianfeng Gu and Shing-Tung Yau. Surface classification using conformal structures. In *ICCV*, pages 701–708. IEEE Computer Society, 2003.

[2] M. Jin, F. Luo, and Xianfeng Gu. Computing general geometric structures on surfaces using ricci flow. *Computer-Aided Design*, 39(8):663–675, August 2007.

[3] Miao Jin, Junho Kim, and Xianfeng David Gu. Discrete surface ricci flow: Theory and applications. In Ralph R. Martin, Malcolm A. Sabin, and Joab R. Winkler, editors, *IMA Conference on the Mathematics of Surfaces*, volume 4647 of *Lecture Notes in Computer Science*, pages 209–232. Springer, 2007.

[4] Miao Jin, Feng Luo, Shing-Tung Yau, and Xianfeng Gu. Computing geodesic spectra of surfaces. In Lévy and Manocha [90], pages 387–393.

[5] Xianfeng Gu and Shing-Tung Yau. Global conformal parameterization. In *Symposium on Geometry Processing*, pages 127–137, 2003.

[6] Xianfeng Gu. *Parametrization for surfaces with arbitrary topologies*. Doctor of philosophy in computer science, Harvard University, Cambridge Massachusetts, December 2002.

[7] Xianfeng Gu. Computing conformal structures of surfaces. *Communications in Information and Systems*, 2(2):121–146, December 2002.

[8] Yalin Wang, Xianfeng Gu, Kiralee M. Hayashi, Tony F. Chan, Paul M. Thompson, and Shing-Tung Yau. Surface parameterization using riemann surface structure. In *ICCV* [91], pages 1061–1066.

[9] Miao Jin, Yalin Wang, Shing-Tung Yau, and Xianfeng Gu. Optimal global conformal surface parameterization. In *IEEE Visualization*, pages 267–274. IEEE Computer Society, 2004.

[10] Lujin Wang, Xianfeng Gu, Klaus Mueller, and Shing-Tung Yau. Uniform texture synthesis and texture mapping using global parameterization. *The Visual Computer*, 21(8-10):801–810, 2005.

[11] Xianfeng Gu, Steven J. Gortler, and Hugues Hoppe. Geometry images. In Tom Appolloni, editor, *SIGGRAPH*, pages 355–361. ACM, 2002.

[12] Yang Wang, Mohit Gupta, Song Zhang, Sen Wang, Xianfeng Gu, Dimitris Samaras, and Peisen Huang. High resolution tracking of non-rigid 3d motion of densely sampled data using harmonic maps. In *ICCV* [91], pages 388–395.

[13] Christopher Carner, Miao Jin, Xianfeng Gu, and Hong Qin. Topology-driven surface mappings with robust feature alignment. In *IEEE Visualization*, pages 543–550. IEEE Computer Society, 2005.

[14] Sen Wang, Yang Wang, Miao Jin, Xianfeng Gu, and Dimitris Samaras. 3d surface matching and recognition using conformal geometry. In *CVPR (2)*, pages 2453–2460. IEEE Computer Society, 2006.

[15] Sen Wang, Yang Wang, Miao Jin, Xianfeng David Gu, and Dimitris Samaras. Conformal geometry and its applications on 3d shape matching, recognition, and stitching. *IEEE Trans. Pattern Anal. Mach. Intell.*, 29(7):1209–1220, 2007.
[16] Xianfeng Gu, Ying He, and Hong Qin. Manifold splines. *Graphical Models*, 68(3):237–254, 2006.
[17] Ying He, Kexiang Wang, Hongyu Wang, Xianfeng Gu, and Hong Qin. Manifold t-spline. In Kim and Shimada [92], pages 409–422.
[18] Xianfeng Gu, Ying He, Miao Jin, Feng Luo, Hong Qin, and Shing-Tung Yau. Manifold splines with single extraordinary point. In Lévy and Manocha [90], pages 61–72.
[19] Xianfeng Gu, Ying He, and Hong Qin. Manifold splines. In Leif Kobbelt and Vadim Shapiro, editors, *Symposium on Solid and Physical Modeling*, pages 27–38. ACM, 2005.
[20] Ying He, Xianfeng Gu, and Hong Qin. Automatic shape control of triangular -splines of arbitrary topology. *J. Comput. Sci. Technol.*, 21(2):232–237, 2006.
[21] Hongyu Wang, Ying He, Xin Li, Xianfeng Gu, and Hong Qin. Polycube splines. In Lévy and Manocha [90], pages 241–251.
[22] Xianfeng Gu, Yalin Wang, Tony F. Chan, Paul M. Thompson, and Shing-Tung Yau. Genus zero surface conformal mapping and its application to brain surface mapping. In Christopher J. Taylor and J. Alison Noble, editors, *IPMI*, volume 2732 of *Lecture Notes in Computer Science*, pages 172–184. Springer, 2003.
[23] Xianfeng Gu, Yalin Wang, Tony F. Chan, Paul M. Thompson, and Shing-Tung Yau. Genus zero surface conformal mapping and its application to brain surface mapping. *IEEE Trans. Med. Imaging*, 23(8):949–958, 2004.
[24] Craig Gotsman, Xianfeng Gu, and Alla Sheffer. Fundamentals of spherical parameterization for 3d meshes. *ACM Trans. Graph.*, 22(3):358–363, 2003.
[25] Yalin Wang, L. Lui, Xianfeng Gu, K. Hayashi, T. Chan, A.W. Toga, P.M. Thompson, and S.-T. Yau. Brain surface conformal parameterization using riemann surface structure. *IEEE Trans. Med. Imaging*, 26(6):853–865, June 2007.
[26] Wei Hong, Xianfeng Gu, Feng Qiu, Miao Jin, and Arie E. Kaufman. Conformal virtual colon flattening. In Ralph R. Martin and Shi-Min Hu, editors, *Symposium on Solid and Physical Modeling*, pages 85–93. ACM, 2006.
[27] Czes Kosniowski. *A First Course in Algebraic Topology.* Cambridge University Press, 1980.
[28] Allen Hatcher. *Algebraic Topology.* Cambridge University Press, 2002.
[29] Shiing-Shern Chern, Wei-Huan Chern, and Kai Shue Lam. *Lectures on Differential Geometry.* World Scientific Publishing Co. Pte. Ltd., 1999.
[30] Thierry Aubin. *A Course in Differential Geometry*, volume 27. American Mathematics Society, 2000.
[31] Lars V. Ahlfors. *Lectures on Quasiconformal Mappings*, volume 38 of *University Lecture Series.* American Mathematical Society, 1966.
[32] Alastair Fletcher and Vladimir Markovic. *Quasiconformal Maps and Teichmuller Theory.* Oxford University Press, 2007.
[33] Otto Forster. *Lectures on Riemann Surfaces*, volume 81 of *Graduate texts in mathematics.* Springer, 1991.
[34] Jürgen Jost. *Compact Riemann Surfaces: An Introduction to Contemporary Mathematics.* Springer Berlin Heidelberg, 2000.
[35] William Goldman. Projective geometry on manifolds. Lecture Notes, 1988.
[36] William P. Thurston. *The Geometry and Topology of 3-manifolds.* Princeton University Press, 1981.
[37] Kenneth Stephenson. *Introduction to Circle Packing: The Theory of Discrete Analytic Functions.* Cambridge University Press, 2005.
[38] Bennett Chow, Peng Lu, and Lei Ni. *Hamilton's Ricci Flow*, volume 3 of *Lectures in Contemporary Mathematics.* American Mathematical Society and Science Press, 2006.

[39] Richard Schoen and Shing-Tung Yau. *Lecture on Harmonic Maps*, volume 2. International Press Incorporated, Boston, 1997.

[40] Richard Schoen and Shing-Tung Yau. *Lecture on Differential Geometry*, volume 1. International Press Incorporated, Boston, 1994.

[41] Michael S. Floater and Kai Hormann. Surface paramterization: a tutorial and survey. In N. A. Dodgson, M. S. Floater, and M. A. Sabin, editors, *Advances in Multiresolution for Geometric Modelling*, pages 157–186. Springer-Verlag, Heidelberg, 2005.

[42] Alla Sheffer, Emil Praun, and Kenneth Rose. *Mesh Parameterization Methods and Their Applications*, volume 2 of *Foundations and Trends in Computer Graphics and Vision*. Now Publishers Inc., 2006.

[43] Xiaotian Yin, Miao Jin, and Xianfeng Gu. Computing shortest cycles using universal covering space. *The Visual Computer*, 2007.

[44] Junfei Dai, Wei Luo, Shing-Tung Yau, and Xianfeng Gu. Geometric accuracy analysis for discrete surface approximation. In Kim and Shimada [92], pages 59–72.

[45] Greg Leibon and David Letscher. Delaunay triangulations and voronoi diagrams for riemannian manifolds. In *Symposium on Computational Geometry*, pages 341–349, 2000.

[46] Tamal K. Dey, Kuiyu Li, and Jian Sun. On computing handle and tunnel loops. In *IEEE Proc. NASAGEM*, 2007.

[47] Edwin E. Moise. *Geometric Topology in Dimension 2 and 3*. Springer-Verlag, 1977.

[48] Shiing-Shern Chern. An elementary proof of the existence of isothermal parameters on a surface. *Proc. of AMS*, 6:771–782, 1955.

[49] Richard Schoen and Shing-Tung Yau. Compact group actions and the topology of manifolds with non-positive curvature. *Topology*, 18:361–380, 1979.

[50] Richard S. Hamilton. Three manifolds with positive ricci curvature. *Journal of Differential Geometry*, 17:255–306, 1982.

[51] Richard S. Hamilton. The ricci flow on surfaces. In *Mathematics and general relativity (Santa Cruz, CA, 1986), Contemp. Math.* , volume 71, pages 237–262, Providence, RI, 1988. Amer.Math.Soc.

[52] Bennett Chow. The ricci flow on the 2-sphere. *Journal of Differential Geometry*, 33(2):325–334, 1991.

[53] John W. Milnor. On the existence of a connection with curvature zero. *Comm. Math. Helv.*, 32:215–223, 1958.

[54] John C. Loftin. *Applications of Affine Differential Geometry to RP(2) Surfaces*. Phd in mathematics, Harvard University, Cambridge, MA, April 1999.

[55] Wei Luo. Error estimates for discrete harmonic 1-forms over riemann surfaces. *Communications in Analysis and Geometry*, 14(5):1027–1035, 2006.

[56] Mario Botsch, Stephan Steinberg, Stephan Bischoff, and Leif Kobbelt. Openmesh – a generic and efficient polygon mesh data structure, 2002.

[57] CGAL. Computational geometry algorithms library (cgal), 1996.

[58] Hugues Hoppe. Progressive meshes. *SIGGRAPH 1996*, pages 99–108, 1996.

[59] Craig Gotsman, Stefan Gumhold, and Leif Kobbelt. Simplification and compression of 3D meshes. In Armin Iske, Ewald Quak, and Michael Floater, editors, *Tutorials on Multiresolution in Geometric Modeling*. Springer, 2002.

[60] Pierre Alliez and Craig Gotsman. Recent advances in compression of 3D meshes. In Neil A. Dodgson, Michael S. Floater, and Malcolm A. Sabin, editors, *Advances in Multiresolution for Geometric Modelling*, pages 3–26. Springer-Verlag, 2005.

[61] G. Vegter and C.K.Yap. Computational complexity of combinatorial surfaces. In *Sympos. Comput. Geom.*, pages 102–111, 1990.

[62] T.K. Dey and H. Schipper. A new technique to compute polygonal schema for 2-manifolds with application to null-homotopy detection. *Discrete Comput. Geom.*, 14:93–110, 1995.

[63] J. Erickson and S. Har-Peled. Optimally cutting a surface into a disk. *Discrete Comput. Geom.*, 31:37–59, 2004.

[64] Éric Colin de Verdiére and F. Lazarus. Optimal system of loops on an orientable surface. *Discrete Comput.Geom.*, 33:507–534, 2005.

[65] J. Erickson and K. Whittlesey. Greedy optimal homotopy and homology generators. In *Proc. ACM-SIAM Sympos. Discrete Algorithms*, pages 1038–1046, 2005.

[66] T. Dey, H. Edelsbrunner, and S. Guha. Computational topology, 1999.

[67] Ulrich Pinkall and Konrad Polthier. Computing discrete minimal surfaces and their conjugates. *Experimental Mathematics*, 2(1):15–36, 1993.

[68] T. Duchamp, A. Certain, A. Derose, and W. Stuetzle. Hierarchical computation of PL harmonic embeddings. Technical report, University of Washington, 1997.

[69] Michael S. Floater. Parametrization and smooth approximation of surface triangulations. *Computer Aided Geometric Design*, 14(3):231–250, 1997.

[70] Mathieu Desbrun, Mark Meyer, and Pierre Alliez. Intrinsic parameterizations of surface meshes. *Computer Graphics Forum (Proc. Eurographics 2002)*, 21(3):209–218, 2002.

[71] Michael S. Floater. Mean value coordinates. *Computer Aided Geometric Design*, 20(1):19–27, 2003.

[72] Steven Haker, Sigurd Angenent, Allen Tannenbaum, Ron Kikinis, Guillermo Sapiro, and Michael Halle. Conformal surface parameterization for texture mapping. *IEEE Trans. Vis. Comput. Graph.*, 6(2):181–189, 2000.

[73] Craig Gotsman, Xianfeng Gu, and Alla Sheffer. Fundamentals of spherical parameterization for 3d meshes. *ACM Trans. Graph.*, 22(3):358–363, 2003.

[74] Emil Praun and Hugues Hoppe. Spherical parametrization and remeshing. *ACM Trans. Graph.*, 22(3):340–349, 2003.

[75] Bruno Lévy, Sylvain Petitjean, Nicolas Ray, and Jérome Maillot. Least squares conformal maps for automatic texture atlas generation. *SIGGRAPH 2002*, pages 362–371, 2002.

[76] Steven J. Gortler, Craig Gotsman, and Dylan Thurston. Discrete one-forms on meshes and applications to 3d mesh parameterization. *Computer Aided Geometric Design*, 23(2):83–112, 2006.

[77] Ke Wang, Weiwei, Yiying Tong, Mathieu Desbrun, and Peter Schröder. Edge subdivision schemes and the construction of smooth vector fields. *ACM Trans. Graph.*, 25(3):1041–1048, 2006.

[78] Matthew Fisher, Peter Schröder, Mathieu Desbrun, and Hugues Hoppe. Design of tangent vector fields. *ACM Trans. Graph.*, 26(3):56, 2007.

[79] Nicolas Ray, Wan Chiu Li, Bruno Lévy, Alla Sheffer, and Pierre Alliez. Periodic global parameterization. *ACM Transactions on Graphics*, 25(4):1460–1485, 2005.

[80] Burt Rodin and Dennis Sullivan. The convergence of circle packings to the riemann mapping. *Ournal of Differential Geometry*, 26:349–360, 1987.

[81] Zheng-Xu He and Oded Schramm. On the convergence of circle packings to the riemann map. *Invent. Math.*, 125:285–305, 1996.

[82] Bennett Chow and Feng Luo. Combinatorial ricci flows on surfaces. *Journal Differential Geometry*, 63(1):97–129, 2003.

[83] Alla Sheffer and Eric de Sturler. Parameterization of faced surfaces for meshing using angle based flattening. *Engineering with Computers*, 17(3):326–337, 2001.

[84] Alla Sheffer, Bruno Lévy, Maxim Mogilnitsky, and Alexander Bogomyakov. ABF++: Fast and robust angle based flattening. *ACM Transactions on Graphics*, 24(2):311–330, 2005.

[85] William P. Thurston. *Geometry and Topology of Three-Manifolds*. Princeton lecture notes, 1976.

[86] Boris A. Springborn. *Variational principles for circle patterns*. PhD thesis, Technical University of Berlin, 2003.

[87] Alexander I. Bobenko and Boris A. Springborn. Variational principles for circle patterns and koebe's theorem. *Transactions of the American Mathematical Society*, 356:659–689, 2004.
[88] Liliya Kharevych, Boris Springborn, and Peter Schröder. Discrete conformal mappings via circle patterns. *ACM Transactions on Graphics*, 25(2):412–438, 2006.
[89] Xianfeng Gu, Sen Wang, Junho Kim, Yun Zeng, Yang Wang, Hong Qin, and Dimitris Samaras. Ricci flow for 3d shape analysis. In *ICCV*. IEEE Computer Society, 2007.
[90] Bruno Lévy and Dinesh Manocha, editors. *Proceedings of the 2007 ACM Symposium on Solid and Physical Modeling, Beijing, China, June 4-6, 2007*. ACM, 2007.
[91] *10th IEEE International Conference on Computer Vision (ICCV 2005), 17-20 October 2005, Beijing, China*. IEEE Computer Society, 2005.
[92] Myung-Soo Kim and Kenji Shimada, editors. *Geometric Modeling and Processing - GMP 2006, 4th International Conference, Pittsburgh, PA, USA, July 26-28, 2006, Proceedings*, volume 4077 of *Lecture Notes in Computer Science*. Springer, 2006.

Index

Q

R

S

T

U

V

Made in the USA
Monee, IL
08 November 2026

68136727R20173